8th Asian International Seminar on Atomic and Molecular Physics 2008

Journal of Physics: Conference Series Volume 185

Perth, Australia
24-28 November 2008

ISBN: 978-1-61738-291-8

8th Asian International Seminar on Atomic and Molecular Physics 2008

Journal of Physics: Conference Series Volume 185

Perth, Australia
24-28 November 2008

ISBN: 978-1-61738-291-8

IOP Institute of Physics

Printed from e-media with permission by:

Curran Associates, Inc.
57 Morehouse Lane
Red Hook, NY 12571

Printed by Curran Associates, Inc. (2010)

Institute of Physics
Dirac House, Temple Back
Bristol BS1 6BE UK

Phone: 44 1 17 929 7481
Fax: 44 1 17 920 0979

techtracking@iop.org

Additional copies of this publication are available from:

Curran Associates, Inc.
57 Morehouse Lane
Red Hook, NY 12571 USA
Phone: 845-758-0400
Fax: 845-758-2634
Email: curran@proceedings.com
Web: www.proceedings.com

TABLE OF CONTENTS

These proceedings arose from the 8th Asian International Seminar on Atomic and Molecular Physics (AISAMP) which was held at the University of Western Australia 24–28 November 2008.

The history of AISAMP (Takayanagi and Matsuzawa 2002) recognizes its origin from the Japan–China meeting of 1985, and the first use of the name 'The First Asian International Seminar on Atomic and Molecular Physics (AISAMP)' in 1992. The initial attendees, Japan and China, were joined subsequently by scientists from Korea, Taiwan, India, Australia and recently by Malaysia, Thailand, Vietnam, Turkey Iran, UK and USA.

The main purpose of the biennial AISAMP series is to create a wide forum for exchanging ideas and information among atomic and molecular scientists and to promote international collaboration. The scope of the AISAMP8 meeting included pure, strategic and applied research involving atomic and molecular structure and processes in all forms of matter and antimatter. For 2008 the AISAMP conference incorporated the Australian Atomic and Molecular Physics and Quantum Chemistry meeting.

The topics for AISAMP8 embraced themes from earlier AISAMP meetings and reflected new interests, in atomic and molecular structures, spectroscopy and collisions; atomic and molecular physics with laser or synchrotron radiation; quantum information processing using atoms and molecules; atoms and molecules in surface physics, nanotechnology, biophysics, atmospheric physics and other interdisciplinary studies.

The implementation of the AISAMP themes, as well as the international representation of research interests, is indicated both in the contents list of these published manuscripts as well as in the program for the meeting.

Altogether, 184 presentations were made at the 8th AISAMP, including Invited Talks and Contributed Poster Presentations, of which 60 appear in the present Proceedings after review by expert referees in accordance with the usual practice of *Journal of Physics: Conference Series* of the Institute of Physics. The support from the IOPCS staff made this publication possible.

The 8th AISAMP was sponsored primarily by the University of Western Australia and Curtin University of Technology, both in Perth, Western Australia, and by *Journal of Physics: Conference Series*. Support was also received from the International Council of Science, ICSU.

Guidance and active participation from colleagues, particularly from the University of Western Australia, and Curtin University, and from the Australian National University and Melbourne University were sources of strength for the actual organization of the conference. Dr Elena Semidelova receives special thanks for her organizing abilities.

We hope that this issue of *Journal of Physics: Conference Series* will be referenced widely and that it will strengthen ties between all scientists and their countries.

Evan Bieske, Stephen Buckman and Jim F Williams
Guest Editors

Dates

Issue 1 (2009)

All papers published in this volume of *Journal of Physics: Conference Series* have been peer reviewed through processes administered by the proceedings Editors. Reviews were conducted by expert referees to the professional and scientific standards expected of a proceedings journal published by IOP Publishing

Journal of Physics: Conference Series **185** (2009) 012001 doi:10.1088/1742-6596/185/1/012001

Effects of variation of hyperfine splitting(structure) in atomic physics.

S. A. Alavi

Department of Physics, Tarbiat Moallem university of Sabzevar, Sabzevar, P. O. Box 397, Iran.

E-mail: alavi@sttu.ac.ir

Abstract. The possible variation of the fundamental constants is currently a very popular research topic. Theories unifying gravity and other interactions suggest the possibility of spatial and temporal variation of physics constants in the universe. Current interest is high because in superstring theories which have additional dimensions compactified on tiny scales any variation of the size of the extra dimensions results in changes in the 3-dimensional coupling constants. String theory also suggest the space to be noncommutative i.e., the space coordinates do not commute with each other. In this paper we study the hyperfine splitting in the framework of the noncommutative quantum mechanics(NCQM) developed in the literature. We show that the energy difference between two excited $F = I + \frac{1}{2}$ and the ground $F = I - \frac{1}{2}$ states in a noncommutative space(NCS) is bigger than the one in commutative case, so the radiation wavelength in NCS_s must be shorter than the radiation wavelength in commutative spaces. We also find an upper bound for the non-commutativity parameter. Since in the very tiny string scale or at very high energy situation the effects of non-commutativity of space may appear so the hyperfine splitting is not constant and changes as energy changes(high energy situation). The results would be of interest both for theoretical and optical spectroscopists.

1. Introduction

Recently there have been much interest in the study of physics in noncommutative spaces(NCS_s). In the usual quantum mechanics, the coordinates and momenta have the following commutation relations :

$$[x_i, x_j] = 0 \qquad [x_i, p_j] = i\hbar\delta_{ij} \qquad [p_i, p_j] = 0, \tag{1}$$

At very short scales, say string scale, the coordinates may not commute and the commutation relations are as follows :

$$[\hat{x}_i, \hat{x}_j] = i\theta_{ij} \qquad [\hat{x}_i, \hat{p}_j] = i\hbar\delta_{ij} \qquad [\hat{p}_i, \hat{p}_j] = 0, \tag{2}$$

where θ_{ij} is an antisymmetric tensor which can be defined as $\theta_{ij} = \frac{1}{2}\epsilon_{ijk}\theta_k$. The hat symbol represents the operators in the NCS.

2. Hyperfine splitting in noncommutative quantum mechanics.

It is shown in [1] that for the Hamiltonian of the type $H(\hat{p}, \hat{x}) = \frac{\hat{p}^2}{2m} + V(\hat{x})$, the noncommutative Hamiltonian $H_{nc} = H_\theta$ can be obtained by a shift in the argument of the potential :

$$x_i = \hat{x}_i + \frac{1}{2}\theta_{ij}\hat{p}_j \qquad \hat{p}_i = p_i. \tag{3}$$

which leads to $H_\theta = \frac{p^2}{2m} + V(x_i - \frac{1}{2}\theta_{ij}p_j)$. The variables x_i and p_i now, satisfy in the same commutation relations as the usual(commutative) case i.e. Eq.(1). The magnetic dipole moment of the nucleus is given by $\vec{M} = \frac{Zeg_N}{2M_N c}I$.

where Ze, M_N, I and g_N are its electric charge, mass, spin and gyromagnetic ratio respectively. In the commutative case, the vector potential due to a point-like magnetic dipole is given by $\vec{A}(\vec{r}) = -\frac{1}{4\pi}(\vec{M} \times \nabla)\frac{1}{r}$.

As shown in [1], their proposal for the non commutative H-atom Hamiltonian can be generalized to other systems, i.e. taking the usual Hamiltonian that is now a function of noncommutative coordinates$(\hat{x}, \hat{p})$, so we take the usual expression for vector potential but now being a function of noncommutative coordinates, accordingly this leads to a expression for the noncommutative Hamiltonian which is the same as the usual Hamiltonian that is now a function of noncommutative coordinates, so $\hat{\vec{A}}(\vec{r}) = -\frac{1}{4\pi}(\vec{M} \times \hat{\nabla})\frac{1}{\hat{r}}$. where the hat symbol represents the same variable in the NCS.
From Eq.(3), we observe that $\frac{\partial}{\partial x_i} = \frac{\partial}{\partial \hat{x}_i}$, so $\nabla = \hat{\nabla}$. $\hat{\vec{M}} = \vec{M}$, because the noncommutativity of space has no effect on spin.
One can derive the magnetic field produced by the nucleus from the vector potential $\hat{\vec{B}} = \nabla \times \hat{\vec{A}}$. Then the perturbative Hamiltonian is given by :

$$H_I = -\vec{M}_e \cdot \vec{B} = \frac{e}{m_e c}\vec{S} \cdot \vec{B} = \frac{Ze^2 g_N}{2m_e M_N c^2}\frac{1}{4\pi}\vec{S} \cdot \left[-\vec{I}\nabla^2\frac{1}{\hat{r}} + \nabla(\vec{I} \cdot \vec{\nabla})\frac{1}{\hat{r}}\right] \tag{4}$$

where $\vec{M}_e = \frac{e}{m_e c}\vec{S}$ is the electron magnetic dipole and m_e is its mass.

We first consider the second term in H_I:

$$\left\langle \left((\vec{S} \cdot \vec{\nabla})(\vec{I} \cdot \vec{\nabla})\frac{1}{\hat{r}}\right)\right\rangle = S_i I_k \left\langle \frac{\partial}{\partial \hat{x}_i}\frac{\partial}{\partial \hat{x}_k}\frac{1}{\hat{r}}\right\rangle = S_i I_k \left\langle \frac{\partial}{\partial x_i}\frac{\partial}{\partial x_k}\frac{1}{\hat{r}}\right\rangle \tag{5}$$

Using Eq.(3), we have :

$$\frac{1}{\sqrt{\hat{r}\hat{r}}} \equiv \frac{1}{\sqrt{\hat{x}\hat{x}}} = r^{-1}\left[1 + \frac{1}{r^2}\left(\frac{1}{4}\vec{L} \cdot \vec{\theta} - \frac{1}{8}\theta_{ij}\theta_{ik}p_j p_k + O(\theta^3)\right)\right] \tag{6}$$

One can show that for the states with $\ell = 0$ in Eq.(5), the contributions of the terms with $i \neq j$ will vanish and for the terms $i = j$ it gives $\frac{1}{3}S_i I_i \left\langle \nabla^2 \frac{1}{\hat{r}}\right\rangle$. We note that the first term in the brackets in Eq.(4) is $-S_i I_i \nabla^2 \frac{1}{\hat{r}}$, so :

$$\langle H_I \rangle = -\vec{M}_e \cdot \vec{B} = \frac{e}{m_e c}\vec{S} \cdot \vec{B} = \frac{Ze^2 g_N}{2m_e M_N c^2}\frac{1}{4\pi}\left[-\frac{2}{3}S_i I_i \left\langle \nabla^2 \frac{1}{\hat{r}}\right\rangle\right] \tag{7}$$

Using perturbation theory we proceed as follows :

$$\left\langle \Psi + \theta\Delta^{(1)}\Psi \left| V(r) + \theta H_I^{(1)} + \theta^2 H_I^{(2)} \right| \Psi + \theta\Delta^{(1)}\Psi \right\rangle \tag{8}$$

where :

$$\theta H_I^{(1)} = -\frac{Ze^2 g_N}{48\pi m_e M_N c^2} S_i I_i \nabla^2 \frac{1}{r^3}\vec{L}.\vec{\theta} \tag{9}$$

$$\theta^2 H_I^{(2)} = \frac{Ze^2 g_N}{96\pi m_e M_N c^2} S_i I_i \nabla^2 \frac{1}{r^3}\theta_{ij}\theta_{ik}p_j p_k \tag{10}$$

and $V = -\frac{Ze^2}{r}$. We set $\theta_3 = 0$ and the rest of the θ-components to zero, which can be done by a rotation or a redefinition of coordinates, so $\vec{L}.\vec{\theta} = L_z\theta$.

Using the first-order perturbation theory for the wave functions, we have :

$$\Psi_n^\theta = \Psi_n + \sum_{k\neq n} \frac{\langle k\ell' j' j_z' \,|\beta\theta L_z|\, n\ell j j_z\rangle}{|E_n^0 - E_k^0|}\Psi_k, \beta = -\frac{Ze^2 g_N}{48\pi m_e M_N c^2} S_i I_i \nabla^2 \frac{1}{r^3} \tag{11}$$

This leads to:

$$\Psi_n^\theta = \Psi_n + \beta\theta \sum_{k\neq n} \frac{j_z\hbar\left(1 \mp \frac{1}{2\ell+1}\right)\delta_{\ell\ell'}\delta_{j_z j_z'}}{|E_n^0 - E_k^0|}\Psi_k \qquad j = \ell \pm \frac{1}{2} \tag{12}$$

The first-order correction terms are $2\left\langle\Psi\,|V(r)|\,\theta\Delta^{(1)}\Psi\right\rangle + \left\langle\Psi\left|\theta H_I^{(1)}\right|\Psi\right\rangle$.

We consider the $\ell = 0$ states, so it is obvious that the second term i.e. the correction to the Hamiltonian to the first order, vanishes: $\theta H_I^{(1)} = 0$.

we note that in the relation $j = \ell \pm \frac{1}{2}$, for $\ell = 0$ states only the upper sign(plus) is acceptable which gives $j = \frac{1}{2}$. But this corresponds to the upper sign(minus) in Eq.(12), which leads to $\Psi_n^\theta = \Psi_n$, and this means that to the first order in θ, there is no corrections to the wave functions i.e. $\theta\Delta^{(1)}\Psi = 0$. One can show the same is true for the second-order term $\theta\Delta^{(2)}\Psi = 0$. Therefore, the noncommutativity of space to the first order has no effect on the hyperfine splitting, so we study the noncommutativity effects to the second order. The third term in Eq.(6), can be written as :

$$\frac{1}{8r^3}(\theta_{ij}\theta_{ik})p_j p_k = \frac{1}{32r^3}(\epsilon_{ij\mu}\epsilon_{ik\nu})p_j p_k \theta_\mu \theta_\nu = \frac{1}{32r^3}(p^2\theta^2 - (\vec{p}.\vec{\theta})^2) \tag{13}$$

Then we have $\left\langle \nabla^2\left(\frac{\theta_{ij}\theta_{ik}p_j p_k}{8r^3}\right)\right\rangle = \frac{1}{8}\theta^2\left\langle\left(p_x^2 + p_y^2\right)\nabla^2\frac{1}{r^3}\right\rangle$.

By symmetry considerations we have $\langle p_x^2\rangle = \left\langle p_y^2\right\rangle = \langle p_z^2\rangle = \frac{1}{3}\langle p^2\rangle$, so:

$$\left\langle H_I^{(2)}\right\rangle_\theta = \frac{Ze^2 g_N}{96\pi m_e M_N c^2}\theta^2 \vec{S}\cdot\vec{I}\left\langle\frac{p^2}{r^5}\right\rangle \tag{14}$$

To calculate the expectation value of $\frac{p^2}{r^5}$, we note that $\frac{p^2}{2m} + V = E$, where $V = -\frac{Ze^2}{r}$, then we have $p^2 = 2mH_0 + 2m\frac{Ze^2}{r}$. Therefore Eq.(10) reads :

$$\theta^2\left\langle H_I^{(2)}\right\rangle_\theta = \frac{Ze^2 g_N}{48\pi M_N c^2}\theta^2 \vec{S}\cdot\vec{I}\left[E_n\left\langle\frac{1}{r^5}\right\rangle + Ze^2\left\langle\frac{1}{r^6}\right\rangle\right] \tag{15}$$

If F is the total spin of the electron and nucleus, then the value of $\frac{\vec{S}.\vec{I}}{\hbar^2}$ is $\frac{1}{2}I$ for $F = I + \frac{1}{2}$ and $\frac{1}{2}(-I-1)$ for $F = I - \frac{1}{2}$. So the expectation value of the Hamiltonian $\langle H_I\rangle_\theta$ to the second order is given by :

$$\langle H_I\rangle_\theta \equiv \Delta E_{NC} = \frac{(Ze^2)^2\hbar^2 g_N}{48\pi M_N c^2}\theta^2\left(-\frac{1}{2}\mu\frac{Ze^2}{n^2\hbar^2}f(5) + f(6)\right) \tag{16}$$

where $f(\alpha) = \left\langle\frac{1}{r^\alpha}\right\rangle$ and $\alpha = \frac{e^2}{\hbar c}$.

for the ground state of hydrogen atom, $n = 1; Z = 1$, and by substituting the values of μ, e, $\hbar$, $f(5)$ and $f(6)$ we found out that the term in the parentheses is positive and therefore the correction due to noncommutativity of space on $\Delta E_{NC}(F = 1 \to F = 0)$ is positive. This leads

us to the fact that the energy difference between the two states in a NCS is bigger than the one in commutative case, so the radiation wavelength in NCS_s spaces must be shorter than the radiation wavelength in commutative spaces.
The magnitude of ΔE_{NC}, for a H-atom can be obtained by substitution of the values of various quantities : $c = 3.0 \times 10^8 ms^{-1}$, $e = 1.602 \times 10^{-19} C$, $m_e = 9.109 \times 10^{-31} kg$, $M_N = 1.673 \times 10^{-27} kg$, $\hbar = 1.055 \times 10^{-34} Js$, $g_N \simeq 5.56$ and the Bohr radius $a_0 = 5.292 \times 10^{-11} m$, which leads us to the result $\Delta E_{NC} = 10^{-28} \theta^2$.
On the other hand one can use the data on the hyperfine splitting to impose some bounds on the value of noncommutativity parameter θ. The best measurement of the hyperfine splitting is for Hydrogen and its relative accuracy is about 10^{-12}[2]. Since the noncommutativity of space has not been detected sofar, the value of ΔE_{NC}, should be of the order of 10^{-12}, so we have $\Delta E_{NC}(F = 1 \rightarrow F = 0) \leq 10^{-12}$, which gives $\theta \leq (10^3 Gev)^{-2}$. This is in agreement with other results presented in the literature, e.g.[1].

References

[1] Chaichian M, Sheikh-Jabbari M M and Tureanu A 2001 *Phys. Rev. Lett* **86** 2716 and 2004 *Eur. Phys. J. C* **36** 251.
[2] Essen L et.al., 1971 *Nature* **229** 110.

Efficient up-conversion by triplet-triplet annihilation

Josie E. Auckett[1], Yuen Yap Chen[1], Tony Khoury[1], Raphaël G. C. R. Clady[1], N. J. Ekins-Daukes[2], Maxwell J. Crossley[1] and Timothy W. Schmidt[1]

[1] School of Chemistry, The University of Sydney, NSW 2006, Australia
[2] Department of Physics and the Grantham Institute for Climate Change, Imperial College London, U.K

E-mail: `t.schmidt@chem.usyd.edu.au`

Abstract. Following experimental determination of kinetic parameters governing the up-conversion of light by triplet-triplet annihilation (TTA-UC), we develop a kinetic model to determine the conditions required for efficient up-conversion. We discuss the assumptions underpinning statistical arguments for an upper limit to TTA-UC and argue that no such limit exists.

1. Introduction

Single threshold photovoltaic convertors suffer from an inability to harvest photons of an energy less than the threshold, restricting their efficiency to about 30%.[1] One possible improvement to the single threshold device is to place an up-converting material behind the cell which can efficiently convert the low sub-threshold photons into usable light.[2] The limiting efficiency of a cell utilizing an up-conversion (UC) system is over 50% with a threshold energy of 2 eV.[3] However, using rare-earth containing phosphors, the best UC quantum efficiency recorded is $\sim 10^{-6}$.[4] Recently, a way to up-convert red light using triplet-triplet annihilation (TTA) in organic molecules has been developed.[5, 6] In TTA-UC, efficiently emitting molecules, "emitters", are placed in their long lived triplet states through collisions with triplet sensitizer molecules. The sensitizers are chosen such that they undergo intersystem crossing from the S_1 state to the T_1 state following absorption of a low energy photon. When two triplet emitter molecules encounter another, a complex is formed which may take on singlet, triplet or quintet spin states. Due to the degeneracies of these states, they are statistically weighted, respectively, 1:3:5, such that there is only a 1/9 statistical chance of producing a singlet encounter complex. When formed, the singlet undergoes internal conversion to a lower energy state whereby one emitter moiety is in its S_1 state and the other is in its ground state (S_0). The excited moiety then promptly fluoresces, yielding up-converted light. The general scheme is given in Figure 1. The limiting efficiency of TTA-UC is not known, but it is widely held that it cannot exceed 11%, the statistical weighting of singlet encounter complexes. However, in 2008 we showed that 28% UC efficiency is possible using a palladium tetrakisquinoxalinoporphyrin as the sensitizer and rubrene as the emitter.[7] Using a pulsed laser source and time-resolved spectroscopy, we measured the rate constants indicated in Figure 1. In this article, we present the kinetic model which emerged from the previous study and discuss how spin statistics and the various kinetic parameters affect the predicted TTA-UC efficiencies.

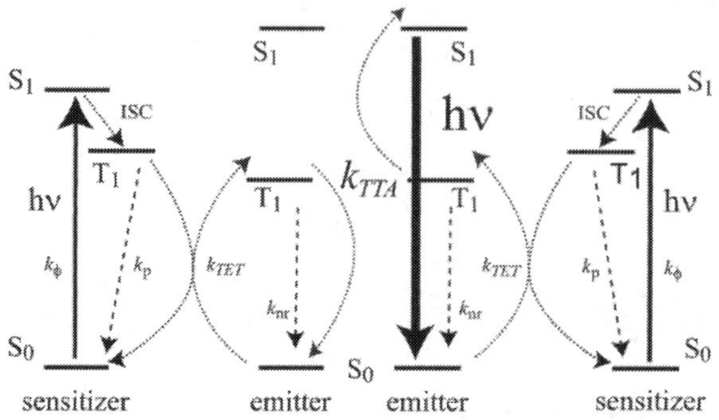

Figure 1. Kinetic scheme of TTA-UC.

2. Kinetic Model of TTA-UC

Under continuous illumination, there is a continuous pumping of population from the S_0 level into the S_1 level of the sensitizer. This rate is given by $k_\phi = \phi\epsilon'$, where ϕ has units of moles of photons $\mathrm{dm^{-2}s^{-1}}$, and the molar extinction coefficient, $\epsilon' = \epsilon \log(10)$, is in $\mathrm{M^{-1}dm^{-1}}$. In the case of solar irradiation, the rate is given by the integral $k_\phi = \int d\lambda \rho(\lambda)\epsilon'(\lambda)$, where ρ is given in moles of photons $\mathrm{dm^{-2}s^{-1}nm^{-1}}$. The S_1 state of the sensitizer lives only $\sim 2\,\mathrm{ps}$ as the molecules undergo intersystem crossing to T_1.[7]

In the following, X represents the sensitizer molecules, and Y represents the emitter species. The kinetics of the triplet energy transfer and triplet-triplet annihilation procedure proceed as

$$\frac{d[^3X]}{dt} = k_\phi[^1X] - k_{TET}[^1Y][^3X] - k_p[^3X] - k_{TTA}\left(2[^3X]^2 + [^3Y][^3X]\right) = -\frac{d[^1X]}{dt}$$

$$\frac{d[^3Y]}{dt} = k_{TET}[^1Y][^3X] - k_{nr}[^3Y] - k_{TTA}\left(2[^3Y]^2 + [^3Y][^3X]\right) = -\frac{d[^1Y]}{dt}$$

where k_{TET} and k_{TTA} represent the rate constants for the triplet energy transfer and triplet-triplet annihilation respectively. The constants k_p and k_{nr} represent the total first-order decay constants for the sensitizer and emitter triplets respectively. The kinetic processes are also shown schematically in Figure 1. These equations are solved to find steady state conditions. The efficiency, e, is given by

$$e = 2\frac{k_{TTA}[^3Y]^2}{k_\phi[^1X]}. \tag{1}$$

In all simulations, kinetic rate constants and concentrations were held at our experimental values: $[X] = 1.16 \times 10^{-4}\,\mathrm{M}$, $[Y] = 2.30 \times 10^{-3}\,\mathrm{M}$, $k_{TET} = 3.33 \times 10^8\,\mathrm{M^{-1}s^{-1}}$, $k_{TTA} = 1 \times 10^8\,\mathrm{M^{-1}s^{-1}}$, $k_{nr} = 9000\,\mathrm{s^{-1}}$, $k_p = 2.5 \times 10^4\,\mathrm{s^{-1}}$ and $k_\phi = 10\,\mathrm{s^{-1}}$ unless specified otherwise.

3. Results and Discussion

Our modelling reveals a threshold behaviour for efficiency as a function of k_{TET} and $[Y]$. Essentially, efficient TTA-UC can proceed if $k_{TET}[^1Y] \gg k_p$, as seen in Figure 2. When this is satisfied, the efficiency comes down to the competition between $k_{nr}[^3Y]$ and $k_{TTA}[^3Y]^2$. We require that $k_{TTA}[^3Y] \gtrsim k_{nr}$.

In Figure 3, a plot of photon absorption rate against photon emission rate is presented. On the double-logarithmic plot, the order of the process is revealed by the slope. Interestingly, the slope of the plot is very sensitive to the value of k_{nr}, the rate of spontaneous decay of emitter triplets. At high values of k_{nr}, say $10^4\,\mathrm{s^{-1}}$, the process is seen to exhibit a quadratic nature

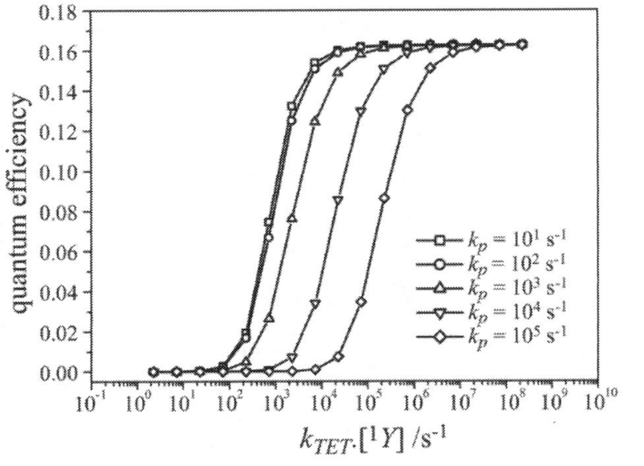

Figure 2. Quantum efficiency of the TTA system over several values of k_p, showing competition between the desired triplet energy transfer process (TET) and alternative sensitizer triplet decay pathways. The quantity $k_{TET}.[^1Y]$ describes the transfer between sensitizer triplets and emitter triplets per second. At very low values, emitter triplet production is negligible and TTA efficiency approaches zero regardless of the value of k_p. At very high values, sensitizer triplet "loss" mechanisms cannot compete with TET; k_p becomes insignificant and efficiency reaches an upper limit set by the values of k_{nr}, k_ϕ and k_{TTA}.

across the ranges of k_ϕ employed here. A value of $k_\phi = 10^4\,\mathrm{s}^{-1}$ corresponds to monochromatic light intensity of about $10\,\mathrm{Wcm}^{-2}$, for the molecules employed here. Typically, sunlight would bring about $k_\phi \sim 10\,\mathrm{s}^{-1}$. The quadratic dependence of the TTA-UC across several orders of magnitude of irradiance is due to the short life of the triplet states. Where the dominant triplet decay mechanism is non-radiative, rather than TTA, the amount of up-converted light will depend quadratically on the concentration of triplets which is controlled linearly by the level of irradiation. However, where the non-radiative processes are cut-off, so that $k_{nr} = 0$, the process is entirely linear. In this case, triplets *always* have the time to react by TTA and thus the TTA-UC output is linearly dependent on the irradiation level, bringing about a slope of unity. At intermediate values of k_{nr}, the process changes smoothly from quadratic to linear as the input light is increased, as opposed to exhibiting a threshold.[8] This behaviour has been recently observed.[9]

It is only where the process deviates from quadratic towards linear that the TTA-UC is significantly efficient. At typical solar excitation rates of $k_\phi \sim 10\,\mathrm{s}^{-1}$, k_{nr} must be below about $1000\,\mathrm{s}^{-1}$ to achieve significant TTA-UC. One contaminant that likely increases k_{nr} is dioxygen. Being a triplet ground state, when it encounters a triplet emitter molecule in solution it can undergo TTA to yield singlet oxygen and a ground state emitter, short circuiting the TTA-UC process. With concentrations of dioxygen as low as $10^{-7}\,\mathrm{M}$ (one atmosphere is $10^{-2}\,\mathrm{M}$), k_{nr} will be enhanced by a contribution on the order of $1000\,\mathrm{s}^{-1}$. As such, dioxygen concentrations must be kept low to achieve efficient TTA. In our experiments concerning TTA-UC kinetics in degassed toluene, the measured k_{nr} value of $9000\,s^{-1}$ is too high to achieve efficient TTA-UC under solar illumination. It is likely that our experiments were affected by trace levels of oxygen, foreshortening the lifetime of the rubrene triplet state. Nevertheless, the high efficiencies obtained under pulsed-laser illumination (28%) and the excellent agreement with our kinetic model suggests that higher efficiencies could be obtained with more careful deoxygenation.[7] The competition between intramolecular and intermolecular processes will be more important still when implementing TTA-UC in solid polymer films.[10, 11] In such environments, the rates k_{TTA} and k_{TET} will be reduced by a factor consistent with the increased viscosity of the environment.

Solving for a steady-state population of triplets, we find that the condition $k_{TTA}[^3Y] \gtrsim k_{nr}$ translates to the condition

$$\sqrt{k_{nr}^2 + 8k_\phi[^1X]k_{TTA}}/k_{nr} \gtrsim 5. \tag{2}$$

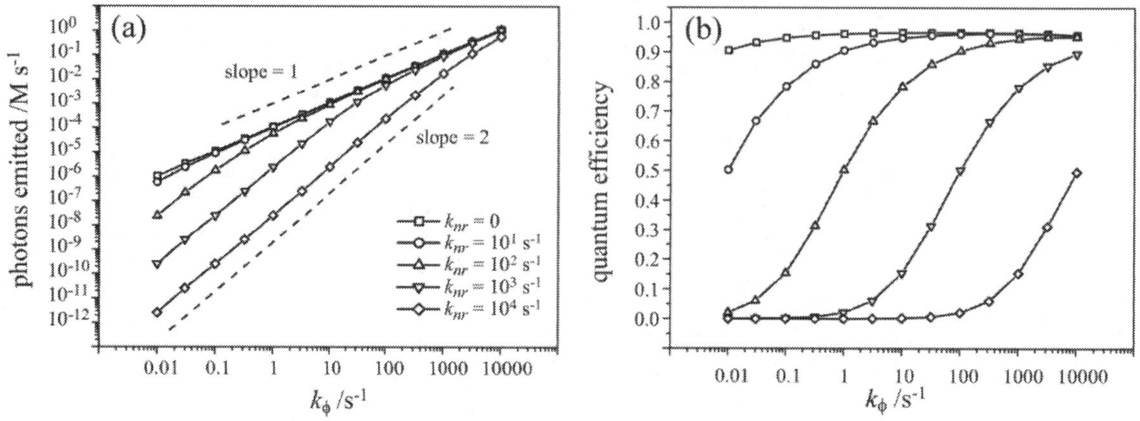

Figure 3. (a) Photon output of the rubrene emitter in the TTA process over several values of k_{nr}, the rate constant for decay of the rubrene triplet state via all pathways other than TTA, including quenching by molecular oxygen. The slopes for $k_{nr} = 10^3$ and $= 10^4\,s^{-1}$ at moderate values of k_ϕ are close to 2, yet $k_{nr} = 0$ has a slope of 1, suggesting that if loss mechanisms of the emitter triplet can be suppressed, such as by careful removal of all oxygen from the system, TTA proceeds with linear dependence on the light intensity.(b) The corresponding quantum efficiencies, defined as the percentage of sensitizer triplet pairs that successfully undergo TTA to place one emitter molecule in its S_1 state. At very high k_ϕ, all curves approach the maximum efficiency given by $k_{nr} = 0$, showing that triplet loss mechanisms cannot compete with TTA when sensitizer triplets are being regenerated very quickly.

4. Final remarks

Our kinetic model predicts that efficient TTA-UC can proceed providing that certain conditions are met: triplet energy must be rapidly transfered to the emitter molecules, and the emitter triplet states must be long lived with respect to the TTA rate. We find no evidence in support of a spin-statistical limitation on TTA. Indeed, our experiments, on which the present model is based, suggest that TTA can proceed with high efficiency. If quintet encounter complexes are to decay non-radiatively, rather dissociating back into triplets, the the question remains: since there is no lower quintet state, to which state does the encounter complex decay? Indeed, if it does decay, then due to the energy-gap law, it is likely to decay to a state which subsequently fluoresces. The only possible quenching process is if the triplet encounter complex decays into the lower triplet state which then dissociates into a triplet and a ground state emitter. In this case the limiting efficiency is 40%.

References

[1] Shockley W and Queisser H J 1961 *J. Appl. Phys.* **32** 510–519
[2] Trupke T, Green M A and Würfel P 2002 *J. Appl. Phys.* **92** 4117–4122
[3] Trupke T, Shalav A, Richards B, Würfel P and Green M 2006 *Sol. Ener. Mat. Sol. Cells* **90** 3327 – 3338
[4] Shalav A, Richards B and Green M 2007 *Sol. Ener. Mat. Sol. Cells* **91** 829 – 842
[5] Baluschev S, Yakutkin V, Miteva T, Wegner G, Roberts T, Nelles G, Yasuda A, Chernov S, Aleshchenkov S and Cheprakov A *New J. Phys.* **10** 013007
[6] Singh-Rachford T and Castellano F 2008 *J. Phys. Chem. A* **112** 3550–3556
[7] Chen Y Y, Khoury T, Clady R, Ekins-Daukes N J, Crossley M J and Schmidt T W 2009 *In Preparation*
[8] Monguzzi A, Mézyk J, Scotognella F, Tubino R and Meinardi F 2008 *Phys. Rev. B* **78** 195112
[9] Yakutkin V, Aleshchenkov S, Chernov S, Miteva T, Nelles G, Cheprakov A and Baluschev S 2008 *Chem. Eur. J.* **14** 9846–9850
[10] Islangulov R R, Lott J, Weder C and Castellano F N 2007 *J. Am. Chem. Soc.* **129** 12652–12653
[11] Monguzzi A, Tubino R and Meinardi F 2009 *J. Phys. Chem. A* **113** 1171–4

Journal of Physics: Conference Series **185** (2009) 012003 doi:10.1088/1742-6596/185/1/012003

Electron-impact ionisation of atoms and ions

I. Bray, D. V. Fursa and A. T. Stelbovics

ARC Centre for Antimatter-Matter Studies, Institute of Theoretical Physics, Curtin University, Perth 6845, Australia

E-mail: I.Bray@curtin.edu.au

Abstract. We report on the recent applications of the convergent close-coupling method to electron impact ionisation of atoms and ions. Total and fully differential cross sections are considered. Some excellent agreement with experiment is identified, but some long-standing discrepancies still remain.

Electron-impact ionisation of atoms and ions is of fundamental and practical interest. On the fundamental side, the interactions involve at least three charged particles. Such problems defied adequate formulation since the inception of Quantum Mechanics. Only very recently a workable formulation has been given [1]. On the practical side, ionisation cross sections are vital in modeling of artificial and astrophysical plasma processes. Electron-impact single ionisation cross sections range from the fully differential, where the energies and angles of the two outgoing electrons are specified, to the total, where all the angles and energy-sharings are integrated over. In this paper we review some recent applications of the convergent close-coupling (CCC) method to electron-impact ionisation. Note that the close-coupling method was originally intended for the calculation of excitation processes, thereby avoiding any issues associated with the boundary conditions in breakup problems with all charged particles. However, an extension of the CCC method to fully differential ionisation cross sections is particularly straightforward, see [2, 3, 4].

We begin by considering the e-Na ionisation problem. The sodium atom is treated as a one active electron above an inert Hartree-Fock core. A small polarisation potential is also added to take into account virtual core excitations. Such a treatment is sufficient to yield very accurate e-Na excitation cross sections [5]. Application to ionisation is a little problematic due to the possibility of ejecting a core electron, a significant contribution at sufficiently high energies. However, fully differential measurements ensure by energy conservation that it is the valence electron that is being ejected. Such measurements were performed by Murray [6] who measured the coplanar symmetric geometry for the case of equal energy outgoing electrons. Owing to the relatively small ionisation threshold of 5.1 eV, excitation of large angular momentum states becomes more probable, requiring some of the largest to date close-coupling calculations to obtain convergence. The details of the convergent close-coupling (CCC) calculations have been given by Bray *et al* [7], and the comparison with experiment is presented in figure 1.

The excellent agreement between the CCC theory and experiment for fully differential ionisation cross sections suggests that the same should hold for total ionisation cross sections. Indeed this is the case for atomic hydrogen, whether in the ground (1s) or excited (2s) state, see figure 2. In fact we are unaware of any significant discrepancies between CCC and experiment for any electron-impact ionisation processes. What the figure shows is that there is nothing particularly special about the choice of the initial state, so long as convergence in the target states used to expand the total wavefunction is obtained.

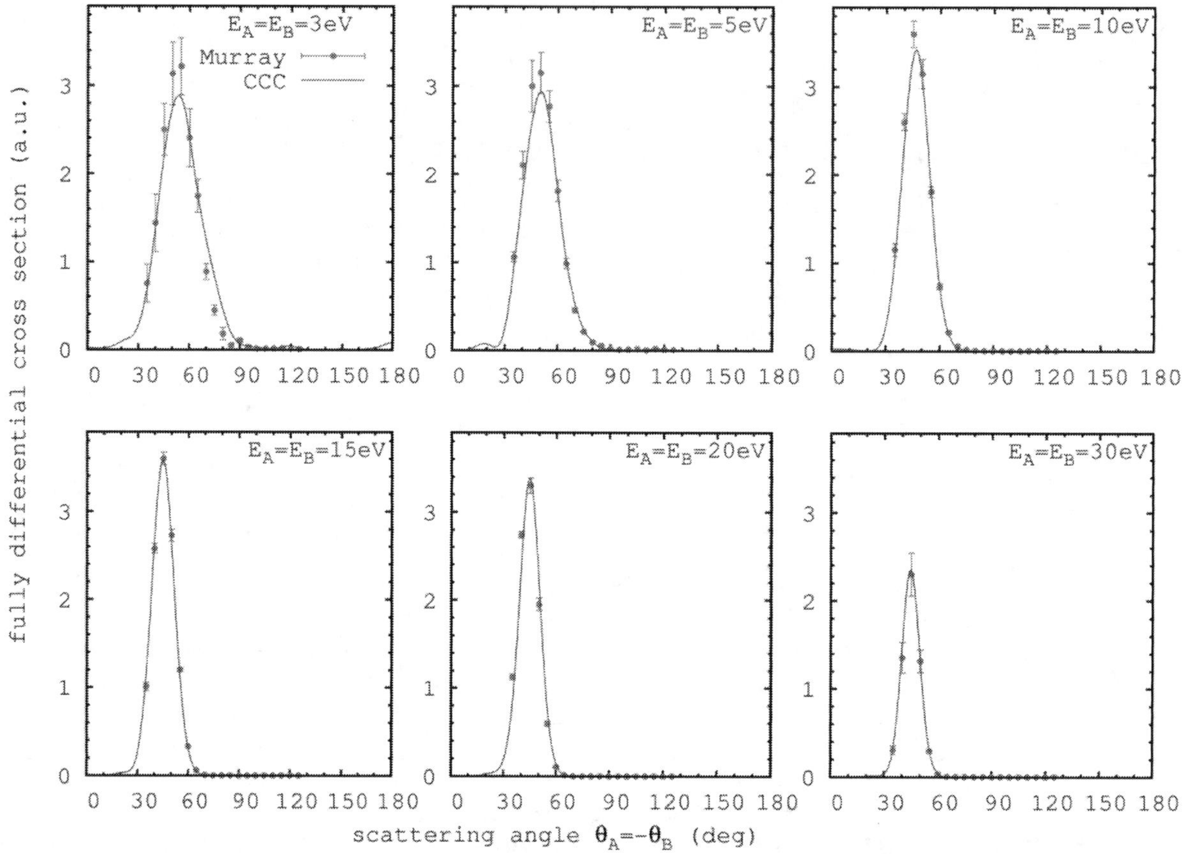

Figure 1. Electron-impact on sodium fully differential equal energy-sharing ionisation cross section in the symmetric geometry. The experimental data and calculations are due to Murray [6] and Bray *et al* [7].

The situation is not quite so simple if we consider helium, the next simplest target. In this case there is a substantial discrepancy between CCC and experiment, though only for the case where the initial state is 2^3S, see figure 3. This discrepancy is a long standing one with other similarly sophisticated theories in agreement with the CCC results [10].

Clearly, a new measurement of the e-He(2^3S) ionisation is desirable. While this has not yet happened a closely related measurement, that of electron impact ionisation of helium-like lithium (Li^+), has been performed for the ground state and a mix of the ground and metastable 2^3S initial states [11]. The CCC theory for He is only marginally modified by the fact the the Li^+ is a singly charged ion. If the theory can yield good agreement with measurements of e-Li^+(2^3S) ionisation then we can be even more confident in the corresponding He results.

By choosing different ways of creating Li^+ Borovik *et al* [11] were able to control whether the target was prepared in a pure ground 1^1S state, or a controlled mixture of the 1^1S and 2^3S states. Since the ionisation cross sections of the metastable states are much larger than for the ground state, and their thresholds are at much lower energies, their signal is unmistakable even for relatively small fractions in the target beam. In figure 4 we compare the experiment for the two cases with the CCC calculations. The left panel shows excellent agreement when the

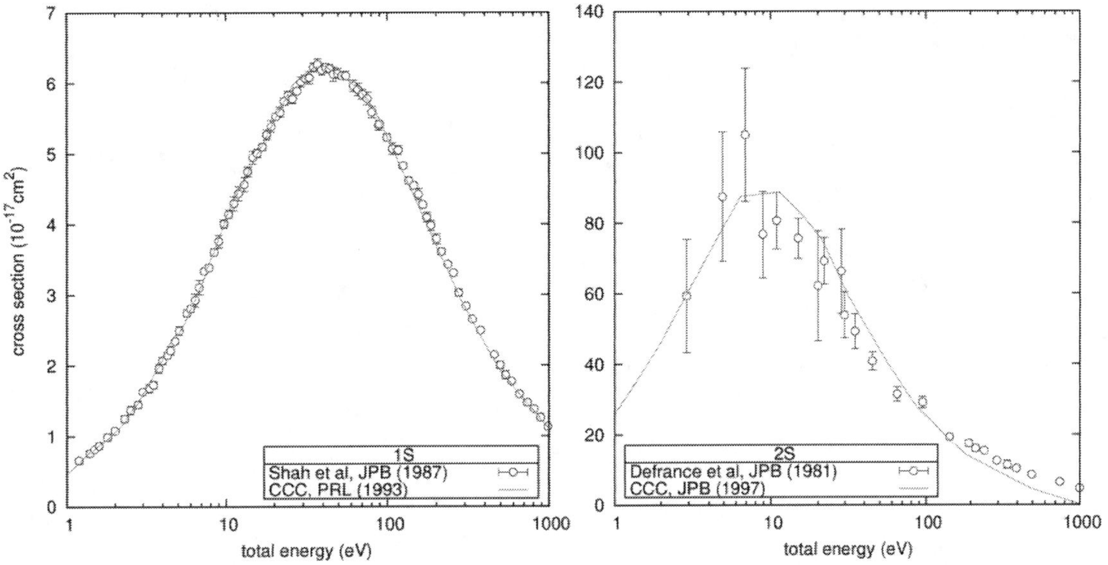

Figure 2. Electron-impact of atomic hydrogen total ionisation cross sections for the ground and 2s states. The measurements are due to Shah *et al* [8] and Dixon *et al* [9]. The calculations are due to Fursa and Bray [10].

beam contains pure Li^+ in the ground state. The right panel shows excellent agreement also, assuming that the beam contains around 13% of the metastable 2^3S state. Such agreement over the broad energy range is only possible if the theory correctly obtains the cross sections for both initial states.

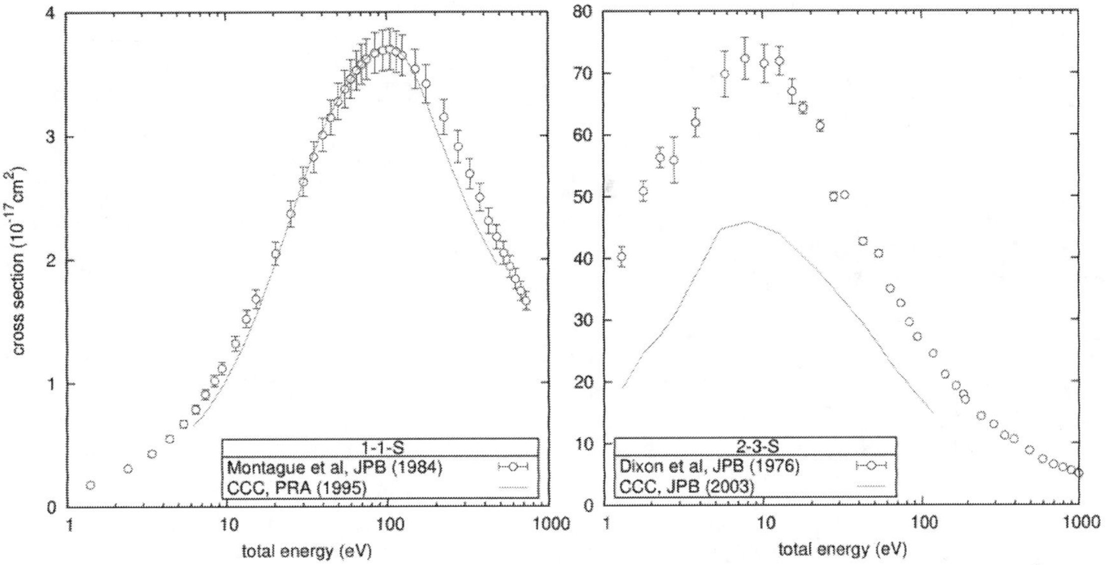

Figure 3. Electron-impact of helium total ionisation cross sections for the ground and 2^3S states. The measurements are due to Shah *et al* [12] and Defrance *et al* [13]. The calculations are due to Bray and Stelbovics [14] and Bartschat and Bray [15].

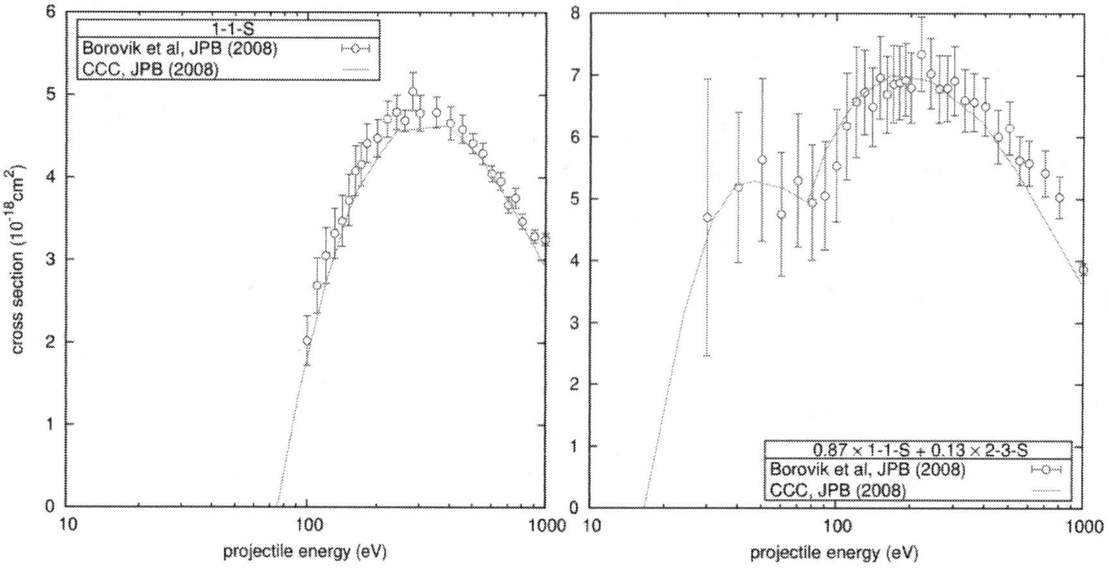

Figure 4. Electron-impact ionisation of helium-like lithium in the ground and a mixture of the ground and 2^3S states. The experiment and calculations have been recently presented by Borovik *et al* [11].

To conclude, the CCC theory has been shown to yield accurate fully differential ionisation cross sections for Na, a target more complicated than H or He. Application to e-$Li^+(2^3S)$ suggests that the CCC theory should also yield correct e-He(2^3S) cross sections. Nevertheless, an independent experimental verification is still desirable.

Acknowledgments

This work was supported by the Australian Research Council and Curtin University. We are grateful for access to the Australian National Computational Infrastructure and its Western Australian node iVEC.

References

[1] Kadyrov A S, Bray I, Mukhamedzhanov A M and Stelbovics A T 2008 *Phys. Rev. Lett.* **101** 230405 (pages 4)
[2] Bray I and Fursa D V 1996 *Phys. Rev. A* **54** 2991–3004
[3] Bray I 2002 *Phys. Rev. Lett.* **89** 273201
[4] Stelbovics A T, Bray I, Fursa D V and Bartschat K 2005 *Phys. Rev. A* **71** 052716(13)
[5] Bray I 1994 *Phys. Rev. A* **49** 1066–1082
[6] Murray A J 2005 *Phys. Rev. A* **72** 062711
[7] Bray I, Fursa D V and Stelbovics A T 2008 *J. Phys. B* **41** 215203
[8] Shah M B, Elliot D S and Gilbody H B 1987 *J. Phys. B* **20** 3501–3514
[9] Dixon A J, Harrison M F A and Smith A C H 1976 *J. Phys. B* **9** 2617–2631
[10] Fursa D V and Bray I 2003 *J. Phys. B* **36** 1663
[11] Borovik Jr A, Müller A, Schippers S, Bray I and Fursa D V 2009 *J. Phys. B* **42** 025203 (10pp)
[12] Shah M B, Elliot D S, McCallion P and Gilbody H B 1988 *J. Phys. B* **21** 2751–2761
[13] Defrance P, Claeys W, Cornet A and Poulaert G 1981 *J. Phys. B* **14** 111–117
[14] Bray I and Stelbovics A T 1993 *Phys. Rev. Lett.* **70** 746–749
[15] Bartschat K and Bray I 1996 *J. Phys. B* **29** L577–L583

Journal of Physics: Conference Series **185** (2009) 012004 doi:10.1088/1742-6596/185/1/012004

Theoretical studies of photoassociation in ultracold metastable helium

D G Cocks and I B Whittingham

School of Mathematics, Physics and Information Technology, James Cook University, Townsville, Australia, 4811

E-mail: daniel.cocks@jcu.edu.au, ian.whittingham@jcu.edu.au

Abstract. Line shifts and line shapes for photoassociation of spin-polarised metastable helium to long-range vibrational states in the $J = 1, 0_u^+$ potential dissociating to the $2s\,{}^3\mathrm{S}_1 + 2p\,{}^3\mathrm{P}_0$ limit are studied using a nonperturbative multichannel calculation valid for arbitrary laser intensities. Asymptotically-free dressed states of the laser plus matter system are used to obtain the S-matrix elements required to generate the photoassociation profiles. Issues associated with the very shallow nature of the potentials that support the excited states are investigated.

1. Introduction

Photoassociation (PA) experiments in which two interacting ultracold atoms (usually ground state alkali atoms or metastable rare gas atoms) are resonantly excited by a laser to a molecular bound state provide a powerful technique to study the dynamics of ultracold collisions [1]. Since the colliding atoms are so cold, the energy of the initial scattering state is well determined and the resonant laser energies corresponding to transitions to various bound states produce a very high resolution spectrum (< 1 MHz). At this level of precision, energy level shifts induced by the PA laser can be significant [2] and an understanding of the dependence of the energy level shifts upon the laser intensity, polarization and frequency is crucial.

Line profiles for PA in metastable helium (see figure 1) have been observed experimentally by Kim *et al* [3]. A short laser pulse was applied to the magnetically trapped gas, the trapping potential turned off and ballistic expansion used to measure the PA-induced temperature profile of the gas for various laser frequencies. Lorentzian fits to the profiles yielded resonance shifts that varied linearly with laser intensity up to 6 W/cm^2. Difficulties in accurately determining the laser intensity incident upon the gaseous cloud led to large error bars.

Previous theoretical investigations of PA have either relied upon perturbative techniques [4, 2] or have assumed that the laser interaction vanishes at large interatomic separation [5]. To properly treat the PA process, one must use laser-dressed states [6] that asymptotically decouple, thus allowing the S-matrix to be correctly defined. This will be shown to be especially important for the weakly bound levels in the present study.

2. Theory

The eigenstate $|\Psi_j\rangle$ of two atoms colliding in the presence of a radiation field satisfies

$$[\hat{H}_{\mathrm{mol}} + \hat{H}_{\mathrm{int}} + \hat{H}_{\mathrm{rad}}]|\Psi_j\rangle = E|\Psi_j\rangle \tag{1}$$

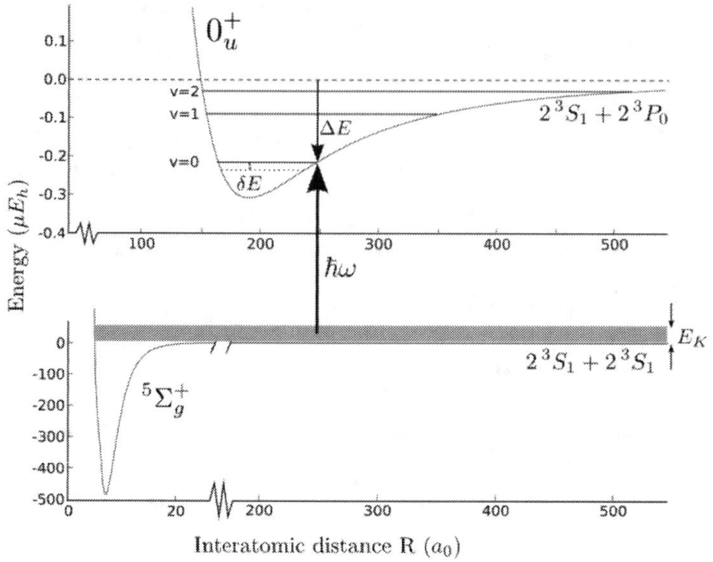

Figure 1. Photoassociation in metastable helium. Two spin-polarized metastable helium atoms ($2s\,^3S_1$+$2s\,^3S_1$) with very low kinetic energy E_K absorb a photon of frequency ω and form a short-lived molecule in the $2s\,^3S_1+2p\,^3P_0\,0_u^+$ excited state. The molecule can then spontaneously decay back to the disassociated metastable state (all other decay paths are spin-forbidden). The energy level shift induced by the laser field is represented by δE and ΔE is the laser detuning energy from the separated atom resonance.

where $\hat{H}_{\rm mol}$ is the total molecular Hamiltonian, $\hat{H}_{\rm rad}$ is the Hamiltonian for a free laser field of angular frequency ω, intensity I and polarization $\boldsymbol{\epsilon}$, and $\hat{H}_{\rm int}=-(\sqrt{I/2\epsilon_0 c})\boldsymbol{d}\cdot\boldsymbol{\epsilon}$ describes the interaction between the laser and the molecular dipole moment $\boldsymbol{d}$. Using the expansion $|\Psi_j\rangle=\sum_i R^{-1}G_{ij}(R)|i\rangle$ in terms of a basis $\{|i\rangle\}$, yields the close-coupled equations

$$-\frac{\hbar^2}{2\mu}\frac{d^2}{dR^2}G_{ij}(R)+\sum_l W_{il}(R)G_{lj}(R)=0 \tag{2}$$

where $W_{il}(R)=\langle i|\hat{H}-E|l\rangle$ and $\hat{H}$ contains all the terms of $\hat{H}_{\rm mol}$ except the radial derivative term.

In choosing the basis, we note that molecular interactions are most naturally described in the molecular frame whereas selection rules for the laser interaction refer to the space-fixed frame. Hence we use the basis $|i\rangle\equiv|\gamma j_1 j_2, j\Omega_j, Jm_J\rangle\otimes|n,\omega,\epsilon\rangle$, where Ω_j is the projection of the total electronic angular momentum j upon the molecular axis and m_J labels the projections of the total angular momentum J in the space-fixed frame. In this basis, W_{il} for the states coupled by σ^- polarised light has the asymptotic $R\to\infty$ form,

$$\begin{bmatrix} -E_K & 0 & 0 & 0 & 0 & 0 & \hbar\Omega_1^0 \\ 0 & -E_K & 0 & 0 & 0 & 0 & 0 \\ 0 & 0 & -E_K & 0 & 0 & 0 & \hbar\Omega_5^{-1} \\ 0 & 0 & 0 & -E_K & 0 & 0 & \hbar\Omega_5^0 \\ 0 & 0 & 0 & 0 & -E_K & 0 & \hbar\Omega_5^1 \\ 0 & 0 & 0 & 0 & 0 & -E_K & 0 \\ \hbar\Omega_1^0 & 0 & \hbar\Omega_5^{-1} & \hbar\Omega_5^0 & \hbar\Omega_5^1 & 0 & -\Delta E-i\frac{\Gamma}{2} \end{bmatrix} \tag{3}$$

where $\hbar\Omega_{2j+1}^{\Omega_j}\equiv\langle i|\hat{H}_{\rm int}|l\rangle$ are the Rabi couplings and Γ is the spontaneous decay width of the excited state.

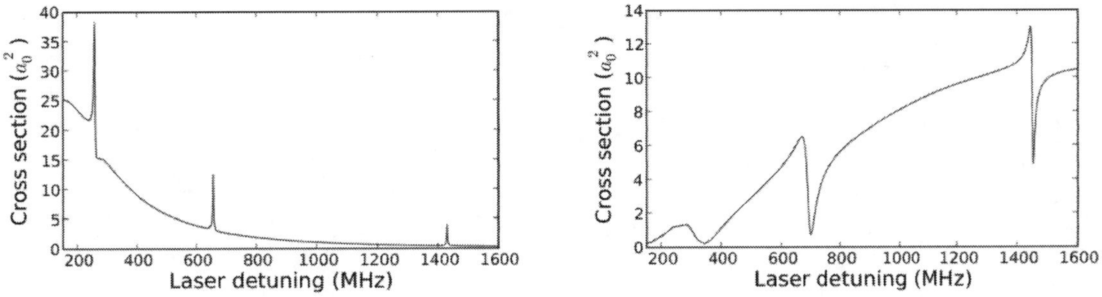

(a) (b)

Figure 2. Spontaneous emission cross section profiles as a function of laser detuning ΔE averaged over degenerate entrance channels for laser intensities of (a) 64 mW/cm^2 and (b) 3.2 W/cm^2.

In this basis the system is not asymptotically uncoupled; in fact the shallow nature of the potential wells (small ΔE), the ultracold temperatures (small E_K) and high laser intensities (large $\hbar\Omega$) can cause the off-diagonal terms to be of the same order of magnitude as the diagonal terms. To properly define the S-matrix, we must transform to a laser-dressed set of basis states $|\beta\rangle = \sum_i U_{\beta i}|i\rangle$ that diagonalise the dressed potential matrix $W^D_{\beta\beta'}$. Although Napolitano [6] was able to analytically form dressed state combinations for the sodium system investigated, the large number of couplings in W_{il} for our system forces us to perform a numerical diagonalisation.

All dressed states potentially include the undressed excited state, resulting in complex energies $E_\beta = -W^D_{\beta\beta'}(R = \infty)$. However, for the current problem, we find only two combinations are formed with the excited state: one a closed channel and the other a 'pseudo-open' channel. The remaining dressed states are degenerate open channels (real E_β) that span a subspace with an arbitrariness that varies with laser frequency. The channel designated as the 'pseudo-open' channel has $\mathrm{Re}(E_\beta) \gg \mathrm{Im}(E_\beta)$, which suggests open-channel behaviour, however the small imaginary component introduces exponential behaviour characteristic of closed channels. In a rigorous treatment, one would be forced to treat this as a closed-channel but for numerical reasons this is not a feasible approach. Napolitano [6] circumvented this problem by allowing a complex total energy, which removed the asymptotic imaginary component but has the side effect of introducing small complex perturbations to the excited states. Although reasonable for the sodium system, the method will introduce additional asymptotic imaginary components in the degenerate open states for the metastable helium system. Instead, we prefer to slightly redefine the scattering matrix, by matching the asymptotic functions to a finite interatomic distance $R = R_{\mathrm{max}}$ instead of $R \to \infty$. Although the particular choice of R_{max} will modify the cross-sections, we do not expect it to modify the profile shape as long as R_{max} is outside the interaction region.

To extract the S matrix elements, we solve the radial equation (2) in the dressed state basis and, for the open and pseudo-open channel scattering states, match to the asymptotic form

$$G_{\beta\beta'}(R) = A_{\beta\beta'}h^-_\beta(R) + B_{\beta\beta'}h^+_\beta(R) \qquad (4)$$

where $h^\pm_\beta \sim 1/\sqrt{(2|k_\beta|)}\exp(\pm ik_\beta R)$ and $k_\beta = \sqrt{2\mu\mathrm{Re}(E_\beta)}/\hbar$. The S-matrix is then $\boldsymbol{S} = -\boldsymbol{B}\boldsymbol{A}^{-1}$ and the cross section for spontaneous photon emission from a particular entrance channel β is obtained from the loss of unitarity of the S-matrix:

$$\sigma^{\mathrm{photon}}_\beta = \frac{\pi}{k^2_\beta}[1 - \sum_{\beta'}|S_{\beta'\beta}|^2]. \qquad (5)$$

Journal of Physics: Conference Series **185** (2009) 012004 doi:10.1088/1742-6596/185/1/012004

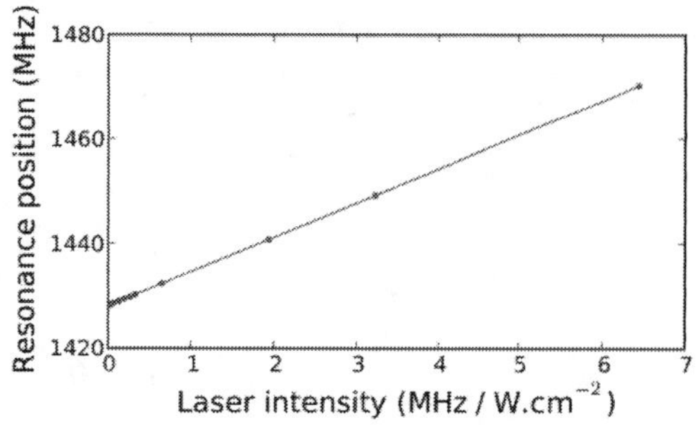

Figure 3. Resonance position of the $v = 0$ vibrational level.

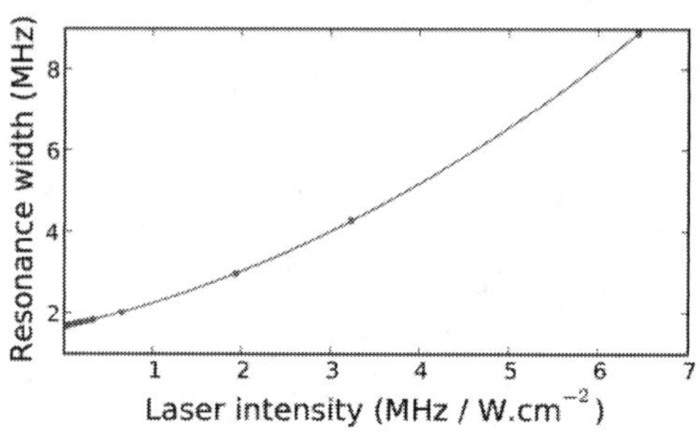

Figure 4. Resonance width for the $v = 0$ vibrational level.

3. Results and discussion

Two sample PA spectra for the $v = 0$ $J = 1, 0_u^+$ vibrational level are shown in figure 2. These spectra exhibit two features not present in previously calculated PA profiles: (i) a non-zero background variation in $\sigma^{\rm photon}$ and (ii) the appearance of interference behaviour at larger laser intensities. The non-zero background appears to arise from the shallow nature of the 0_u^+ potential well as artificially deepening the well by two orders of magnitude completely removes the background. The interference feature can be matched perfectly by a Fano profile [7] of the form

$$A_{\rm bg} - A_{\rm res}\frac{(\epsilon + q)^2}{1+\epsilon^2}, \qquad \epsilon = \frac{E - E_{\rm res}}{\Gamma_{\rm res}/2} \tag{6}$$

where $E_{\rm res}$ and $\Gamma_{\rm res}$ are the position and width respectively of the resonance. The experimental measurements [3] do not show such interference and are fitted by standard Lorentzian profiles. A detailed comparison between actual experimental profiles and theoretical predictions needs further investigation.

The dependence of the resonance position and width on laser intensity is shown in figures 3 and 4 respectively. Our calculations indicate that the line shift has a linear dependence and the line width a quadratic dependence. Perturbative calculations of energy level shifts [2] predict a linear dependence and our result of 6.446 MHz / W.cm^{-2} agrees closely with our perturbative result of 6.439 MHz / W.cm^{-2} obtained using the F-operator technique [8] adapted to the laser-matter interaction. Details of this perturbative calculation, together with a more comprehensive account of the nonperturbative calculations presented here, will be the subject of a forthcoming paper.

References

[1] Weiner J, Bagnato V S, Zilio S and Julienne P S 1999 *Rev. Mod. Phys.* **71** 1-85
[2] Portier M, Moal S, Kim J, Leduc M, Cohen-Tannoudji C and Dulieu O 2006 *J. Phys.* B **39** S881-904
[3] Kim J, Moal S, Portier M, Degué J, Leduc M and Cohen-Tannoudji C 2005 *Europhys. Lett.* **72** 548-54
[4] Simoni A, Julienne P S, Tiesinga E and Williams C J 2002 *Phys. Rev.* A **66** 063406
[5] Bohn J L and Julienne P S 1999 *Phys. Rev.* A **60** 414-25
[6] Napolitano R 1998 *Phys. Rev.* A **57** 1164-75
[7] Fano U 1961 *Phys. Rev.* **124** 1866-78
[8] Beams T J, Peach G and Whittingham I B 2006 *Phys. Rev.* A **74** 014702

Journal of Physics: Conference Series **185** (2009) 012005 doi:10.1088/1742-6596/185/1/012005

Post collision interaction effects in the excitation of zinc atoms

D. Cvejanović, S. A. Napier, J. F. Williams, L. Pravica, S. N. Samarin, A. D. Sergeant, P. Guagliardo and P. Wilkie

ARC Centre for Antimatter-Matter Studies, Centre for Atomic, Molecular and Surface Physics, Physics Department, University of Western Australia, Perth 6009 WA, Australia

E-mail: `jfw@cyllene.uwa.edu.au`

Abstract.
Post collision interaction of scattered and ejected electrons in the decay of the $3d^94s^24p$ autoionizing states was observed to contribute to the population of the $4snd\,^3D_{1,2,3}$ states of zinc. Energies of PCI structures and percentage contributions to the emission cross sections have been measured for the $n = 4, 5, 6$ states. Also, the effects of PCI were observed in an energy loss at 13.4 eV and affected both the positions and the width of the energy loss structures. In the energy loss study a strong excitation of the $3d^94s^24p\,^3P_1$ state was observed in the near threshold regime.

1. Introduction

Photon intensity measurements of the emission lines from the electron impact excited states of zinc show remarkable structures in the autoionizing region above 10 eV. Our experiments have shown that the two phenomena of formation of temporary negative ions and Post Collision Interaction (PCI), both governed by strong electron correlations, can significantly affect the cross section when excitation of an inner 3d electron becomes possible [1]. In particular we investigated the population of the $4snd\,^3D_{1,2,3}$ states of zinc via the decay of the $3d^94s^24p$ autoionizing states when excited near their excitation thresholds. That process is made possible by the post collision Coulomb interaction of the slow scattered and ejected electrons while still in field of the Zn^+ ion. The interaction, which involves the exchange of energy and momentum, can lead to the capture of one, usually the slower scattered, electron into a higher energy state.

Here we observe the electron capture into three different excited states of the same orbital and spin momentum, i.e. the $4snd^3D_{1,2,3}$ $n = 4, 5, 6$ states, by measuring the intensity of the decay photons. Also, in the presence of the PCI and depending on the amount of energy exchanged, an autoionizing state can decay by ejection of one electron but with a kinetic energy different from their spectroscopic values. Due to energy conservation, the scattered electron energy would be changed appropriately. This process we observe in the energy loss spectrum at incident energies close to the threshold of the $3d^94s^24p$ autoionizing states. The PCI effect was observed previously in electron impact excitation of noble gases [2, 3] for which different outcomes were discussed.

2. Experiment

The apparatus was based on crossed electron and atom beams. Briefly, an electrostatic 127° analyzer selected a portion of the energy spread of electrons emitted from a tungsten cathode. Using aperture lenses electrons were focused into the interaction region where they were crossed with a Zn atomic beam emanating from a thermally heated oven. Two types of measurements were performed to investigate the electron impact excitation of zinc. First, decay photons, emitted in the direction perpendicular to the incident electron beam, were wavelength selected using an interference filter and detected by a photomultiplier tube. The incident electron energy was scanned and the intensity of the detected photons recorded using a PC-based system. The energy scale was calibrated on known positions of negative ion resonances in zinc [1]. Photon intensities were corrected for background to separate those photons which were not related to the electron impact excitation of zinc. Also, using the same electron impact spectrometer, electrons scattered on zinc atoms at a variable angle θ were energy analyzed using a 127° cylindrical analyzer and detected using a channel electron multiplier.

3. Results and discussion

Photon intensity measurements of the unresolved fine structure emission lines from the $4snd\,^3\mathrm{D}_{1,2,3}$, $n = 4, 5, 6$ states of zinc show remarkable structures in the autoionizing region above 10 eV [1]. The structures, superimposed on a smoothly varying cross section, are identified on the basis of the specific energy dependence of negative ion resonances and of the PCI effects. To show the structures in more detail a smoothly varying cross section is subtracted and the percentage contributions of the remaining structures are shown in figure 1.

At the low energy end of the spectrum for each of the three $^3D_{1,2,3}$ states, the sharp structures labeled a,b and c$''$ appear at the same incident electron energy. These structures correspond to the $3d^94s^24p^2$ negative ion resonances observed in the excitation of different neutral states of zinc, [4, 5, 1], in electron transmission and angle differential elastic scattering [6]. The broad structures observed at higher energies are associated with the PCI effect in the decay of the $3d^94s^24p$ autoionizing states excited close to their respective thresholds. They move on an energy scale with increase of the n quantum number of excited electron in observed channel, i.e. with decrease of neutral excited state binding energy, as can be seen in the three spectra in figure 1. In the decay of the autoionizing state the ejected electron and the slow scattered electron, while still in the field of the residual ion, interact and exchange energy and angular momentum. Various outcomes of this interaction were discussed [2] for the noble gasses He, Ne and Ar. Here we observe the situation where one of the outgoing electrons loses more energy than its kinetic energy leading to recapture in the $4snd\,^3D_{1,2,3}$ states of the neutral Zn atom. PCI appears in the cross section as an interference of direct excitation and PCI mediated population of these states via excitation of one of the autoionizing states. In the excitation of the triplet D-states in zinc this leads to significant enhancement of the cross section as can be seen in figure 1.

It follows from the nature of the post collision interaction that the energies of autoionizing states associated with observed PCI effects have to be close to, but not higher than, the onset of the observed structures as indicated in figure 1(a)-(c). Of the many autoionizing states with the $3d^{10}4p^2$ and $3d^94s^24p$ electron configurations situated in the energy region below the observed PCI structures, we consider as likely contributors to excitation of the $4snd^3\mathrm{D}_{1,2,3}$ states of zinc, those excited with the largest cross sections in the relevant energy region. Extensive spectroscopic information, including the energies and relative intensities of both the optically allowed and optically forbidden states, was obtained by Back *et al* [7]. We used here the ejected electron spectrum at an electron impact energy of 16 eV and ejected electron angle of 60° later interpreted by Mansfield [8]. This ejected electron spectrum at kinematical conditions close to our experiment indicates appreciable excitation of the states with $3d^94s^24p$ electron configuration. The only exception is the $3d^{10}4p^2\,^1S$ state at an excitation energy of 11.50 eV.

We use observed ejected electron intensities to gain an insight into which states are most likely to contribute to the observed PCI. Below the onset of the first PCI structure at 11.64 eV, by far the strongest contribution comes from the $3d^9 4s^2 4p\,^3P_1$ (11.186 eV) autoionizing state decaying with the ejection of 1.793 eV electrons. At higher energies, but below the onset of the second PCI process at 12.68 eV, a group of five more autoionizing states was observed. Their spectroscopic energies and corresponding ejected electron energies are shown in figure 2 (b) and (c).

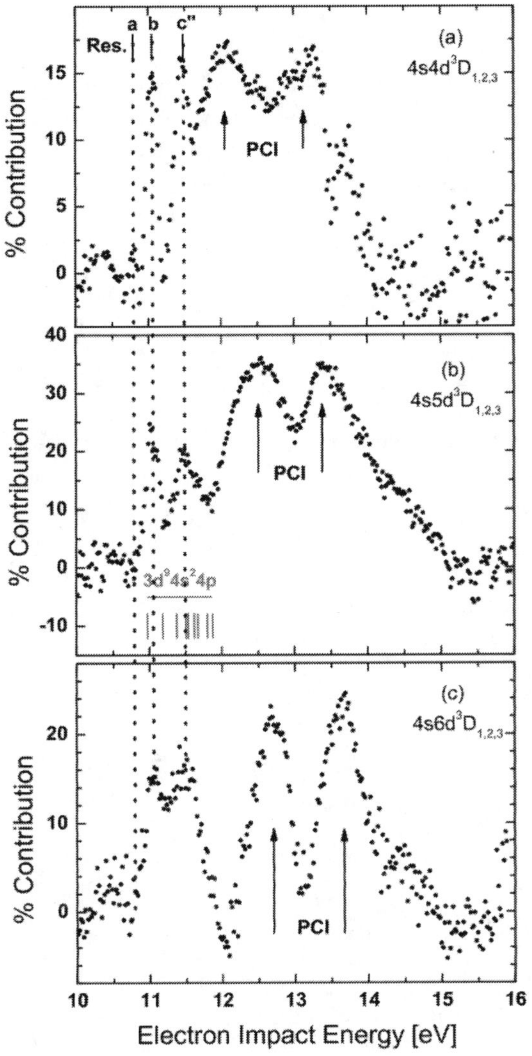

Figure 1. Resonance and PCI structures observed in photon excitation functions.

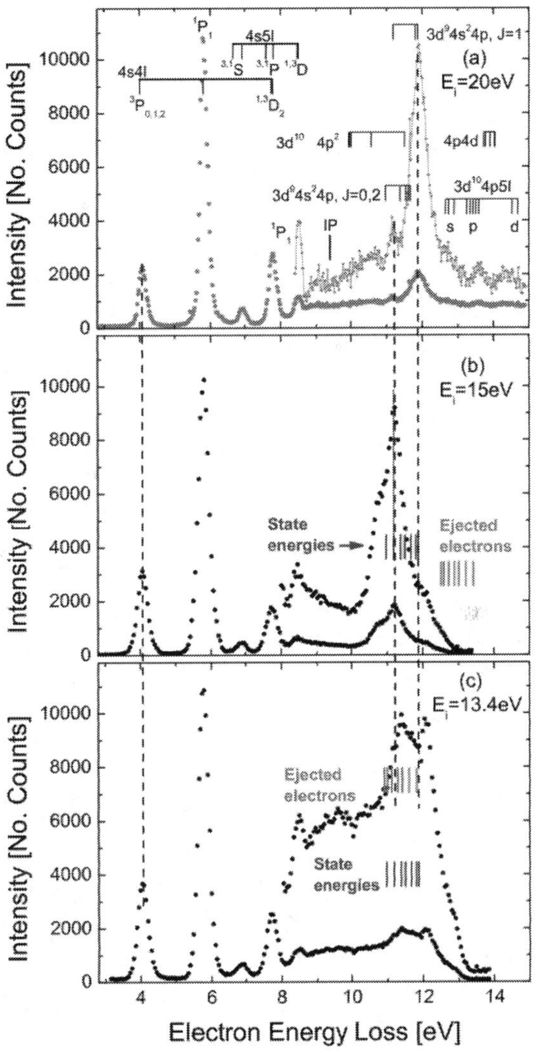

Figure 2. Electron energy loss at θ=40° E_i=20, 15 and 13.4 eV.

Providing an adequate energy resolution is available, additional information on autoionizing states, their near threshold excitation and decay, and also their relative cross sections, can be obtained from energy loss spectra like those shown in figure 2. The three spectra were measured at the same scattering angle of 40°, but at three different electron impact energies, E_i=20, 15 and 13.4 eV. The choice of energy permits the separation or simultaneous detection of both the scattered and ejected electrons from the process of autoionization while approaching the excitation threshold. The three spectra in figure 2 include also the states below the ionization threshold and the energy resolution, low background signal and energy loss scale are indicated.

The 20 eV energy loss spectrum in figure 2(a) is measured far from excitation thresholds where the excitation of optically allowed transitions dominates the spectrum. At an incident electron energy of 15 eV, closer to excitation thresholds, the relative intensities of autoionizing states have changed, the optically forbidden $3d^94s^24p\,^3P_{0,1,2}$ states are strongly excited. The J=1 state is the most prominent in agreement with ejected electrons intensities from Back *et al* [7] and is almost certainly responsible for the lower PCI peak. The spectrum still displays well defined energy loss features in the autoionizing region. In contrast the energy loss spectrum at 13.4 eV, while being still of the same quality below the ionization threshold at 9.39 eV, shows an apparent fuzziness in the region of autoionizing states between 11 and 13 eV.

For interpretation of energy loss spectra in figure 2(a)-(c) we consider the kinetic energy of the detected electron (the residual energy of the scattered electron) and evaluate where ejected electrons may appear in our energy loss spectrum. The positions of ejected electron peaks, calculated using their spectroscopic energies, are shown also on spectra in figure 2(b) and (c). As can be seen, the energy loss peaks corresponding to scattered and ejected electrons will overlap at incident electron energy of 13.4 eV. The kinetic energy of ejected electrons has a fixed spectroscopic value at some distance from threshold, but not near the threshold due to PCI, and the same is true for the scattered electrons after excitation of an autoionizing states susequently decaying by autoionization. For both types of structures, the energy loss and ejected electron peaks are shifted and broadened, and the spectrum appears substantially different and appears to have lower energy resolution.

In contrast to PCI observed in noble gases, where ejected electrons have considerable kinetic energy and gain energy, while scattered electron is slow and loses energy, a specific situation occurs in zinc. In zinc, both the scattered and the ejected electron have relatively small energies and spend a substantial time in the vicinity of the residual ion while interacting. As a consequence, high values of angular momentum can be transferred [9]. An order of magnitude estimate [1, 3, 9] shows that for electron energies in zinc a transfer of $\Delta J \leq 2.6\ \hbar$ is possible explaining the strong effect on excitation of the D states. Also due to similar energies electrons are expected to be more correlated both in the escape and in re-capture processes.

Acknowledgments
This work was supported by the Australian Research Council and University of Western Australia, SAN was supported by an Australian Postgraduate Award.

4. References

[1] Napier S A, Cvejanovic D, Williams J F and Pravica L 2008 *Phys. Rev. A* **78** 032706
[2] King G C, Read F H and Bradford R C 1975 *J. Phys. B: At. Mol. Phys.* **8** 2210
[3] van Ittersum T, Heideman H G M, Nienhuis G and Prins J 1976 *J. Phys. B: At. Mol. Phys.* **9** 1713
[4] Napier S A, Cvejanovic D, Williams J F and Pravica L 2007 *J. Phys. B: At. Mol. Opt. Phys.* **40** 1323
[5] Pravica L, Cvejanovic D, Williams J F and Napier S A 2007 *Phys. Rev. A* **75** 030701(R)
[6] Napier S A, Cvejanovic D, Burrow P D, Williams J F, Michejda J A and Pravica L 2009 *To be published*
[7] Back C G, White M D, Pejčev V and Ross K J 1981 *J. Phys. B: At. Mol. Phys.* **14** 1497–1507
[8] Mansfield M W D 1981 *J. Phys. B: At. Mol. Phys.* **14** 2781–2792
[9] Fano U 1974 *J. Phys. B: At. Mol. Phys.* **7** L401–4

Studies of whispering gallery modes of single, coated microspheres by steady-state and time-resolved laser fluorescence microscopy

Venkata Ramanaiah Dantham and Prem Ballabh Bisht*

Indian Institute of Technology, Madras, Chennai – 600 036, India

*E-mail: bisht@iitm.ac.in

Abstract. Dye coated single silica and barium titanate microcavities exhibit the whispering gallery modes (WGMs) in their fluorescence spectra. These modes are extremely sensitive to the size, shape and refractive index of the microcavity as well as of the coating thickness (mantle). Theoretical simulations on WGM have been carried out based on the Aden and Kerker model of Lorentz-Mie theory. The effects of the mantle thickness and the surrounding medium on WGMs have been studied. The variation in the decay times of dye molecules with the microcavity size and the mantle thickness have been studied.

1. Introduction

Small optical resonators, of micrometer dimensions with large Q value have diverse technological and scientific applications such as biosensing [1-4], low threshold lasing [5] and enhanced energy transfer [6-10] in the visible region. While undergoing total internal reflection, the electromagnetic wave propagates along the surface to create the standing wave field, known as the whispering gallery mode (WGM). If an atom is embedded or placed in the vicinity of an optical microcavity, its radiative behavior is significantly altered. The Q value and cavity induced rate alteration can be much higher in spherical microcavities than in planar cavities because of the 3D optical confinement.

Chew *et al.* have proposed the theoretical expressions for the radiative rates of atoms in the vicinity of the microsphere by using classical theory [11] while Klimov *et al.* have given these expressions by using quantum theory [12]. However, work on coated microspheres is scarce in the literature [13, 14].

2. Experimental

The standard laser dyes rhodamine B (RB) and N, N'-bis (2, 5-di-tert-butylphenyl)-3, 4:9, 10-Perylenebis (dicarboximide) (DBPI) (Aldrich) were used as received. The solvents (methanol and benzene) were obtained from Qualigens Fine Chemicals and Central Drug House Ltd respectively. Silica (Duke Scientific Corp.) and barium titanate (Mo-sci Corp.) microspheres were immersed in the dye solutions (10^{-4}M) for more than 10 hours. These microspheres were then filtered and dried under ambient conditions for a few hours. After completely drying, the microspheres were transferred to a glass slide for the measurement.

Single microspheres were selected with the help of the epifluorescence microscope (NIKON Eclipse E-400). The excitation source was a He-Ne laser (543.5 nm). The spectra were measured with a combination of a PC controlled spectrograph (Spectrapro 2750, PI Acton) and CCD detector

Journal of Physics: Conference Series **185** (2009) 012006 doi:10.1088/1742-6596/185/1/012006

(PIXIS, PI Acton). Decay time measurements were done with the technique of time-correlated single photon counting (TCSPC) technique [15].

3. Theoretical aspects

The separation between the adjacent WGMs (or free spectral range, FSR) for a coated sphere is approximately given by [14]

$$\Delta\lambda = \left(\frac{\lambda^2}{2\pi m_2 r_1}\right)\tan^{-1}\left[\left(\left(\frac{m_1}{m_2}\right)^2 - 1\right)^{\frac{1}{2}}\right] \Big/ \left(\left(\frac{m_1}{m_2}\right)^2 - 1\right)^{\frac{1}{2}} \quad (1)$$

where λ and r_1 represent the wavelength of the light and the radius of the microsphere, respectively. m_1 and m_2 are the relative refractive index of the core and mantle with respect to the surrounding medium (m_{med}), respectively (figure 1). Aden and Kerker have [16] solved the problem of scattering from a layered sphere using Mie – Debye theory. The scattering cross section is given by

$$\sigma_s = \frac{2\pi}{|k|^2}\sum_{n=1}^{\infty}(2n+1)\left(|a_n|^2 + |b_n|^2\right) \quad (2)$$

where k is the complex propagation constant. a_n and b_n are the scattering coefficients.

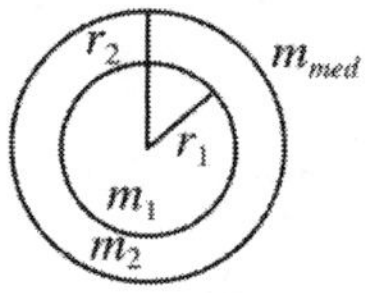

Figure 1. Schematic of the coated microsphere. Here r_1-r_2 gives the thickness of the mantle (see text for details).

4. Results and discussion

A small portion of the fluorescence spectrum of a RB coated single silica microsphere of diameter ~ 8 μm is shown in figure 2 (panel A). The observed spectrum matches reasonably well with that obtained from the simulations as indicated in panel B. The mode assignment is done by using the procedure given elsewhere [17].

4.1. Effect of refractive index of the microspheres

The separation between the adjacent WGMs (FSR) decreases with increase in the refractive index of the microspheres. This is shown in figure 3. The WGMs in the fluorescence spectrum of RB coated single barium titanate microsphere ($n \sim 2.1$) are much sharper than those observed for the case of silica microsphere ($n \sim 1.45$) coated with the same dye. The Q values for the barium titanate microspheres are approximately double those of the silica microspheres.

4.2. Effect of refractive index of the surrounding medium and the mantle thickness

Fluorescence spectra of the dye DBPI coated single silica microsphere of size ~20 μm surrounded by water ($m_{\text{med}} \sim 1.33$) and air ($m_{\text{med}} \sim 1$) are shown in figure 4. The separation between the WGMs is large in water medium as compared to that in air. This is because the separation decreases with the relative refractive index of the microsphere (equation (1)). The full width at half maximum (FWHM) of the WGMs is found to be larger in water medium as compared to that in air. The Q value decreases by a factor of 2.6 from the case of water. The ratio of the peak fluorescence to the background fluorescence is 1.9 in air as compared to 1.1 in water medium. The observed Q value for a microsphere is given by [18]

$$\frac{1}{Q} = \frac{1}{Q_0} + \frac{1}{Q_a} \quad (3)$$

where Q_0 represents the quality factor associated with the real value of the refractive index while Q_a is associated with the imaginary refractive index of the microsphere due to the self-absorption. For

Journal of Physics: Conference Series **185** (2009) 012006 doi:10.1088/1742-6596/185/1/012006

thin coatings of the order of a few tens of nm, Q is approximately equal to Q_0. For thicker coatings (obtained from the concentrated solutions) the quality factor decreases due to appearance of Q_a. This can be seen in figure 5. The Q value as well as the ratio of peak intensity to the background fluorescence decrease nearly by a factor of 3 as the concentration increases from 10^{-4} M to 10^{-2} M.

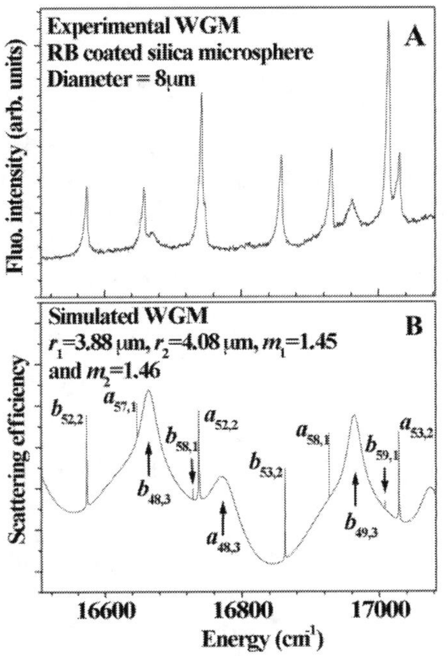

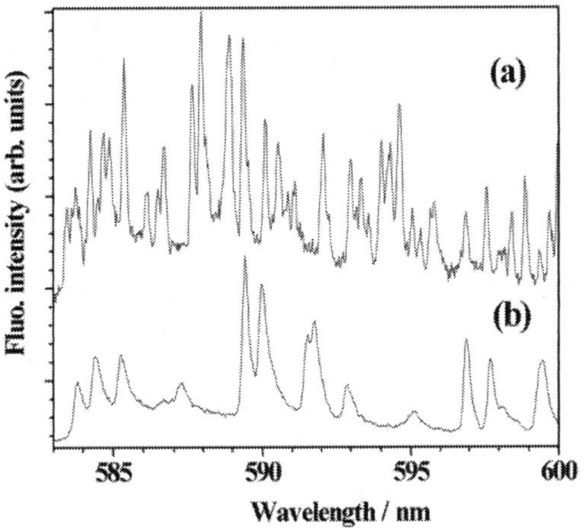

Figure 3. Observed WGMs in the fluorescence spectra of RB coated (a) single barium titanate (m_1 = 2.1) microsphere and (b) silica (m_1=1.45) microsphere of diameter ~ 10 μm.

Figure 2. Fluorescence spectrum of RB coated silica microsphere sitting on a glass plate (panel A). Panel B shows the calculated scattering efficiency for a coated microsphere.

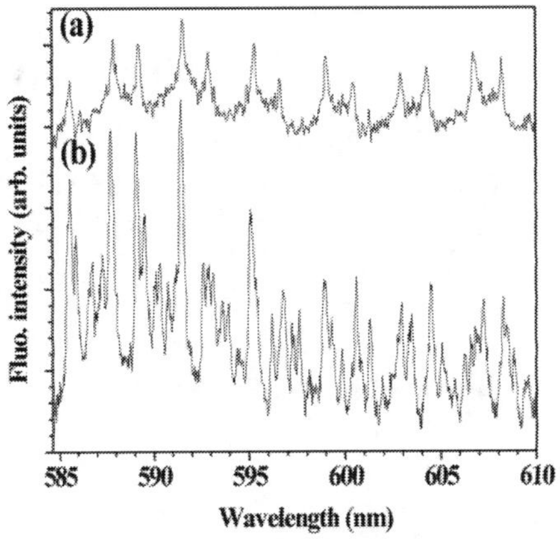

Figure 4. Observed WGMs in the fluorescence spectra of DBPI coated single silica microsphere of diameter ~ 20 μm kept in (a). water (b). air.

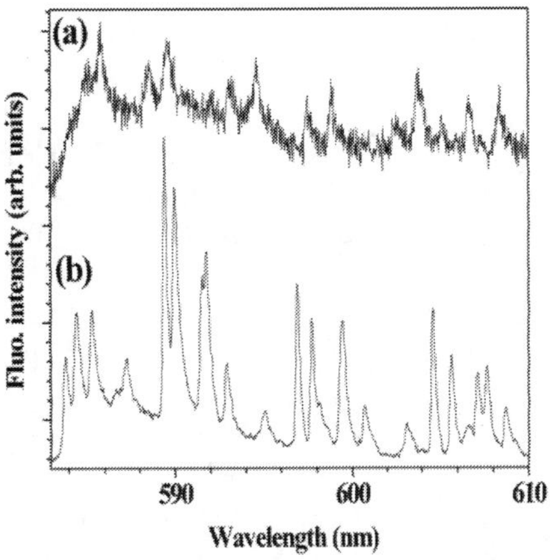

Figure 5. Observed intensities of WGM in the fluorescence spectra of single silica microsphere of diameter 10 μm coated with two different concentrations of dye RB (a). 10^{-2}M and (b). 10^{-4}M.

Journal of Physics: Conference Series **185** (2009) 012006 doi:10.1088/1742-6596/185/1/012006

4.3. Cavity effects on the decay times of the RB molecules coated on silica microspheres

The decay profiles of microspheres coated with different concentrations (10^{-3} M and 10^{-4} M) of RB were recorded. The fluorescence intensity (I(t)) exhibits a bi-exponential behaviour

$$I(t) = B_1 e^{-t/\tau_1} + B_2 e^{-t/\tau_2} \tag{4}$$

where B_1 and B_2 are the amplitudes of the fluorescence lifetime components τ_1 and τ_2 respectively. The fluorescence decay for a coated microsphere of 25 μm diameter exhibits two components viz., 1.3 ns and 3.2 ns, respectively. The fractional intensity of the slow component dominates here. On decreasing the size to 18 μm, the corresponding decay times become 1.6 ns and 3.8 ns, respectively with approximately same fractional intensities. This decrease in the decay time is observed only for the microsphere coated with a lower concentration (10^{-4} M) where the coating thicknesses are of the order of a few tens of nm. Since the quantum yield of RB is close to unity, and there is no other nonradiative process that can appear on reducing the size, this behaviour indicates towards the inhibition of the radiative rate for the smaller sized microcavities. For the microcavity coated with higher concentration (10^{-3} M), the decay times do not change appreciably with the size and are close to the decay times of the single crystal of 8 μm x 13 μm size.

Conclusions

Fluorescence spectra of dyes coated on single microspheres exhibit WGMs. The FWHM of the WGMs are controlled by the coating thickness, refractive index of the microsphere and the medium. Single silica microspheres coated with low concentrations show inhibition of the decay rate at smaller sizes due to thin coatings. High concentration coatings fail to show this effect due to presence of the imaginary refractive index.

Acknowledgement

We thank Department of Science and Technology (DST), New Delhi for financial support.

References

[1] Sandeep P and Bisht P B 2005 *Chem. Phys. Lett.* **415** 15-19
[2] Vollmer F, Braun D, Libchaber A, Khoshsima M, Teraoka I and Arnold S 2002 *Appl. Phys. Lett.* **80** 4057-59
[3] Arnold S, Khoshsima M, Teraoka I, Holler S and Vollmer F 2003 *Opt. Lett.* **28** 272-274
[4] Armani A M, Kulkarni R P, Fraser S. E. Flagan R C and Vahala K J 2007 *Science* **317** 783-787
[5] Spillane S M, Kippenberg T J and Vahala K J 2002 *Nature* (London) **415** 621-623
[6] Druger S D, Arnold S and Folan M 1987 *J. Chem. Phys.* **87** 2649-59
[7] Agarwal G S and Gupta S D 1998 *Phys. Rev. A* **57** 667-670
[8] Pandey K K and Hirayama S J 1996 *Photochem. and Photobio. A: Chem.* **99** 165-175
[9] Kiraz A, Doğanay, Kurt A and Demirel A L 2007 *Chem. Phys. Lett.* **444** 181-185
[10] Fujiwara H, Sasaki K and Mahasura H 2005 *Chem. Phys. Chem.* **6** 2410-16
[11] Chew H 1987 *J. Chem. Phys.* **87** 1355-60
[12] Klimov V V, Ducloy M and Letokhov V S 1996 *J. Mod. Opt.* **43** 2251-67
[13] Shibata S, Ashida S, Segawa H and Yano T 2006 *J. Sol-Gel Sci. Techn.* **40** 379-384
[14] Wang J, Wang M, Liu L, Hao W, Hou B and Lu Y 2007 *J. Lumin.* **122** 949-950
[15] Sandeep P and Bisht P B 2005 *J. Chem. Phys.* **123** 204713-7
[16] Aden A L and Kerker M 1951 *J. Appl. Phys.* **22** 1242-46
[17] Dantham V R and Bisht P B 2008 *J. Opt. Soc. Am. B.* (in press)
[18] Chýlek P, Lin H B, Eversole J D and Campillo A J 1991 *Opt. Lett.* **16** 1723-25

Electron emitted in transfer ionization in He^{2+} on helium collisions

W. T. Feng[1,2], X. Ma[1], S. F. Zhang[1,2], X. L. Zhu[1], S. Y. Xu[1,2], B. Li[1,2], S. C. Yan[1,2], L. J. Meng[1,2], D. C. Zhang[1], D. B. Qian[1], H. P. Liu[1]

[1]Institute of Modern Physics, Chinese Academy of Sciences, Lanzhou, 730000, China

[2]The Graduate University of the Chinese Academy of Sciences,Beijing, 100049, China

E_mail: x.ma@impcas.ac.cn

Abstract. A complete experiment for He^{2+} on helium collisions was performed over a relatively broad energy range by using a reaction microscope in Lanzhou in order to clarify the emission properties. In the experiment the momenta of recoil ions, the emitted electron, and the charge state of the scattered projectiles were recorded in coincidence in event-by-event mode. By reconstruction of the momentum vector of electrons, the momentum distributions of the continuum electrons in the scattering plane were obtained. The distributions show characteristics of the electrons originated from transfer ionization in the collisions.

1. Introduction

The electron emission into the continuum in slow ion-atom collisions is of great interest [1, 2]. Transfer ionization is a two-electron processes where one target electron is captured into the bound state of the projectile and the other target electron is emitted into the continuum states. Several papers showed recently the emission characteristic of the continuum electron in single ionization and transfer ionization by total and differential energy spectra [3, 4] as well as by momentum imaging methods [5, 6]. We performed a complete experiment for He^{2+} on helium collisions over a relatively broad energy range (70 – 400 keV) by electron momentum imaging in order to clarify the emission properties. Here we report on the momentum distributions of the continuum electrons projected in the scattering plane for 100keV and 400keV He^{2+} impact on helium atoms.

2. Experiment and results

The experiment was performed using the reaction microscope in Lanzhou (ReMiLa) [7, 8]. It was built on the atomic beam line of 320 kV platform for research with highly charged ions at the institute of modern physics [9]. The helium target beam is provided by a two-stage supersonic gas-jet, which is cooled to liquid nitrogen temperature. The background pressure in collision chamber is maintained to 6.5×10^{-8} mbar when target gas is supplied, the target density is estimated to be about $10^{11}-10^{12}$ atom/cm^2 at collision zone. The charged products are extracted by a weak electric field D2 of 3 V/cm which is perpendicular to the gas-jet and the projectile beam as shown in figure 1. A 3-dimension focusing design is employed to recoil ion optics to reduce the influence on the recoil time and the position resolution due to the extended collision zone. Separate delay-line micro-channel plate detectors with an effective diameter of 75 mm are used for both the recoil ion and the electron detection. The scattered beam is analyzed by an electric deflector and projected to the third position sensitive detector. All the timing and position signals are handled by front end electronics and then fed into the PXI data acquisition system, and recorded and stored in event-by-event mode. A complex of Helmholtz coils was installed to produce magnetic field which confines most of the electrons traveling along spiral trajectories to the detector. According to its flight time and the position on the detector the three dimensions of momentum can be reconstructed. See details in ref [7, 10].

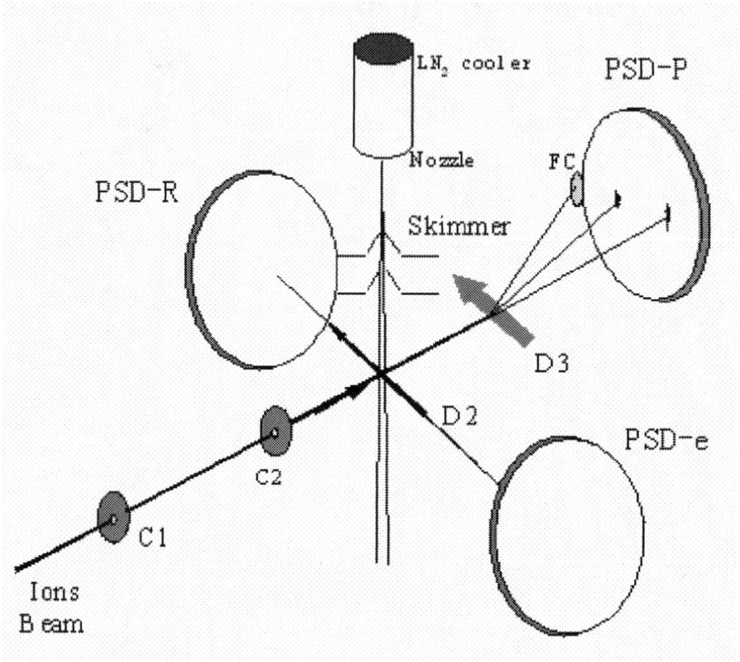

Figure 1. Schematic view of the experimental setup

In the laboratory frame, the projectile incident direction is defined as the z-axis, the atomic jet direction and the recoil ion extraction direction are chosen as –y-axis and x-axis, respectively. The details of the electron distribution pattern are integrated in the laboratory frame due to the rotation symmetry of the collision about z-axis. In our analysis, an ion-collision frame was defined where the scattering plane is constructed by the incident direction of the projectile (Z-axis) and the direction of recoil ion momentum (X-axis). The measured data were transformed from the laboratory frame to the ion-collision frame event-by-event. For easy comparison the electron velocity is normalized to the projectile velocity. The momentum distributions of the continuum electrons projected onto the scattering plane (XOZ plane) for He^{2+} on helium collisions at two values of incident energy are shown in figure 2. The abscissa in figure 2 is the normalized longitudinal velocity (parallel to Z/z-axis) and the ordinate is the normalized transverse velocity (X-axis). The recoil ion is scattered upwards in the figure. The right and the left plots show the electron momentum distributions for 100keV and 400keV He^{2+} on helium collisions, respectively. It is obvious from the left plot that the electrons with lower longitudinal momenta ($V_{e,Z}/V_P < 0.5$ a.u.) prefer to be emitted in the opposite direction of the recoil ions, and there is a minimum area in the vicinity of $0.2 < V_{e,Z}/V_P < 0.8$. Similar results were also observed by Frankfurt group [6], the overall emission pattern was explained by the molecular orbital promotion picture reasonably, and the pattern reflects the symmetry of bound σ or π orbitals. On the contrary, the electrons with lower longitudinal momenta ($V_{e,Z}/V_P < 0.5$ a.u.) prefer to follow the recoil ions for higher collision energy (400 keV, V_P = 2 a.u.) as shown in the right plot, where most events are accumulated in the neighbourhood of $0 < V_{e,Z} / V_P < 0.5$ and $0 < V_{e,X} / V_P < 0.2$. These results are not reported in literatures up to now and detailed data analysis is being carried out.

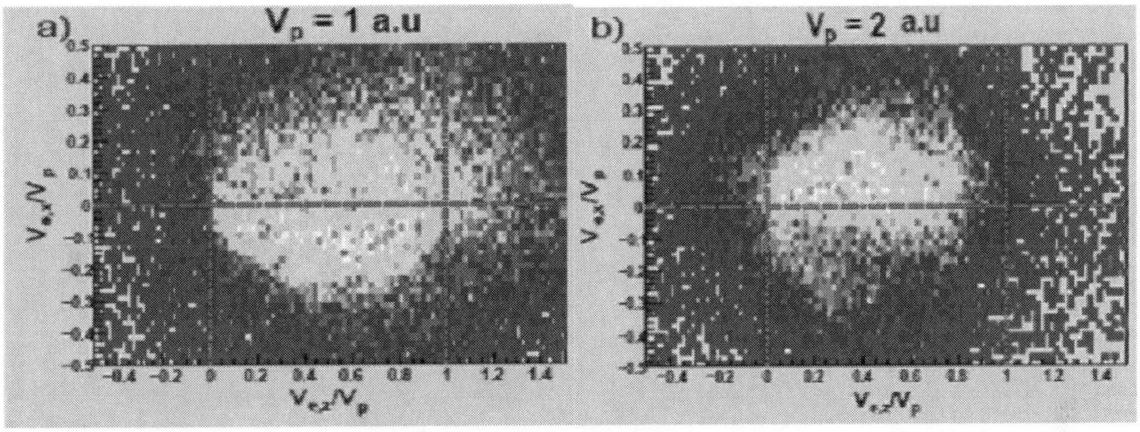

Figure 3. The momentum distributions of continuum electron in the scattering plane for He^{2+} on helium collisions at incident energies of 100 keV (left) and 400 keV (right), respectively.

This work was supported by the National Nature Science Foundation of China under grant No. 10434100 and 10674140.

References

[1] M. A. Abdallah, C. L. Cocke, W. Wolff, et. al., Phys. Rev. Lett. 81 (1998) 3627
[2] R. Dörner, H. Khemliche, M. H. Prior, et. al., Phys. Rev. Lett. 77, 4520 (1996)
[3] S. Martinez, G. Bernardi, P. Focke, et al, J. Phys. B 35(2002)2261
[4] S. Martinez, G. Bernardi, P. Focke, et al, J. Phys. B 41(2002)145204
[5] L. Ph. H. Schmidt, M. Schoeffler, K. E. Stiebing et. al., Phys. Rev. A 76(2007) 012703
[6] X. Ma, X. L. Zhu, H. P. Liu, et al, Sciences in China (Series G) 51 (2008)755
[7] X. Ma, X. L. Zhu, B. Li, et. al., J. Phys. B, Conf. Series 88 (2008)012019
[8] L. Ph. H. Schmidt, F. Afaneh, M. Schöffler, et. al., Physica Scripta T110(2004) 379
[9] X. Ma, H. P. Liu, et. al., to be published in J. Phys. Conf. serious
[10] J. Ullrich, R. Moshammer, et. al., Rep. Prog. Phys. 66 (2003)1463.

Journal of Physics: Conference Series **185** (2009) 012008 doi:10.1088/1742-6596/185/1/012008

Electron scattering from atomic gallium

D. V. Fursa and I. Bray
ARC Centre for Antimatter-Matter Studies
Physics Department, Curtin University, Perth 6845, Australia

E-mail: d.fursa@curtin.edu.au

Abstract. Theoretical study of electron scattering from the $4P_{1/2}$ ground state of gallium atoms is reported. We have used relativistic convergent close-coupling method to obtain cross sections for elastic scattering and excitations of the $4P_{3/2}$, $5S_{1/2}$, $5P_{1/2,3/2}$ and $4D_{3/2,5,2}$ states for incident electron energies ranging from 0.1 to 500 eV. Comparison of our results with recent theoretical estimates that have been used in plasma modeling shows significant discrepancies for all considered transitions.

1. Introduction

Accurate electron-atom collision data is of primary importance for plasma physics modeling. Often the number of required transitions is very large and realistically can only be provided by theoretical calculations. With development of sophisticated scattering methods, such as the convergent close coupling (CCC) [1] method and R-matrix (RM) method [2, 3], accurate collision data can be obtained for a large number of scattering systems. Recent developments in the CCC [4] and RM [5] methods allow for accurate calculations of target atoms that are strongly affected by relativistic effects. Gallium is one such target, with the ground state being a P-state that is strongly affected by spin-orbit interaction resulting in fine structure splitting of 0.1 eV.

Gallium has recently attracted considerable attention in relation to the design of new mercury-free fluorescent light sources. Light emission from gallium-iodide discharge plasma has been studied by Adamson *et al* [6] and Smith *et al* [7]. In particular, Adamson *et al* [6] have presented detailed plasma physics modeling that relied on electron-atom collision data which were calculated using a rather simple technique, a combination of the similarity function method and the Born-Ochkur approximation. Given the scarcity of e-Ga collision data and very large number of transitions required it was a reasonable approach atthe time.

The purpose of this paper is to examine the accuracy of such data by comparing with the results of an advanced theoretical method. We have performed relativistic CCC (RCCC) calculations of electron scattering from gallium for incident electron energies ranging from 0.1 eV ($4P_{3/2}$ threshold) to 500 eV. Selected cross sections for elastic scattering and excitation of a number of low-lying states from the $4P_{1/2}$ ground state are presented and compared with estimates of Adamson *et al* [6].

2. Theoretical method

The RCCC method has been described by Fursa and Bray [4]. Hence we give only the most important aspects here and details specific to the e-Ga scattering system. We model the Ga atom as a quasi one-electron atom with one active electron above a frozen inert Dirac-Fock

$[Ar]3d^{10}4s^2$ core. The core orbitals are obtained from the GRASP package [9]. The set of Ga target states is obtained via diagonalization of Ga frozen-core quasi one-electron Hamiltonian in a relativistic Sturmian basis (Dirac L spinors [10]). Specifically, we have used 50 L spinors with exponential fall-offs 2.5 for $s_{1/2}$, 2.4 for $p_{1/2,3/2}$, 2.2 for $d_{3/2,5/2}$ and $f_{3/2,1/2}$ target symmetries. Similarly to the nonrelativistic CCC method [11, 12] we include phenomenological one-electron and two-electron potentials [8] in order to account more accurately for core-valence electron correlations. The resulting set of Ga target states provide an accurate representation of the Ga low-lying bound states and a finite square-integrable representation of the infinite number of high-lying bound states and the continuum. Table 1 demonstrates very good agreement of the ionization energies obtained in the RCCC structure calculations with the experimental values (NIST) [13] for the states included in the present study. Good agreement was also found the oscillator strength for the optically allowed $4P_{1/2} - 5S_{1/2}$ and $4P_{1/2} - 4D_{3/2}$ transitions, see Table 2.

The set of Ga target states is used to perform the multichannel expansion of the e-Ga scattering system total wave function and formulate a set of coupled Lippman-Schwinger equations for the T-matrix [4]. The calculations have been performed in two models. For incident electron energies above 2 eV we have performed calculations using the 75-state model that includes 11 $S_{1/2}$, 11 $P_{1/2,3/2}$, 11 $D_{3/2,5/2}$, and 10 $F_{5/2,7/2}$ states. At energies below 2 eV we have done calculations using a smaller 26-state model that includes 4 $S_{1/2}$, 4 $P_{1/2,3/2}$, 4 $D_{3/2,5/2}$, and 3 $F_{5/2,7/2}$ states.

Table 1. Ionization energies of the low-lying states of gallium. The present RCCC results compared to experiment (NIST) [13].

State	RCCC	NIST
$4P_{1/2}$	6.00	6.00
$4P_{3/2}$	5.90	5.90
$5S_{1/2}$	2.89	2.93
$5P_{1/2}$	1.87	1.90
$5P_{3/2}$	1.86	1.89
$4D_{3/2}$	1.66	1.69
$4D_{5/2}$	1.66	1.69

Table 2. Oscillator strengths for transitions involving the ground state of gallium. The present RCCC results compared to experiment (NIST) [13].

Transition	RCCC	NIST
$4P_{1/2} - 5S_{1/2}$	0.12	0.123
$4P_{1/2} - 4D_{3/2}$	0.334	0.29

3. Results and discussion

Figs. 1-5 present the results of the RCCC calculations and comparison with the theoretical estimates of Adamson *et al* [6] (SFM, obtained by digitising their Fig. 17) We find very large discrepancies both in shape and magnitude for the elastic scattering cross section, Fig. 1, and for the first excited $4P_{3/2}$ state cross section, Fig. 2. For the latter cross section, the difference at the cross maximum is about a factor of six. We note that excitation of the $4P_{3/2}$ level from the ground $4P_{1/2}$ level is a transition between fine structure levels. Such a transition can be expected to be described reliably only by theoretical methods that take into account the relativistic effects sufficiently accurately.

Excitations of the $5S_{1/2}$ and $4D_{3/2}$ states are optically allowed transitions. At high energies the excitation cross section is well described by the Bethe-Born formula [14] and, therefore, is determined by the excitation energy and optical oscillator strength value. Good agreement between RCCC and experiment for Ga structure (see Tables 1 and 2) should lead to sufficient

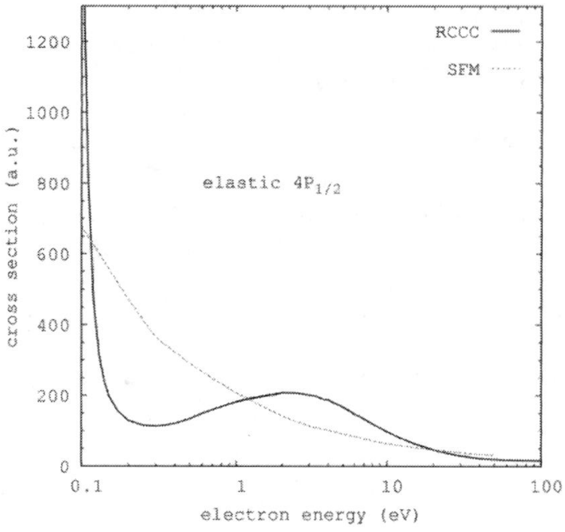

Figure 1. Cross sections for elastic electron scattering on the $4P_{1/2}$ ground state of gallium atom. Present RCCC calculations are described in the text. The theoretical estimates of Adamson *et al* [6], labelled SFM, are a combination of the similarity function method and Born-Ochkur approximation.

accuracy and reliability of the RCCC results for these transitions. The excitation of the $5S_{1/2}$ state is presented in Fig. 3. It shows that both RCCC and SFM converge to the same high-energy limit, but there is still a relatively large discrepancy in the cross section shape and its magnitude near the cross section maximum.

Cross section for the excitation of the optically allowed $4D_{3/2}$ state is much larger than the corresponding excitation cross section for the optically prohibited $4D_{3/2}$ state as follows from Fig. 4. The RCCC cross section for the fine structure combined $4D$ level agree well in the shape, but differ substantially in magnitude with the SFM results.

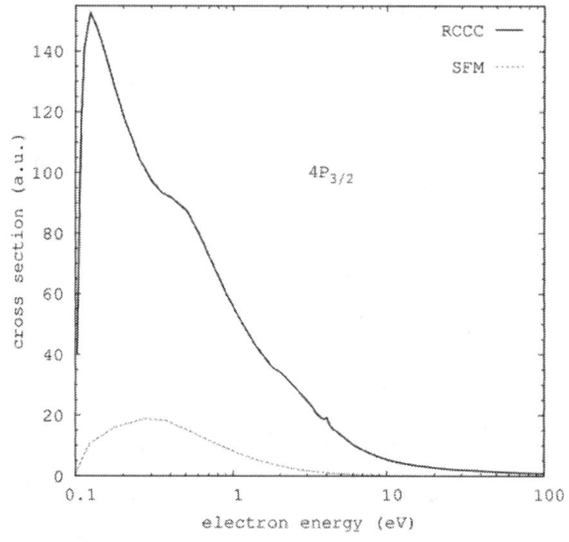

Figure 2. Cross sections for electron impact excitation of $4P_{3/2}$ state from the $4P_{1/2}$ ground state of gallium atom. The calculations as in Fig. 1.

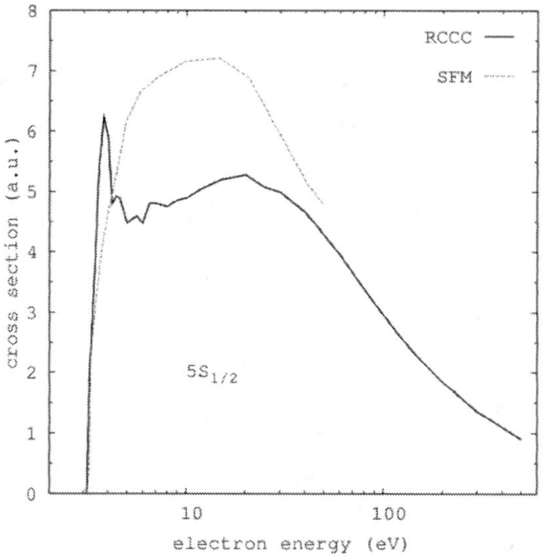

Figure 3. Same as for Fig. 2 but for excitation of $5S_{1/2}$ state.

Fig. 5 presents the RCCC results for the excitation of the optically prohibited $5P_{1/2}$ and $5P_{3/2}$ states and cross section for the fine structure combined $5P$ level. The latter cross section

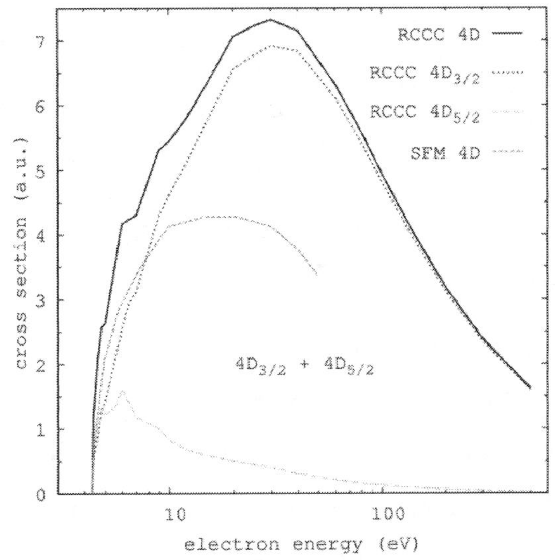

Figure 4. Same as for Fig. 2 but for excitation of $4D_{3/2}$ and $4D_{5/2}$ states. The sfm results are for combined $4D$ level.

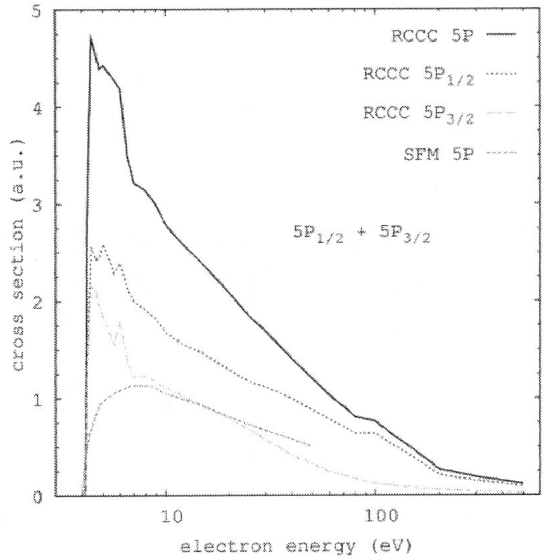

Figure 5. Same as for Fig. 2 but for excitation of $5P_{1/2}$ and $5P_{3/2}$ states. The sfm results are for combined $4D$ level.

is compared with the SFM estimate. We find very large discrepancies in the shape and magnitude for this transition.

In conclusion, we find very large discrepancies between the RCCC and SFM results for elastic scattering and optically prohibited excitations with smaller, but still significant discrepancies for the optically allowed transitions. We expect that incorporation of accurate collision data in plasma modeling calculations might lead to significantly more reliable modeling outcomes.

Acknowledgments

This work was supported by the Australian Research Council and Curtin University. We are grateful for access to the Australian National Computational Infrastructure and its Western Australian node iVEC.

References

[1] Bray I, Fursa D V, Kheifets A S and Stelbovics A T 2002 *J. Phys. B* **35** R117–R146
[2] Bartschat K, Hudson E T, Scott M P, Burke P G and Burke V M 1996 *J. Phys. B* **29** 115–123
[3] Zatsarinny O and Bartschat K 2005 *Phys. Rev. A* **71** 022716
[4] Fursa D V and Bray I 2008 *Phys. Rev. Lett.* **100** 113201 (pages 4)
[5] Zatsarinny O and Bartschat K 2008 *Phys. Rev. A* **77** 062701 (pages 7)
[6] Adamson S, Astapenko V, Chernysheva I, Chorkov V, Deminsky M and Demchenko G 2007 *J. Phys. D: Appl. Phys.* **40** 3857–3881
[7] Smith D J, Michael J D, Midha V, Cotzas G M and Sommerer T J 2007 *J. Phys. D: Appl. Phys.* **40** 3842–3856
[8] Maslov M, Brunger M J, Teubner P J O, Zatsarinny O, Bartschat K, Fursa D, Bray I and McEachran R P 2008 *Phys. Rev. A* **77** 062711 (pages 8)
[9] Dyall K G, Grant I P, Johnson C T, Parpia F P and Plummer E P 1989 *Comp. Phys. Comm.* **55** 425–456
[10] Grant I P and Quiney H M 2000 *Phys. Rev. A* **62** 022508
[11] Bray I 1994 *Phys. Rev. A* **49** 1066–1082
[12] Fursa D V and Bray I 1997 *J. Phys. B* **30** 5895–5913
[13] Ralchenko Y, Kramida A E, Reader J and NIST ASD Team 2008 NIST atomic spectra database (version 3.1.5)
[14] Inokuti M 1971 *Rev. Mod. Phys.* **43** 297–347

Journal of Physics: Conference Series **185** (2009) 012009 doi:10.1088/1742-6596/185/1/012009

Time-resolved reaction imaging by intense few-cycle laser pulses and laser high-order harmonics

M Fushitani[1,2], A Matsuda[1], E J Takahashi[1,2], and A Hishikawa[1,2,3]

[1] Institute for Molecular Science, National Institutes of Natural Sciences, Myodaiji, Okazaki, Aichi 444-8585, Japan
[2] The Graduate University for Advanced Studies (SOKENDAI), Myodaiji, Okazaki, Aichi 444-8585, Japan
[3] PRESTO, Japan Science and Technology Agency, Kawaguchi, Saitama 332-0012, Japan

E-mail: fusitani@ims.ac.jp, hishi@ims.ac.jp

Abstract. We discuss two novel approaches to visualize ultrafast dynamics of polyatomic molecules using few-cycle intense laser pulses and the high-order harmonics in the soft X-ray region, respectively. First, we present the real-time Coulomb explosion imaging of ultrafast isomerization of $C_2D_2^{2+}$ by using intense few-cycle laser pulses (<10 fs, ~1 PW/cm^2, ~800 nm) to show that the hydrogen migration proceeds recurrently in competition with the molecular dissociation. Secondly, we present a new experimental setup for the real-time inner-core excitation imaging, focusing on the generation and separation of the light source using the 59th order harmonic pulses (~91 eV) and their characteristics.

1. Introduction

Understanding molecular dynamics in chemical reaction is one of the central subjects of chemistry. Among others, ultrafast laser spectroscopy has proved to be a powerful technique to unveil complex dynamics of large polyatomic molecules [1]. In its conventional pump-probe scheme, the evolution of the nuclear wavepacket in an excited electronic state created by the pump laser pulse is projected onto a reference electronic state by the time-delayed probe pulse. The pump-probe signal reflects the motion of the nuclear wave packet and thus can be converted to the changes in the instantaneous molecular structure when the potential surfaces of the excited and reference electronic states are well known. Here, we discuss two novel approaches to visualize directly how molecules change their structure in the course of reactions by using few cycle intense laser pulses and the high-order harmonics in the soft X-ray region with an unprecedented time resolution.

2. Real-time Coulomb explosion imaging of recurrently migrating hydrogen in $C_2D_2^{2+}$

Molecules exposed to intense laser fields (~1 PW/cm^2) promptly eject several electrons and undergo rapid bond breakings (called Coulomb explosion) on the repulsive Coulombic potential energy surface of the highly charged states. Since the momenta of the resultant fragment ions reflect sensitively the geometrical structure of the target molecule, the Coulomb explosion provides direct access to the instantaneous structure in the course of reaction processes, as demonstrated for dissociating diatomic [2] and triatomic molecules [3].

Intramolecular hydrogen migration is important in various chemical processes as the large mobility of hydrogen often determines the main route of intra- and inter-molecular reactions [4]. Here we study the acetylene (DCCD) ↔ vinylidene (D_2CC) isomerization of $C_2D_2^{2+}$ by the time-resolved Coulomb explosion imaging [5]. A pair of 9-fs intense laser pulses (0.13 PW/cm^2) is used in a pump-probe scheme where the pump pulse doubly ionizes C_2D_2 and initiates the isomerization of acetylene dication via hydrogen migration and the time-delayed probe pulse induces the three-body Coulomb explosion from the triply charged states, $C_2D_2^{3+} \rightarrow D^+ + C^+ + CD^+$. The momentum angle θ_{12} between D^+ and C^+ fragments as a function of the pump-probe time delay Δt (figure 1) clearly shows that D atom in acetylene dication migrates recurrently [5]: One of the D atoms shifts from one carbon site to the other (D_2CC) in 90 fs and then moves back to the original site (DCCD) by 280 fs.

3. Laser high-order harmonics for time-resolved inner-core excitation imaging

Laser high-order harmonics have novel features such as i) high photon energy up to several keV, ii) extremely short pulse duration in the sub-fs regime, iii) high-quality spatiotemporal coherence, iv) high photon flux comparable to synchrotron radiation and v) simple and precise synchronization with other laser light sources [6]. Because of these aspects, laser high-order harmonic pulses are of great interest as a potential light source for time-resolved spectroscopy of ultrafast dynamics that could not be elucidated in real time by conventional techniques.

Figure 2(a) shows the schematic of our new time-resolved reaction imaging system. It consists of a laser high-order harmonics source, a soft X-ray spectrograph, a beam line equipped with a time delay

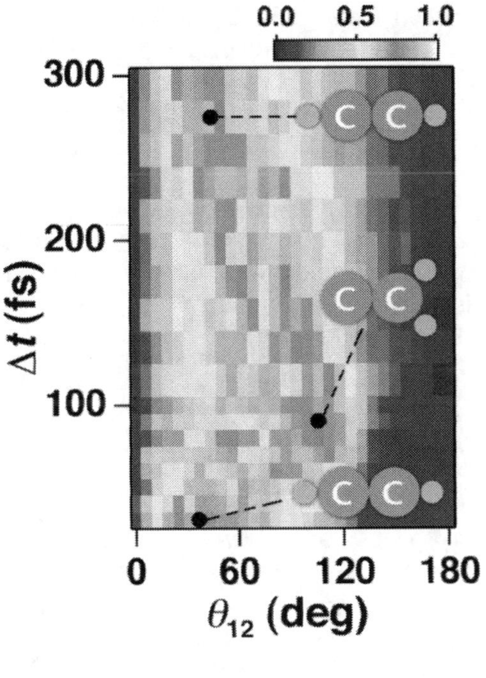

Figure 1 Angular distribution between momenta of D^+ and C^+ from the three-body Coulomb explosion, $C_2D_2^{3+} \rightarrow D^+ + C^+ + CD^+$, as a function of the pump-probe time delay, showing the recurrently migrating D atom.

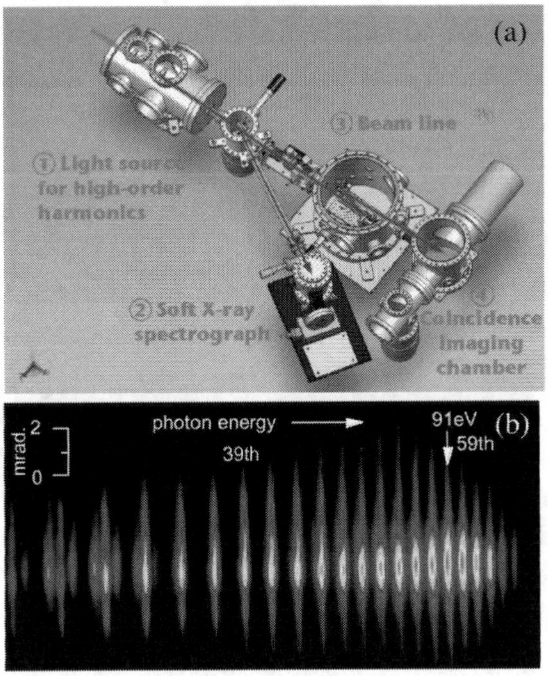

Figure 2 (a) Schematic of time-resolved reaction imaging system by inner-core excitations. (b) Two-dimensional spectrum of laser high-order harmonics for 130 Torr of Ne. Side bands appearing below the 33rd order harmonic are due to the second order diffraction by the grating.

line as well as a pair of dielectric mirrors to select one of the harmonic orders, and an electron-ion coincidence imaging chamber.

3.1. Generation of laser high-order harmonics

Output from a Ti:sapphire laser system (~800 nm, ~30 fs, 1 kHz, 2 mJ/pulse) was focused with a plano-convex lens (f = 600 mm) into a cylindrical Ne cell placed in a high vacuum chamber. The spectral distribution and the divergence of the generated high-order harmonics were characterized by an XUV/soft X-ray spectrometer equipped with a flat-field imaging grating and a MCP/phosphor image intensifier. Figure 2(b) shows a two-dimensional spectrum of laser high-order harmonics at 130 Torr of Ne. The intensity of the fundamental pulse was set to 1 PW/cm^2 at the focus. The spectrum exhibits a flat intensity distribution (plateau) up to the 63rd order harmonic and a rapid decrease (cutoff) from the 65th order.

It is shown [7] that the phase mismatch between the fundamental and high-order harmonics pulses is mainly caused by (i) a phase shift due to absorption of soft X-ray pulses, (ii) dispersion of IR pulses by a nonlinear medium, and (iii) a phase delay of fundamental pulses around the focus (Gouy phase shift). Here, we change the density of a nonlinear medium to minimize the total phase mismatch. Figure 3 shows the intensity and divergence for the 59th order harmonic as a function of Ne gas pressure, P_{Ne}, between 50 and 180 Torr. As P_{Ne} increases, the intensity reaches its maximum while the divergence exhibits the minimum value of 1.1 mrad around P_{Ne} = 130 Torr, indicating that the phase mismatch is minimized at this pressure. The photon flux is evaluated to be 3.4×10^9 photons/s by using a photodiode for soft X-rays under the phase matching condition.

3.2. Beam line

A set of coaxial mirrors is placed in the beam line to independently reflect the fundamental and high-order harmonic pulses; the outer mirror is coated with silver to reflect the fundamental pulse, while a Mo/Si multilayer mirror is adopted as the inner mirror to selectively reflect the 59th order harmonic. The latter is mounted on a piezo transducer stage to control the time delay between the fundamental and 59th order harmonic pulses. Both pulses are focused on a sample beam in a coincidence chamber by a concave Mo/Si multilayer mirror (f = 650 mm).

The reflectance of the Mo/Si multilayer mirrors was measured as a function of photon energy at BL8B1 of UVSOR synchrotron radiation facility at IMS. While the reflectance is constant at the peak (~60 %), the peak position shifts to higher energies as the incident angle increases. At the incident angle of 5 degrees, the reflectance peak and the full width at half maximum (FWHM) of a single

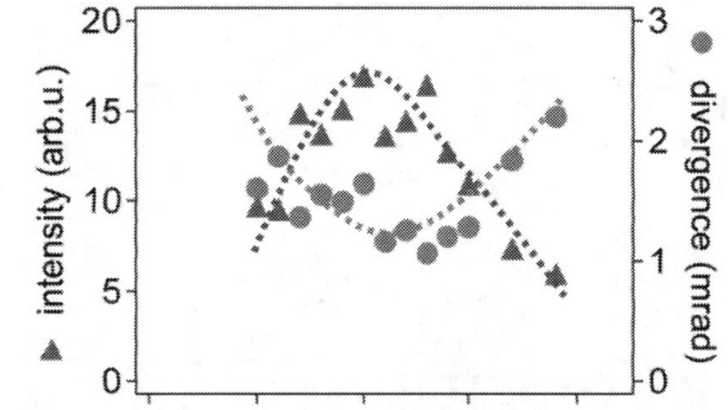

Figure 3 Ne pressure dependence of the intensity (triangle) and divergence (circle) of the 59th order harmonic. The dotted curves are shown to guide to the eye.

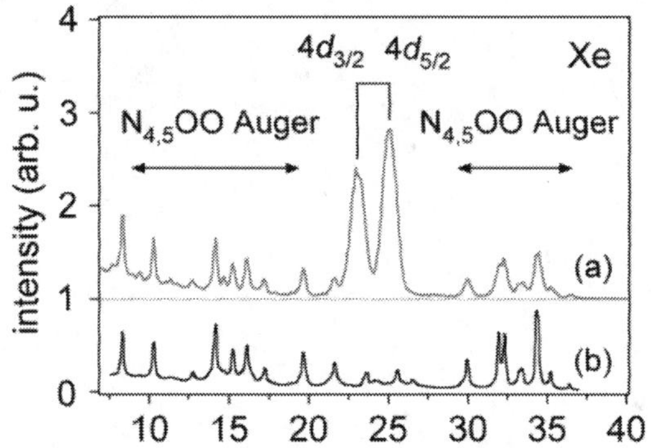

Figure 4 (a) Photoelectron spectrum of Xe 4d inner-core states by exciting with the 59th order harmonic ($h\nu$ = 92.5 eV) from the beam line. (b) High resolution Auger electron spectrum ($h\nu$ = 110 eV) from ref. [8].

Mo/Si mirror is 91 eV and 2.8 eV, respectively.

3.3. Photoelectron spectrum of Xe 4*d* inner-core states

The 59th order harmonic pulse from the beam line was used to ionize Xe from the 4*d* inner-core states. Figure 4(a) shows a photoelectron spectrum of Xe measured by a magnetic bottle photoelectron spectrometer [9]. The broad doublet observed around 24 eV in figure 4(a) are assigned to photoelectrons from the spin-orbit sublevels ($4d_{5/2}$ and $4d_{3/2}$), while other sharp peaks are attributed to the associated Auger electrons (see figure 4(b)). From the binding energies of the 4*d* spin-orbit states ($4d_{5/2}$: 67.5 eV, $4d_{3/2}$: 69.5 eV), photon energy of the 59th order harmonic pulse used was determined to be 92.5 eV. It should be noted that doublet peaks from the 4*d* states associated with neighboring orders (i.e. 57th and 61st) of harmonics are not observed, which clearly demonstrates that the Mo/Si mirror pair selectively reflects the 59th order harmonic to produce a single isolated pulse in the time domain.

The linewidths of the 4*d* photoelectron peaks are much broader than those of Auger electron peaks because the former widths reflect the spectral bandwidth of the 59th order harmonic while the latter ones are determined by the lifetimes of the 4*d* hole states. From the FWHM of the $4d_{5/2}$ peak, 1.1 eV, in figure 4(a), the pulse duration of the 59th order harmonic is estimated to be 1.8 fs in the Fourier-transform limit. A broader spectrum with the FWHM of 2.1 eV corresponding to 0.87 fs in the Fourier transform limit has been obtained by few-cycle intense laser pulses (12 fs, 1.1 mJ, ~800 nm). Characterization of the temporal profile of the harmonic pulse is in progress.

4. Summary

Two novel approaches to visualize ultrafast dynamics of polyatomic molecules are presented. First, we discussed the real-time Coulomb explosion imaging of ultrafast hydrogen migration process in $C_2D_2^{2+}$ by using intense few-cycle laser pulses (<10 fs, ~1 PW/cm^2, ~800 nm). The temporal evolution of the momenta of the fragment ions produced by the three-body explosion, $C_2D_2^{3+} \rightarrow D^+ + C^+ + CD^+$, shows that the migration proceeds recurrently in competition with the molecular dissociation. Secondly, we presented a real-time inner-core excitation imaging system constructed at IMS, focusing on the generation and separation of the light source using the 59th order harmonic pulses (~91 eV) and their characteristics.

Acknowledgements

We are grateful to Mr. Yano and Mr. Horigome in IMS for their support for the design of the inner-core reaction imaging system. We would like to thank the staff of the UVSOR facility for the reflectance measurements of the Mo/Si mirrors and Dr. Hikosaka and Prof. Shigemasa at IMS for their assistance in the photoelectron measurements.

References

[1] Zewail A H 2000 *Angew. Chem. Int. Ed.* **39** 2586.
[2] Ergler T, Rudenko A, Feuerstein B, Zrost K, Schroter C D, Moshammer R and Ullrich J 2005 *Phys. Rev. Lett.* **95** 093001.
[3] Hishikawa A, Ueyama M and Yamanouchi K 2005 *J. Chem. Phys.* **122** 151104.
[4] Schaefer III H F 1979 *Acc. Chem. Res.* **12** 288.
[5] Hishikawa A, Matsuda A, Fushitani M and Takahashi E J 2007 *Phys. Rev. Lett.* **99** 258302.
[6] Brabec T and Krausz F 2000 *Rev. Mod. Phys.* **72** 545.
[7] Tamaki Y, Itatani J, Obara M and Midorikawa K 2000 *Phys. Rev. A* **62** 063802.
[8] Carroll T X, Bozek J D, Kukk E, Myrseth V, Sæthre L J, Thomas T D and Wiesner K 2002 *J Electron Spectrosc. Relat. Phenom.* **125** 127.
[9] Hikosaka Y, Aoto T, Lablanquie P, Penent F, Shigemasa E and Ito K 2006 *Phys. Rev. Lett.* **97** 053003.

Journal of Physics: Conference Series **185** (2009) 012010 doi:10.1088/1742-6596/185/1/012010

Anti-Resonant Interferometric Nonlinear Spectroscopy (ARINS) Study of Metal Nanocluster-Glass Composites.

Binita Ghosh[1], Purushottam Chakraborty[1], B P Singh[2] and T Kundu[2]
[1] Saha Institute of Nuclear Physics, 1/AF Bidhannagar, Kolkata 700064, India
[2] Physics Department, Indian Institute of Technology-Bombay, Mumbai 400076, India

E-mail: binita.ghosh@saha.ac.in

Abstract. Linear and nonlinear optical properties of copper and gold nanoclusters in fused silica glasses synthesized by 200 KeV Cu^+ and 1.5 MeV Au^+ ion implantation at a dose of 3×10^{16} ions/cm^2 have been studied. UV-Vis spectroscopy has revealed prominent linear absorption bands at characteristic surface plasmon resonance (SPR) frequency signifying appreciable formation of copper and gold colloids in glass matrices. Third-order optical properties of the nanocluster-glass composite materials have been studied by Z-Scan and Anti Resonant Interferometric Nonlinear Spectroscopy (ARINS) techniques. The sign of the nonlinear refraction is readily obtained from Z-scan signature. The ARINS technique utilizes the dressing of two unequal-intensity counter-propagating pulsed beams with differential nonlinear phases, which occur upon traversing the sample. This difference in phase manifests itself in the intensity-dependent transmission. The nonlinear refractive index, nonlinear absorption coefficient and the real (and imaginary) parts of the third-order optical susceptibility have been extracted from ARINS data. Results of the investigation of the nonlinear refraction using the above two techniques and the possible mechanisms responsible for the nonlinear optical responses are presented in the current paper.

1. Introduction

There is a considerable interest in finding materials having appreciable optical nonlinearities. Nanomaterials have drawn great attention because of their distinguishable linear and nonlinear optical properties, which differ remarkably from bulk materials. Metal nanoparticles possess interesting optical properties. For example, nanoparticles of alkali metals and noble metals like copper, silver and gold show a broad absorption band in the visible region of the electromagnetic spectrum [1, 2], which is substantially different from the flat absorption of the corresponding bulk metal in this region. The optical absorption spectrum of such materials shows a resonance band, which is due to the surface plasmonic excitation. The metal nanocluster composite glasses (MNCGs) exhibit a resonant third-order optical nonlinearity, which can be enhanced by operating at wavelengths near the Surface Plasmon Resonance (SPR) band of the MNCGs. Glasses possess macroscopic inversion symmetry, allowing only odd-order nonlinearity. Introducing metal nanoclusters in glass matrices can further enhance the third-order nonlinear optical response of the glass matrices. Ion implantation has been shown to produce high-density metal colloids in glasses and other materials [3]. The high precipitate volume fraction and small size of nanoclusters in glasses lead to the generation of third-order susceptibility much greater than those for metal-doped solids [4].

The present work reports a systematic study on the optical properties of Cu^+ and Au^+ implanted fused silica glasses. We have demonstrated that Cu and Au nanoclusters of various sizes can be safely created in as-implanted silica glass samples; evidenced by the pronounced optical absorption bands. Nonlinear optical measurements of these metal nanocluster-glass composites have been made by Z-Scan and Anti-Resonant Interferometric Nonlinear Spectroscopy (ARINS) techniques and the results of the investigation on the nonlinear refraction and the possible mechanisms responsible for nonlinear responses have been discussed.

Experiment:
Optical grade fused silica glass specimens about 1mm thick were used as substrates. Implantation was performed with Au^+ ions using 1.7 MeV Tandem Accelerator and for Cu^+ ions using a high current isotope separator-cum-ion implanter (Danfysik). The implantation energies were chosen as 200 keV and 1.5 MeV for Cu^+ and Au^+ ions, respectively with a dose of 3×10^{16} ions/cm^2. Following implantation the samples were heat treated at 400^{0}C for 1 hour in Ar atmosphere. Rutherford Backscattering Spectrometry (RBS) has been done to study the depth distribution and actual concentration of Au and Cu atoms. The RBS spectra were taken with 2 MeV He^+ ions using a surface barrier detector (resolution ~ 15 KeV) kept at an angle of 165°. Optical absorption of the implanted samples was measured in the wavelength range 300 – 700 nm using UV-Vis Spectroscopy. Nonlinear optical properties of the implanted samples have been investigated using Z-Scan and Anti-Resonant Interferometric Nonlinear Spectroscopy (ARINS) techniques.

The working principle [5] and the experimental details of the Z-Scan technique were described in our earlier works [6, 7]. The Anti-Resonant Interferometric Nonlinear Spectroscopy (ARINS) technique proposed by Lee and Hughes (LH) [8] is a simple, sensitive, single beam technique based on an anti-resonant ring (Sagnac) interferometer for simultaneously measuring the real and imaginary contributions to optical nonlinearity and it utilizes the dressing of two unequal-intensity counter-propagating pulsed beams with differential nonlinear phases that occur upon traversing the sample. This difference in phase manifests itself in the intensity-dependent transmission. Photo-detection of the transmission of the ARINS yields spatially and temporally integrated response. Figure 1 shows the schematic diagram of ARINS setup. We used a Ti-Sapphire laser of pulse width 70fs-100fs and tunability of 700-850 nm. The repetition rate was 100MHz. A 50-50 beam-splitter divides the incoming pulsed beam into two counter-propagating pulses having a π phase difference. The pulses propagating in the clockwise (CW) direction are reflected by an uncoated flat with 6° wedged rear-surface, while those propagating in the counter-clockwise (CCW) direction are reflected by a high-reflectivity mirror.

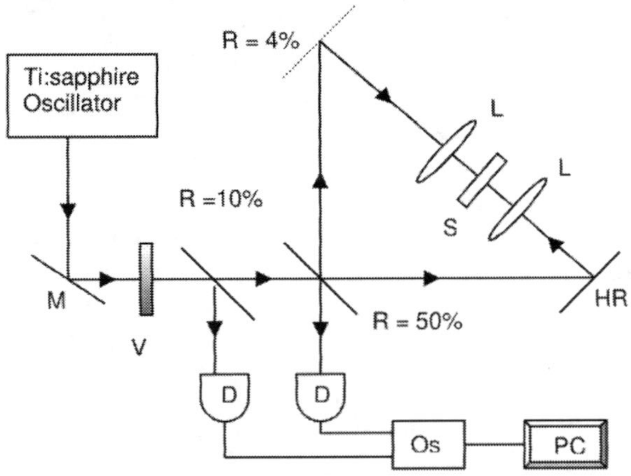

Figure 1. Schematic diagram of ARINS setup

Thus this technique is based on the dressing of two unequal intensity counter-propagating pulses. These two pulses again recombine at the beam splitter to yield ARINS transmission, $|E_{out}|^2 \propto |E_{cw} + E_{ccw}|^2$, where E_{cw} and E_{ccw} are the optical fields travelling in clockwise and counter--clockwise directions, respectively. The two counter-propagating fields traversing the ring acquires linear as well as intensity-dependent nonlinear phase shifts. Since both the fields traverse the same optical path through the ring and encounter the same interactions with the optical elements, linear interactions would affect their amplitude and phase in identical manner. For an exactly 50% beam splitter, in absence of any nonlinear interactions, the two returning fields will be having the same amplitude and phase difference between them and consequently, will interfere destructively at the beam splitter to yield zero transmission. In this 'balanced' condition all the input power is reflected back to the incident direction. Measurement against this dark background provides the basis for the improved sensitivity essential for measuring relatively weak signals. Any small deviation from the ideal splitting ratio (δ) results in a leakage from the ARINS and is responsible for the background that limits the sensitivity of the measurement. If the sample under investigation exhibits nonlinear response, the two unequal intensity counter-propagating pulses will undergo different phase changes after passing through the sample. Their superposition on the beam splitter will result in the intensity-dependent transmission of the ARINS, which is related to the nonlinear response of the sample.

Results and discussion:

Figures (2a) and (2b) show the linear optical absorption bands for Au^+ and Cu^+-implanted glass samples, respectively. As evidenced from the figures the absorption bands appeared at around 490 nm and 580 nm for Au^+ and Cu^+ implanted samples, respectively. Cluster sizes estimated from the Mie theory [9] have been found to be within 3-5 nm in the present cases. These are, however, rough estimates based on the assumption that the clusters are spherical.

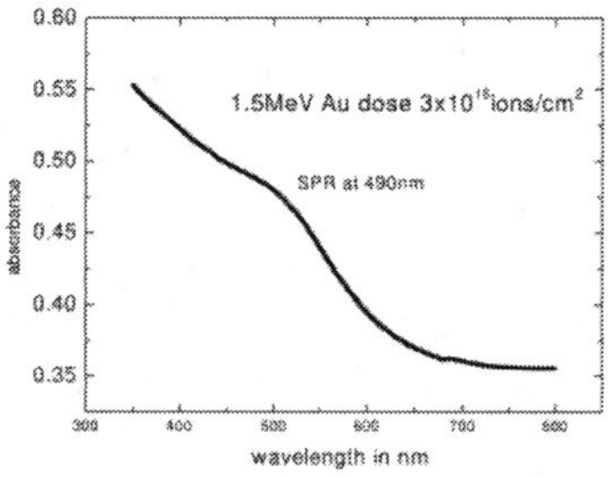

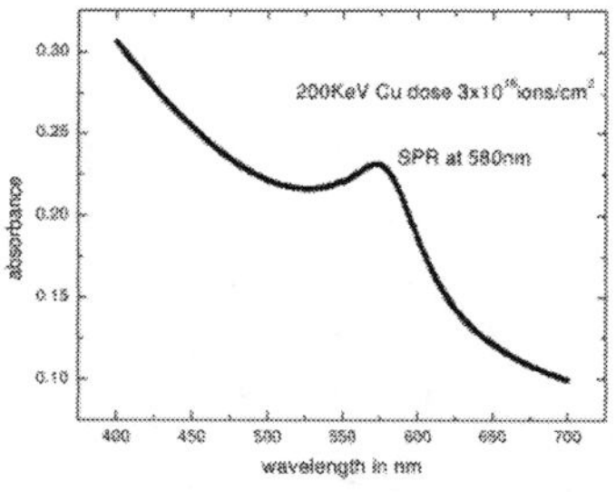

Fig.(2a) Optical absorption spectra of Au^+ sample **Fig.(2b)** Optical absorption spectra of Cu^+ sample

In order to study the nonlinear properties of this effective medium, a pulsed light beam with instantaneous electric field strong enough to generate third-order nonlinear effects is allowed to fall onto the sample. Two-photon absorption coefficient given by β is related to the imaginary part of $\chi^{(3)}$.

$$\beta = \frac{2\pi}{\lambda \eta_0^2 \varepsilon_0 c} \mathrm{Im}\left[\chi^{(3)}\right] \tag{1}$$

where n_0 is the linear refractive index of the material. The nonlinear refractive index of the material is related to the real part of $\chi^{(3)}$ by the relation [10]

$$\eta_2 = \frac{1}{2\eta_0^2 \varepsilon_0 c} \mathrm{Re}\left[\chi^{(3)}\right] \tag{2}$$

The general form for the electric field of a Gaussian beam is given by

$$E(Z,r,t) = E_0(t)\exp\left(\frac{-r^2}{w^2(Z)}\right)e^{i\Phi(Z,t)} \tag{3}$$

where $E_0(r,t)$ is the incident electric field. For the weaker CW beam, $\alpha(I)=\alpha$ and $n(I)=n_0$. For the stronger CCW beam, $\alpha(I)=\alpha+\beta I$ and $n(I)=n_0+n_2 I$.

The ARINS leakage is given by $\left|E_{out}(r,t)\right|^2 = \left|E_{cw}^t(r,t)+E_{ccw}^t(r,t)\right|^2$ (4)

The measured quantity in this experiment is the transmitted pulse energy

$$W = 2nc\varepsilon_0 \int_{-\infty}^{\infty}\int_0^{\infty} E_{out}^2\, 2\pi r\, dr\, dt \tag{5}$$

Defining $I_{out} = 2(\ln 2)^{1/2} W/\pi^{3/2}\tau w^2(Z)$ which has the dimensions of intensity, we get [8,11]

$$I_{out} = 2nc\varepsilon_0 R I_{in} \exp(-\alpha L)\times\left[4\delta^2 + \beta L\delta I_{in} + \left\{\frac{\beta^2}{16}+\left(\frac{kn_2}{2}\right)^2\right\}L^2 I_{in}^2\right] \tag{6}$$

The ARINS output transmission I_{out} is a cubic polynomial of incident intensity I_{in} and the coefficients are directly related to η_2 and β, as shown in the expression. In this equation α is the linear absorption coefficient, L is the sample thickness, R is the reflectivity of the low-reflectivity mirror and δ represents the small deviation from the ideal splitting ratio of the beam splitter. For δ = 0, $I_{out} \propto I_{in}^3$, in which case the simultaneous evaluation of n_2 and β becomes difficult. However, a small nonzero δ makes the analysis much simpler as each term of the polynomial can be evaluated directly. To verify our formulation and demonstrate the application of ARINS, we measured the real and imaginary contribution to the third-order nonlinearity of our samples. As it is clear from the above equation, ARINS cannot give any unique value of the nonlinear parameters (η_2 and β) as far as sign of these parameters are concerned. Z-scan provides us exactly the sign of the parameters [6,7,12]. Combining the output of both the experiments we have calculated the values of η_2 and β. Figures 3a and 3b show the ARINS data for the Au^+ and Cu^+ implanted samples, respectively. Observations were carried out at the wavelength of 737.18 nm. The ARINS output shows a clear curvature indicating the nonlinear signature of the samples. The values of the nonlinear parameters are shown in the table below.

Sample	η_2 cm^2/W	β cm/W	**Re** $\chi^{(3)}$	**Im** $\chi^{(3)}$	**Abs** $\chi^{(3)}$
1.5MeV $Au^+(1\times10^{17})$	-1.45×10^{-11}	-2.6×10^{-8}	1.457×10^{-8}	1.53×10^{-10}	1.456×10^{-8}
200KeV $Cu^+(3\times10^{16})$	-3.65×10^{-15}	-6.4×10^{-12}	3.66×10^{-12}	3.74×10^{-14}	3.66×10^{-12}

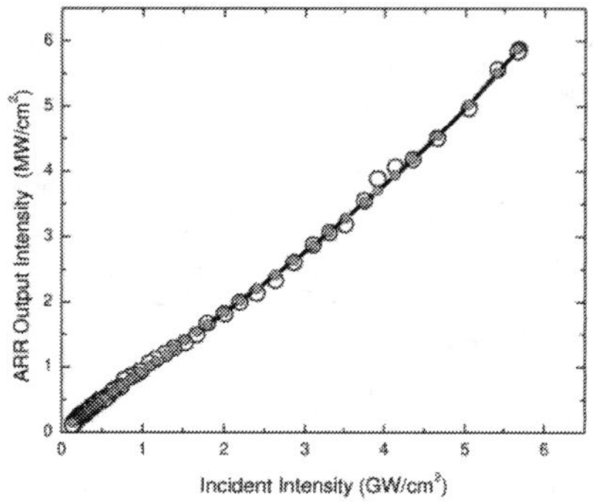

Fig (3a): ARINS data for Au sample

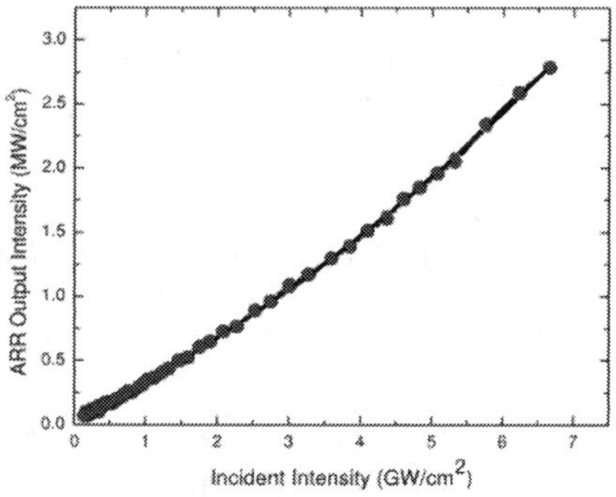

Fig (3b): ARINS data for Cu sample

In absence of any nonlinearity in the sample, the 2nd and 3rd term in Eq.(6) become zero, reducing the equation to a linear one, which when plotted should give a straight line. Figure 4 shows the linear ARINS output for the fused silica substrate, in parity with the above stated reason. Therefore, it can clearly be stated that nonlinearities in this case arise entirely due to the presence of metal ions within the fused silica substrates.

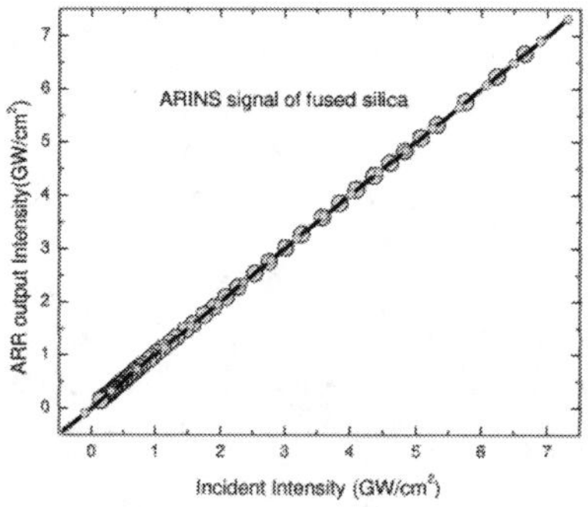

Fig.(4): ARINS data for fused silica substrate

As the interacting wavelength is far away from the resonance in Au and Cu sample (nonresonant condition), the values of the nonlinear parameters obtained from the ARINS calculation is different from the Z-scan part. Much better and accurate values can be extracted by this ARINS technique once we use the resonant wavelength to pump the sample. This part of the work is still in progress and better results are awaited. However the sign of the nonlinearity remains same in both the cases. The scatter in the z-scan data is about $\pm 4\%$, which obviously limits the sensitivity of the technique. Considering the beam waist at the focal point (9.95 μm) and peak power (19 kW) of the laser pulse used in the Z-Scan experiment, the peak intensity at the focal point was estimated to be 12 GW/cm^2, which could be relatively strong to produce significant sample heating. This implies that the nonlinearity measured by Z-Scan have a great extent of thermal contribution.

The third-order nonlinear optical response of the present composite material mainly originates from electronic effects in metal nanoclusters. These electronic contributions are due to both intraband and interband transitions. The first one corresponds to transitions within the conduction band and the second one to transitions from the upper levels of the filled *d* band to the levels above the Fermi level in the conduction band. In the case where the clusters are excited by ultra-short laser pulses, a hot electron phenomenon may superimpose on the pure electronic nonlinear contributions to $\chi^{(3)}$ [13].

Conclusion:
Enhanced nonlinear optical responses have been observed in the MNCGs, synthesized by metal implantation followed by thermal treatments. The sign of the nonlinear absorption and refractive index obtained from Z-Scan measurements have been used in choosing the unique nonlinear parameters extracted from ARINS, enabling us to determine the absolute value of nonlinear susceptibility $\chi^{(3)}$.

Reference:
[1] U. Kreibig and M. Vollmer, *Optical Properties of Metal Clusters* Springer, Berlin, (1995)
[2] A. P. Alivisatos, *J. Phys. Chem.* **100**, 13226 (1996)
[3] F. Gonella, P. Mazzoldi, in: Hari Singh Nalwa (Ed.), *Handbook of Nanostructured Materials and Nanotechnology*, Vol. **4**, Academic Press, (2000).
[4] R.H. Magruder, D.L. Kinser, J.E. Wittig, R.A. Zuhr, *Proc. SPIE Int. Soc Opt. Eng.* **180,** 1761 (1993)
[5] M. Sheik-bahae, A.A. Said, T.H. Wei, D.J. Hagan, E.W. Van Stryland, *IEEE J. Quant. Electron* . **26**, 760 (1989)
[6] B. Ghosh, P. Chakraborty, B. Sunderavel and C. Vijayan, *Nucl. Instrum. Meth. in Phys. Res. B* **266**, 1356 (2008).
[7] B. Ghosh, P. Chakraborty, C. Vijayan, P C Deshmukh, S. Mahapatra and P. Mazzoldi, *Materials Letters* **61**, 4512 (2007)
[8] H.W.H. Lee, R.S. Hughes, *Opt. Lett.* **19**, 1708 (1994)
[9] G. Mie, *Ann. Phys*. (Leipz) **25**, 377 (1908)
[10] Baolong Yu, Guosheng Yin, Congshan Zhu, Fuxi Gan, *Optical Materials* **11**, 17 (1998)
[11] Parinda Vasa, B.P.Singh, Praveen Taneja, Pushan Ayub, *Optics Communications* **233**, 297 (2007)
[12] M. Sheik-Bahae, D.J Hagan and E.W. Ban Stryland, *Phys. Rev. Lett.* **65**, 96 (1990)
[13] F. Hache, D. Ricard, C. Flytzanis, U.Kreibig, *Applied Phys. A* **47**, 347 (1998)

Journal of Physics: Conference Series **185** (2009) 012011 doi:10.1088/1742-6596/185/1/012011

Laser assisted (e, 2e) process of the helium atom

S Ghosh Deb and C Sinha[1]

Department of Theoretical Physics, Indian Association for the Cultivation of Science, Jadavpur, Kolkata – 700 032, India

E-mail: chand_sin@hotmail.com

Abstract. The dynamics of the electron impact multiphoton ionization of a Helium atom in the presence of an intense laser field $(n\gamma e,\ 2e)$ is studied theoretically. Significant modifications are noted in the laser assisted cross sections with respect to the field free ones. The results are compared with the recent kinematically complete experiments for high incident energies (1 KeV).

1. Introduction

Laser assisted (LA) excitation and ionization of atoms by charged particles plays a very dominant role in many applied areas such as fusion plasma, plasma heating, high power gas lasers, semiconductor physics etc [1]. Further, the LA electron-atom collision allows experimental observations [2-4] of different multiphoton (MP) phenomenon at relatively moderate field strengths. Despite such immense importance, the detailed dynamics of such LA processes is far from being well understood and is now a subject of great challenge.

A number of theoretical works have been put forward for the LA (e, 2e) process of the H and He atoms both prior [5-7] to and following [8-10] the recent kinematically complete experiment of Hörr et al [2].

2. Theory

The following laser assisted (LA) multiphoton ionization process of Helium atom in its ground state is studied for exchange of n photons with frequency ω and field strength ε_0:

$$e^-(E_i,\ k_i) + He\ (1s) \pm n\gamma\ (\omega,\ \vec{\varepsilon}_0) \rightarrow e^-(E_1,\ \vec{k}_1) + e^-(E_2,\ \vec{k}_2) + He^+ \quad (1)$$

The laser field is chosen to be monochromatic, spatially homogeneous, linearly polarized and is represented by $\vec{\varepsilon}(t) = \vec{\varepsilon}_o \sin(\omega t + \delta)$ corresponding to the vector potential in the Coulomb gauge $\vec{A}(t) = \vec{A}_0 \cos(\omega t + \delta)$ with $\vec{A}_0 = c\,\vec{\varepsilon}_0/\omega$ and δ is the initial phase of the laser field. The laser polarization is chosen to be parallel to the projectile momentum.

The prior form of the transition matrix element for the process (1) is given as[10]:

$$T_{if} = -i \int_{-\infty}^{\infty} dt \left\langle \Psi_f^- \middle| V_i \middle| \psi_i \right\rangle \quad (2)$$

where V_i is the perturbation potential in the initial channel,

[1] To whom any correspondence should be addressed.

$$V_i = -\frac{2}{r_1} + \frac{1}{r_{12}} + \frac{1}{r_{13}} \quad (3)$$

r_1, r_2, r_3 are the position vectors of the projectile electron and the two target electrons (2, 3) respectively, $r_{12} = r_1 - r_2$ and $\vec{r}_{13} = \vec{r}_1 - \vec{r}_3$, ψ_i is the asymptotic initial state wavefunction and Ψ_f^- is the final state wavefunction satisfying the incoming wave boundary condition. It is evident from Eq. (2) that the perturbation V_i vanishes asymptotically for $r_1 \to 0$, r_2, r_3 finite. In the present model the role of the external laser field is to modify the projectile – electron as well as the ground and residual target atom / ion wavefunctions.

Certain distinctive features of the present model are as follows:

(1) Projectile residual target nucleus interaction in the final channel is chosen to be the Coulomb Volkov (CV) wavefunction [7].The approximate CV solution chosen for the scattered and ejected electron presents the advantage of containing the fields in all orders (incoherent in the plane wave Volkov solution), although it completely decouples the projectile-laser and the projectile-screened ion interaction. For low laser intensity or high laser frequency, the coupling of the slow electron with the laser field is much weaker than the Coulomb interaction so that a perturbative treatment can be used.

(2) Target dressing is taken into account both in the initial and final channel in the framework of first order time dependent perturbation theory, which is quite legitimate since the present laser field strength $\varepsilon_0 \ll$ a.u. of field strength $(5\times10^{11}$ V/m).

(3) Apart from the simple CV, a more refined wavefunction, the Modified Coulomb Volkov (MCV) is considered [6,10] for the dressed ejected electron. For low energy ejected electron and / or for strong laser field with low frequency the CV wavefunction is not adequate, a modification is needed [6].

(4) Correlation between the scattered and the ejected electrons is taken into account in the final channel so that the asymptotic three-body boundary condition is satisfied which is essential for a reliable estimate of an ionization process.

The expression for the triple differential cross-section (TDCS) for n photon transfer is given as:

$$\frac{d^3\sigma}{d\Omega_1\, d\Omega_2\, dE_2} = 2\frac{k_1\, k_2}{k_i}\left| T_{if}^{\,n} \right|^2$$

3. Results & Discussions

The triple differential cross-sections (TDCS) are computed for the process (1) with the kinematics chosen mainly in accordance with the experiment [2].

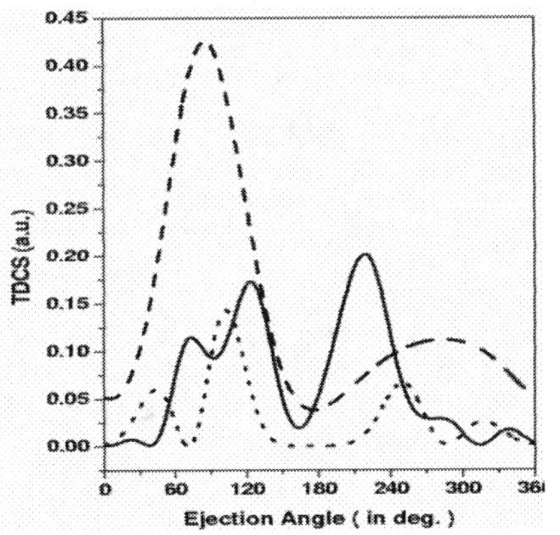

Figure 1. TDCS for $l = 1$ as a function of ejected electron angle (θ_2). The incident energy is $E_i = 1000\ eV$, the ejected electron energy $E_2 = 3.7\ eV$ and the scattering angle is $\theta_1 = 6.4^0$. Field strength $\vec{\varepsilon}_0 = 5\times10^9\ V/m$, laser photon energy $\omega = 1.17\ eV$. Dashed line: FF results, dotted line: CV results and solid line: MCV results

Figure 1 reveals that for single photon transfer ($l = 1$), the CV results are strongly suppressed compared to the FF ones. For the MCV case, on the other hand, the binary peak is suppressed while the recoil peak is enhanced with respect to the FF results.

Figure 2 compares the CV multiphoton TDCS (MTDCS) with the experiment [2], where the present binary peak is normalized to the experimental one. The present results qualitatively follows the experimental behaviour except in the recoil region although the absolute value of the present results overestimates the experiment both in the binary and the recoil region.

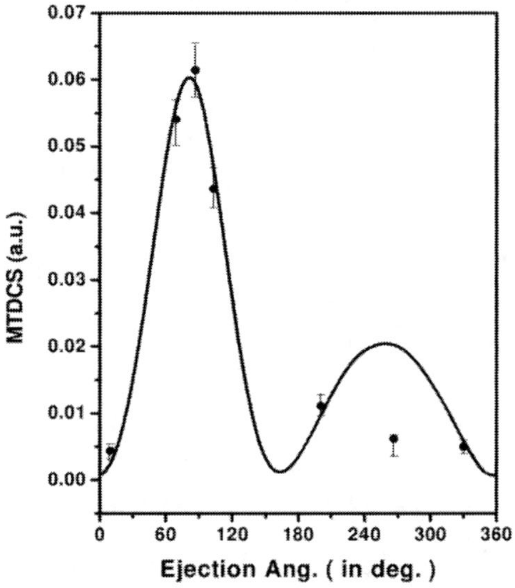

Figure 2. Variation of the CV MTDCS ($\sum_{l=-3}^{3} TDCS$) against the ejected electron angle (θ_2) for same parameters as in figure 1.

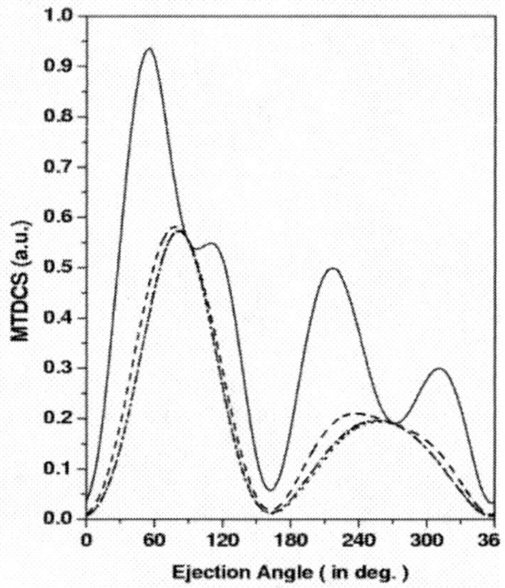

Figure 3. MTDCS versus θ_2 for different laser frequencies for same parameters as in figure 1. Solid line: MCV & Dotted line: CV for $\omega = 1.17\ eV$. Dashed line: MCV & Dashed dot: CV for $\omega = 2\ eV$. The dotted & dashed dot line coincide and is indistinguishable.

Figure 3 exhibits the behaviour of the MTDCS w.r.t. the laser frequency (ω) at $E_i = 1000\ eV$ and it reveals that the magnitude of the MTDCS decreases with increasing laser frequencies as is expected physically. For higher laser frequency, the electron-laser coupling parameter α_0 ($= \varepsilon_0 / \omega^2$) decreases resulting in lesser laser modifications of the cross-section from the field free values. Since the MCV results are much enhanced compared to the FF ones, the former decreases with increasing laser frequency.

Mathematically, the above feature could also be explained from the properties of the Bessel functions [11, 12] occurring in the transition amplitude T_{if}. The argument of the Bessel functions($(\vec{k}_1 + \vec{k}_2 - \vec{k}_i).\vec{\alpha}_0)$) decreases for higher laser frequency i.e. for lower value of α_0. Hence, with decreasing value of the argument $J_0(x)$ increases while $J_\ell(x)$ (for $\ell \neq 0$) decreases

Journal of Physics: Conference Series **185** (2009) 012011 doi:10.1088/1742-6596/185/1/012011

so that when the individual contributions due to different ℓ values (e.g. $0, \pm1, \pm2, \pm3$) are added in the summed TDCS (MTDCS) the net effect is that the summed cross-section diminishes.

Further, the wavefunction of the ejected electron occurring in ψ_f of Eq. (2) contains a term $k_2.\vec{\alpha}_0$ (vide Eq.5 of [10]). This also explains the above behaviour of the TDCS w.r.t frequency.

For low frequency e.g. $\omega = 1.17\ eV\ (\alpha_0\ \sim\ 5.25)$, there is significant deviation between the CV and the MCV results while the deviation diminishes with increasing frequency / decreasing α_0 (electron–laser coupling) (vide figure 3) so that for $\alpha_0 \sim 0.64$ the two (CV & MCV) almost coincide.

4. Conclusions:

A strong enhancement is noted in the present multiphoton binary peak intensity, in qualitative agreement with the experiment. The deviation between the present CV and MCV multiphoton TDCS becomes negligible for higher frequency / lower field strength (i.e. $\alpha_0 < 1$).

5. References:

[1] Dörner R. et al 2002 *Adv. At. Mol. Opt. Phys.* **48** 1

[2] Höhr C, Dorn A, Najjari B, Fischer D, Schröter C D and Ullrich J 2005 *Phys. Rev. Lett.* **94** 153201 (see also the other references cited therein in this context)

[3] Höhr C, Dorn A, Najjari B, Fischer D, Schröter C D and Ullrich J 2007 *Journal of Electron Spectroscopy* **161** 172 (see also other references cited therein in this context)

[4] Sakhelashvili G, Dorn A, Höhr C, Ullrich J, Kheifets A S, Lower J and Bartschat K 2005 *Phys.Rev. Lett.* **95** 033201

[5] Zarcone M, Moores D L and McDowell M R C 1983 *J. Phys. B: At. Mol.Opt. Phys.* **16** L11 (see also other references cited therein in this context).

[6] Martin P, Veniard V, Maquet A, Francken P and Joachain C J 1989 *Phys. Rev. A.* **39** 6178 (see also other references cited therein)

[7] Zangara R, Cavaliere P, Leone C and Ferrante G 1982 *J. Phys. B.* **15** 3881 (see also other references cited therein in this context)

[8] Ghosh Deb S, Roy S and Sinha C 2007 *Arxiv:Physics.atoms-Ph* 0802.1197 v1

[9] Pan J, Li S M, Berakder J 2007 *Optics Letters* **32** 585 (see also other references cited therein)

[10] Chattopadhyay A and Sinha C 2005 *Phys. Rev. A.* **72** 053406

[11] Abramowitz M and Stegun I A 1970 *Handbook of Mathematical Functions* (New York: Dover) pp 359,390

[12] Watson G N 1944 *A Treatise on the Theory of Bessel Functions* 2nd edition (Cambridge: Cambridge University press) p 360

Numerical Study of Contrast Degradation of BEC Interferometers and Countermeasures

G. Gopi Krishna[1,2], Masatoshi Umetsu[2], Chien-Nan Liu[3] and Shinichi Watanabe[2]

[1]Adarsha Junior College, Giddalur, Andhra Pradesh, India
[2]Department of Applied Physics and Chemistry, University of Electro-Communications, 1-5-1 Chofugaoka, Chofu-shi, Tokyo 182-8585, Japan
[3]Department of Physics, Fu-Jen Catholic University, Taipei 24205, Taiwan, The Republic of China

E-mail: `shin@pc.uec.ac.jp`

Abstract. Despite promising features, the BEC-based atomic interferometer needs to overcome momentum recoil caused by the internal atom-atom interaction so as to serve for high precision measurements. We visualize the momentum recoil using the Husimi distribution function and study the BEC's dynamics in phase space. In so doing, we discuss possible countermeasures for contrast degradation in the manner of ref.[1] as well as hitherto unexpected interesting dynamical features.

1. Introduction

The Bose-Einstein condensates is believed to have promising capacities for precision inertial measurements owing to their high brilliance and high coherence. Proof-of-principle experiments have been indeed demonstrated for BEC-based atom interferometer schemes in recent years[2, 3, 4, 5]. Crutial atom optical elements such as mirrors, beam-splitters, and optical lattice gratings have been also studied as part of coherent matter-wave manipulation of cold atoms (see [6] and references therein for non-condensates). Exploring various applications of the BEC-based atom interferometer is obviously very important. However, the atom-atom interaction among 10^4 or even a greater number of atoms causes some hitches like degradation of the interference fringe contrast on account of dephasing as well as decoherence in the course of collisions. It is thus also important to explore remedial measures with a view to applications in precision inertial measurement.

The fringe contrast is dictated by both the amplitude and relative phase of the interfering components. Thus, its study motivates a representation of the dynamics in terms of the phase and amplitude in an appropriate phase space. In quantum electronics, three alternative quasi-probability distributions[7], namely the Glauber-Sudarshan P-function, the Wigner function, and the Husimi Q-function have been studied and applied for analysis. These functions serve as a tool for monitoring the phase and amplitude information, thus also provide insight into the phase space dynamics of the split BEC components, and into the operation of the atom interferometer. Of the three, the positive-definite Husimi function is what we employ for analysis in the present paper, namely

$$P_\alpha(z,p;t) = \frac{1}{2\pi\hbar}\frac{1}{\sqrt{\pi\alpha}}\left|\int_{-\infty}^{\infty} \mathrm{e}^{-\frac{(q-z)^2}{2\alpha}-ipq/\hbar}\Psi(q,t)\,dq\right|^2, \tag{1}$$

defined at an arbitrary instance t where $\Psi(q,t)$ represents the solution of the time-dependent Gross-Pitaevskii equation (2) below, and α sets the phase-space resolution as appropriate for each specific case. We shall describe our theoretical approach, and then discuss the momentum recoil in phase space. We examine two alternative countermeasures, namely momentum-shift and time-delay separately. An interesting quantum feature due to some local oscillations in the colliding BEC components will be mentioned at the end as a preview of the ongoing investigation.

2. Theoretical Approach

As for the equation of motion, we use the Gross-Pitaevskii equation, namely the mean field approximation to numerically study the dynamics of a condensate split by the Bragg diffraction in a 1D Mach-Zehnder-type atom interferometer. It must be thus understood that major features of the coherent components are presumed to be largely retained during the process despite possible inelastic transitions left out here. We have

$$i\hbar\frac{\partial\Psi(z,t)}{\partial t} = \left(-\frac{\hbar^2}{2m}\frac{\partial^2}{\partial z^2} + f_1(t)\times\frac{1}{2}m\omega_0^2 z^2 + f_2(t)\times\hbar\Omega\cos(\tilde{k}\;z-\tilde{\omega}\;t) + g_{1\mathrm{d}}N|\Psi|^2\right)\Psi(z,t), \tag{2}$$

where $g_{1\mathrm{d}} = -2\hbar^2/ma_{1\mathrm{d}}$ is the effective coupling strength deduced from the 3D s-wave scattering length[8]. Here the energy-momentum conservation reads $\tilde{k} = k_1 + k_2 \simeq 2k$ and the detuning $\tilde{\omega} = \omega_2 - \omega_1$. This equation is applied to the specific case of the 1-dimensional ^{87}Rb atom, presuming a 1D trap and the Bragg diffraction laser field under an actual experimental condition. Here $f_1(t)$ and $f_2(t)$ give the time profile of the externally controlled fields. A typical time diagram for the $\pi/2$-π-$\pi/2$ pulse sequence is shown for the case of trap-off in Fig. 1 with the spatial evolution of the split BEC components in parallel. Here the trap is applied for confining the atoms and setting up the initial BEC, but is turned off once the interferometer sequence begins. The signal is picked up by measuring the population of either of the two momentum states $|0\hbar k\rangle$ or $|2\hbar k\rangle$ after merging them by the second $\pi/2$ pulse where $\hbar k$ represents the momentum of a single photon. The contrast is the dynamic range of the signal, namely $v = \dfrac{I_{\max} - I_{\min}}{I_{\max} + I_{\min}}$ where I represents the signal amplitude. It should be noted that in other types of interferometer such as in the Horikoshi-Nakagawa experiment, the trap field is exploited for its isochronism and as a mirror (π-pulse) in order to reduce contrast degradation.

3. Contrast Degradation and Compensation

3.1. Exceptional Contrast Recovery in the case of Trap-On ($f_1(t) \equiv 1$)

Let us first show in Fig. 2, the contrast of an *imperfectly* split BEC in the case of trap-on ($f_1(t) \equiv 1$), "imperfectly" in the sense that the π pulse is applied while the split components still overlap spatially. What happens during this short period is sufficiently rich and proves instructive for our purpose. Let us also note that this corresponds to the actual time span used in earlier proof-of-principle experiments. The parameter of importance here is $T = T_2 + T_1 = 2T_1$. This specific case shows an unexpected recovery of contrast for $N = 8000$, *i.e.* for stronger nonlinearity, after the decline to nearly zero. A somewhat weaker rise is seen near the shoulder in the case of $N = 3000$ too. Let us examine the phase space dynamics that accounts for this puzzling phenomenon. Fig. 3 shows what happens for $N = 8000$ shortly before the third pulse is applied by formally applying the $\pi/2$-pulse unitary matrix to the $|0\hbar k\rangle$ and $|2\hbar k\rangle$ components. The corresponding time t is marked in Fig. 2 by a circle on the local maximum of the contrast. Fig. 4 on the other hand corresponds to the local minimum marked by another circle. As a function of T, there is spatial shift in the relative position coordinates of the components, allowing them to eventually acquire a larger amplitude overlap in phase space, the trap counterbalancing the recoil (see Fig. 2 for the two values of N). This feature of contrast recovery manifests itself as a linear protrusion in Fig.3 of reference [9]. The relative phase of the two components plays a key role in defining finer structures, a fact to be brought attention to for the trap-off case below, *i.e.* $f_1(t) \equiv 0$ at $t > 0$.

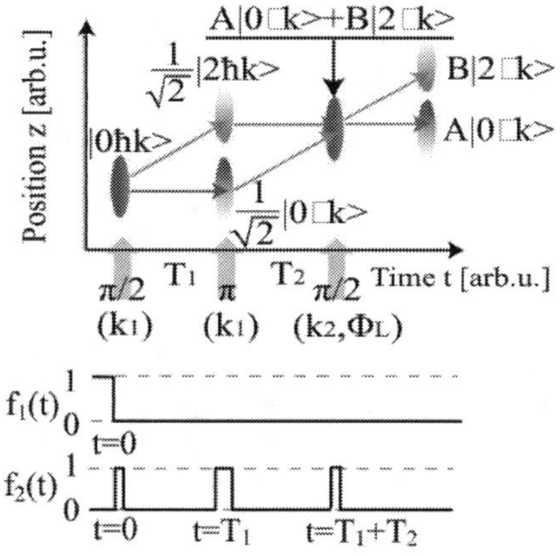

Figure 1. Position vs time for 1D Mach-Zehnder interferometer (above). Time sequence for the trap $f_1(t)$ and the laser pulse sequence $f_2(t)$ for the trap-off case (below).

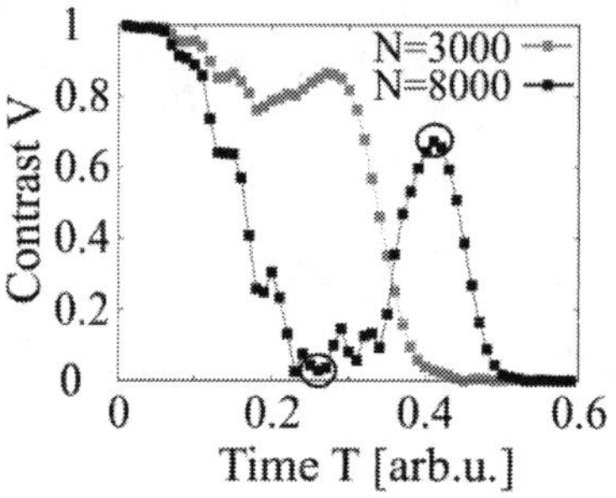

Figure 2. Contrast vs T for an imperfectly split initial stage with trap on. Strong contrast recovery for $N = 8000$ and a small hump around the shoulder for $N = 3000$. Here $T = T_1 + T_2 = 2T_1$.

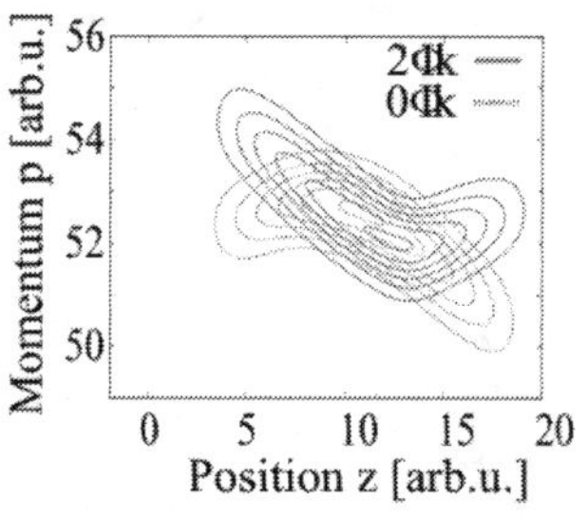

Figure 3. Husimi distribution shown with contours near local maximum at $T \simeq 0.4$ marked by circle in Fig.2. Observe relatively large overlap.

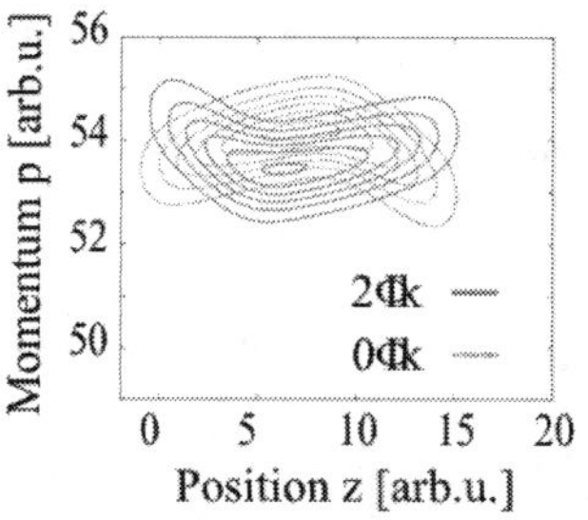

Figure 4. Husimi distribution shown with contours near local minimum at $T \simeq 0.2$ marked by circle in Fig.2. Observe relatively small overlap.

3.2. Countermeasures in the case of Trap-Off ($f_1(t) \equiv 0,\ t > 0$)

Compensating for the momentum recoil by shifting the two components in phase space can counteract the contrast degradation to some extent. One way to shift in phase space is to impart momentum difference $\Delta k = \tilde{k}_2 - \tilde{k}_1$ (see Eq. (2) for $\tilde{k}_1$ and $\tilde{k}_2$). The time-delay approach is another remedial measure but is somewhat nontrivial for the trap-on case[1] and requires an additional consideration. Here we present the case of trap-off. The Δk scheme leads to a considerable improvement as in Fig. 5, and the ΔT scheme to some improvement as in Fig. 6. In principle, one may apply the ΔT and Δk schemes in combination to achieve an even better amplitude overlap of the BEC components.

Let us now consider the relative phase between the split components. Treating once again the time period where the BEC is *imperfectly* split, we map the contrast using color coding in Fig. 7 as a function of T and N, the latter being the number of particles. Consider a section at some fixed T. Then as a function of N, we notice the contrast declines rapidly due to the increase in momentum recoil. However, there appear residual oscillations in contrast, namely an interference fringe with respect to N. The Husimi distribution function (not shown) indicates that each time the contrast attains a local minimum, there emerges a unit of circulation in phase space resembling a vortex in a rotating 2D BEC. The local circulations pertain to local oscillations in coordinate space so that an increase in vorticity by unity corresponds to a creation of vibrational quantum. A further investigation of this phenomenon is currently under way.

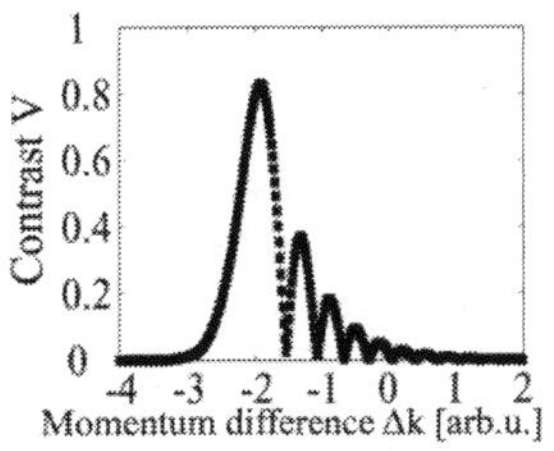

Figure 5. Δk contrast scheme: Trap-off.

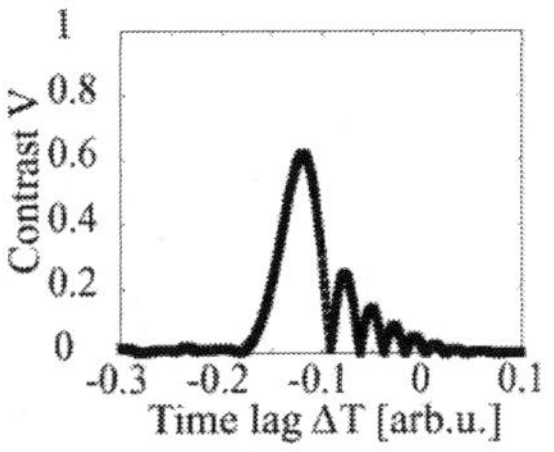

Figure 6. ΔT contrast scheme: Trap-off.

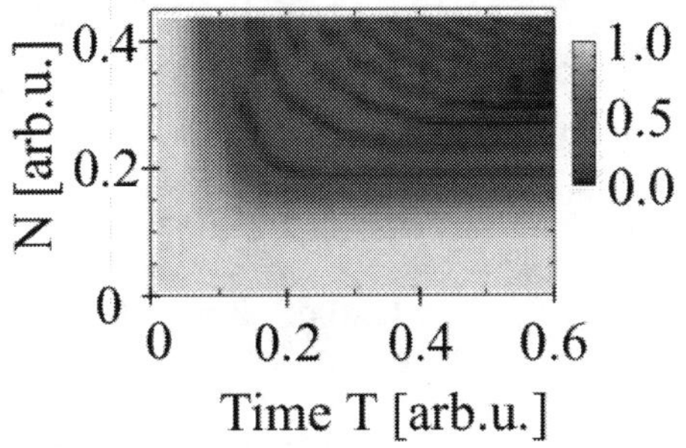

Figure 7. Contrast color-mapped as a function of T and N. Observe fringes in the low contrast regions. Trap is off.

4. Conclusions

We studied the effect of momentum recoil of the BEC components that takes place at the moment of splitting as well as merging. The contrast degradation caused by such recoil can be remedied at least partially by the time-delay and/or the momentum shift scheme(s). BEC-based atomic interferometers are still at their infancy in the sense that they have not been applied for high precision measurements in real systems. There are many technical issues to be cleared both theoretically and experimentally.

This work was partially supported by the JSPS Bilateral joint program between the U.S. and Japan. An early stage of this work was supported in part by a Grant-in-Aid for Scientific Research from the Ministry of Education, Culture, Sports, Science and Technology, Japan, and also in part by the 21st Century COE program on "Coherent Optical Science".

References

[1] Chien-Nan Liu, G. Gopi Krishna, Masatoshi Umetsu, and Shinichi Watanabe, Phys. Rev. A **79**, 013606 (2009)
[2] Y. Torii, *et al*, Phys. Rev. A **61**, 041602(R) (2000).
[3] Y. J. Wang, *et al*, Phys. Rev. Lett. **94**. 090405 (2005).
[4] O. Garcia, B. Deissler, K. J. Hughes, J. M. Reeves, and C. A. Sackett, Phys. Rev. A **74**, 031601(R) (2006).
[5] Y. Shin, *et al*, Phys. Rev. Lett. **92**, 050405 (2004); Y. Shin, et al, Phys. Rev. A **72**, 021604(R) (2005).
[6] See, for instance, the following article and references therein. Hilmar Oberst, Yoshihisa Tashir, Kazuko Shimizu, and Fujio Shimizu, Phys. Rev. A **71**, 052901 (2005).
[7] See, for instance, Chapter 8 of "Quantum Interference and Coherence: Theory and Experiments" by Zbigniew Ficek and Stuart Swain, (Springer Science, Inc., New York, 2004)
[8] M. Olshanii and V. Dunjko, arXiv:cond-mat/0505358.
[9] James A. Stickney, Rudra P. Kafle, Dana Z. Anderson, and Alex A. Zozulya, Phys. Rev. A **77**, 043604 (2008)

Journal of Physics: Conference Series **185** (2009) 012013 doi:10.1088/1742-6596/185/1/012013

Storage Modes in a Low Energy Charged Particle Storage Ring for Atomic and Molecular Experiments

P Hammond[1], T J Reddish[2], A J Alderman[1], D R Tessier[2] and F H Read[3]

[1] School of Physics, CAMSP, University of Western Australia, Perth WA6009, Australia

[2] Department of Physics, University of Windsor, Windsor N9B 3P4, Canada

[3] School of Physics & Astronomy, University of Manchester, Manchester M13 9PL, UK.

E-mail: hammond@physics.uwa.edu.au

Abstract. Specific storage modes have been identified and explored both experimentally and computationally for a new type of low energy charged particle storage ring for use in Atomic & Molecular Physics experiments [Tessier *et al*, *Phys. Rev. Lett.*, **99**, 253201, (2007)]. These modes offer the possibility of storing monoenergetic charged particle beams injected into the ring. Alternatively, some of the storage modes identified may be appropriate for energy filtering a stored charged particle beam within the ring. The ring is entirely electrostatic and has been shown experimentally to be capable of storing low energy electrons with storage lifetimes in the region 50 microseconds (~150 orbits) that are vacuum pressure limited. The ring is constructed from two 180° hemispherical deflector analyzers which are interconnected by two identical cylindrical lens stacks. Each lens stack is primarily controlled by 3 voltages. The storage modes observed occur at specific combinations of these voltages and are consistent with a multiple orbit charged particle optics model of the apparatus explored using both a matrix approach and trajectory integration in a 3 dimensional electrostatic model.

1. Introduction

Historically charged particle storage rings have been achieved using magnetic lenses with the development goal to reach ever higher kinetic energies for the particles circulating in the ring as exemplified by cyclotrons, synchrotrons and tokomak's. Electrostatic storage systems have been a much later development and are typically used for low kinetic energy charged particles [1].

The design aim for the electrostatic charged particle storage system to be described here has been not only to achieve passive storage of charged particles, but also provide flexible setting of the kinetic energy and energy filtering of the stored beam to facilitate collision studies in atomic and molecular physics. Essential to achieving these aims is a charged particle optics model in which the predicted design performance closely matches the observed behaviour of the storage ring. This article briefly describes the progress made in developing and exploring the charged particle optics for an electrostatic storage ring which has already been demonstrated to achieve, when used for electrons, storage times in excess of 150 μs [2].

2. Apparatus

The apparatus is shown schematically in figure 1. It consists of two electrostatic hemispherical deflector analysers (HDA's) interconnected by two identical cylindrical lens systems. In the present apparatus these consist of a field free region located midway between the HDA's and separated from the HDA's by a cylindrically symmetric three element lens. Electrons are introduced into the race-track shaped ring using an external pulsed electron beam injected through HDA 2 while the voltages of the analyzer hemispheres are both briefly held at voltage V_1. Lens 4 transfers the electrons into the source region held at a voltage V_3. During the ~200 ns time period taken by the clockwise traveling electrons to transit around the ring to reach the entrance of HDA 2, the voltages of HDA 2 are rapidly slewed to match the voltages of HDA 1 and allow electrons to return to Lens 4 and re-enter the source region. In the usage of the apparatus to be described here, the source and target regions are held at the same voltage V_3 and each of the 4 lenses are identical in geometrical structure and each has a middle lens element held at the same voltage V_2. The orbit length for a median electron trajectory around the complete storage ring is 0.64 m, with typical orbit times of ~330 ns for electron kinetic energies averaging 20 eV [2].

In the target region electrons in the stored beam are crossed with a gas jet aligned normal to the beam. A channel electron multiplier views the centre of the target region and detects scattered electrons. The yield of scattered electrons is recorded as a function of time from the electron injection pulse using a time to amplitude convertor with output pulse heights analysed using a computer based pulse height analyzer.

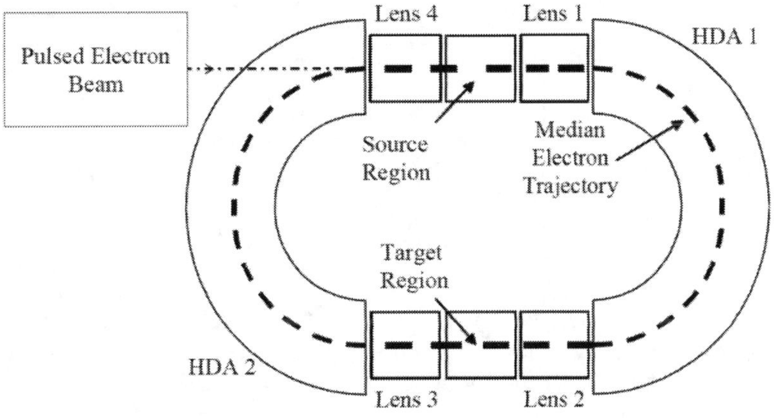

Figure 1. Schematic layout of the low-energy charged particle storage ring configured for electrons.

3. Charged Particle Optics

A charged particle trajectory from the source region passes through three optical components in the transit of one half-orbit of the ring to reach the target region. In the absence of aberrations, each of these components has a unique transfer matrix; m_l for Lens 1, m_h for HDA 1 and m_2 for Lens 2 to give an overall source to target region transfer matrix of $M_{st} = m_2 m_h m_l$. In terms of characteristic lengths associated with thick lenses:

$$M_{st} = \frac{1}{f_1 f_2}\begin{pmatrix} f_1 f_2 - 2K_1 K_2 & 2K_1(f_1 f_2 - K_1 K_2) \\ -2K_2 & f_1 f_2 - 2K_1 K_2 \end{pmatrix} \tag{1}$$

where $K_1 = P - F_1$, $K_2 = Q - F_2$ and focal lengths f_1, f_2 and mid-focal lengths F_1, F_2 of Lens 1. The position of the source and the entrance of HDA 1 with respect to the reference plane of Lens 1 are given by P and Q respectively and are not necessarily the positions of conjugate objects and images. In the specific spatial and voltage geometry under consideration here, the target region is held at the same voltage V_3 as the source region, and the voltages on HDA 1 are the same as HDA 2, Hence in voltage terms Lens 2 is equivalent to Lens 1 but traversed in the opposite direction and the focal lengths of the two lenses are interrelated. This enables m_2 to be expressed in terms of Lens 1 focal properties.

The transit from the target region back to the source region is, in terms of transfer optics, the same as that from source to target since Lens 3 is equivalent to Lens 1 and Lens 4 is equivalent to Lens 2. Hence M_{st} is the transfer matrix for each half orbit of the system. M_{st} can be viewed as a phase space rotation matrix which, for particular numbers of half-orbits H causes $(M_{st})^H$ to rotate phase space through an angle π (in which case trajectories map onto the same trajectory coordinates after $2H$ half orbits) or through an angle 2π (in which case the same coordinates are achieved after H half-orbits). This leads to a general expression for the relationship between focal and mid-focal lengths which needs to be satisfied to achieve stable orbits:

$$\frac{K_1 K_2}{f_1 f_2} = \sin^2\left(\frac{m\pi}{2H}\right) \tag{2}$$

where m is an integer and $0 < m < H$ [3].

Specific focal and mid-focal lengths of three element electrostatic lenses are set by specific voltage ratios V_3/V_1 and V_2/V_1. In the present apparatus the voltages V_1 and V_3 are selected to achieve particular pass energies in the HDA's and collision energy in the target region, respectively. It is therefore V_2 which can be adjusted to vary the focal properties of the lenses to enable equation 2 to be satisfied for a particular m/H ratio. To test equation 2 we have used the parameterisation of focal lengths given by Harting and Read [4] to deduce the mapping from V_2 to m/H for specific V_3/V_1 values.

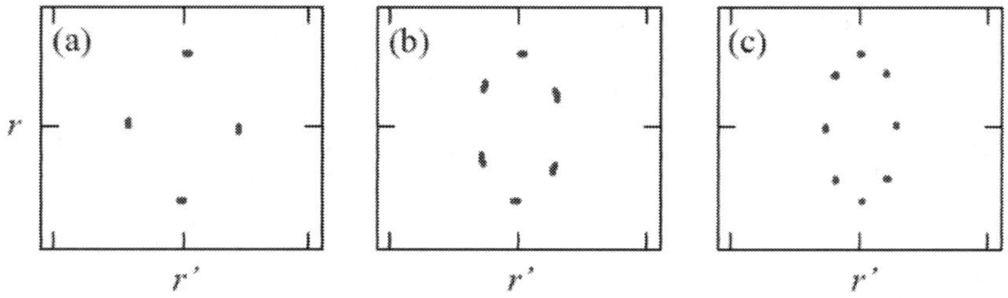

Figure 2. The trajectory coordinates (r, r') in the source and target regions for $V_3/V_1 = 1$ and V_2 voltages derived from the parameterised focal length data [4] and predicted to produce (H,m) modes (a) (2,1) $V_2 = 92.2$ V (b) (3,1) $V_2 = 80.1$ V (c) (4,1) $V_2 = 74.3$ V

In figure 2 are shown the trajectory coordinates (r, r') in the source and target regions resulting from using the focal and mid-focal lengths in a matrix model for 120 half orbits at the V_2 values predicted to achieve specific (H,m) modes in equation 2, namely (2,1), (3,1) and (4,1). When m is odd it is expected phase space will rotate through π after H half-orbits and result in $2H$ phase space "dots" as seen in figure 2. A trajectory integration model using CPO3D [5], in which the apparatus is modelled and which implicitly includes the effects of aberrations, reproduces figure 2 phase space diagrams using V_2 voltages within 3% of those calculated to satisfy equation 2.

4. Experimental Results and Discussion

Experimental results from the storage ring for $V_3/V_1 = 1$ are shown in figure 3. The logarithmic yield of scattered electrons from the target region is plotted as a function of storage time and applied voltage V_2 (recorded in 2 V increments). (H,m) modes of low H and $m = 1$ and 2 are marked on the figure at the predicted voltages from equation 2. Observed structures lie approximately 2 V above the predicted values. Very little yield is observed for voltages above 100 V; the reasons for this are unclear. The observation of reduced storage time near the (5,2) mode may be indicative of instability arising from HDA aberrations for even-*m* modes. Some evidence for this behaviour has been observed in the trajectory integration modelling, and to a lesser extent in an aberration modified matrix model.

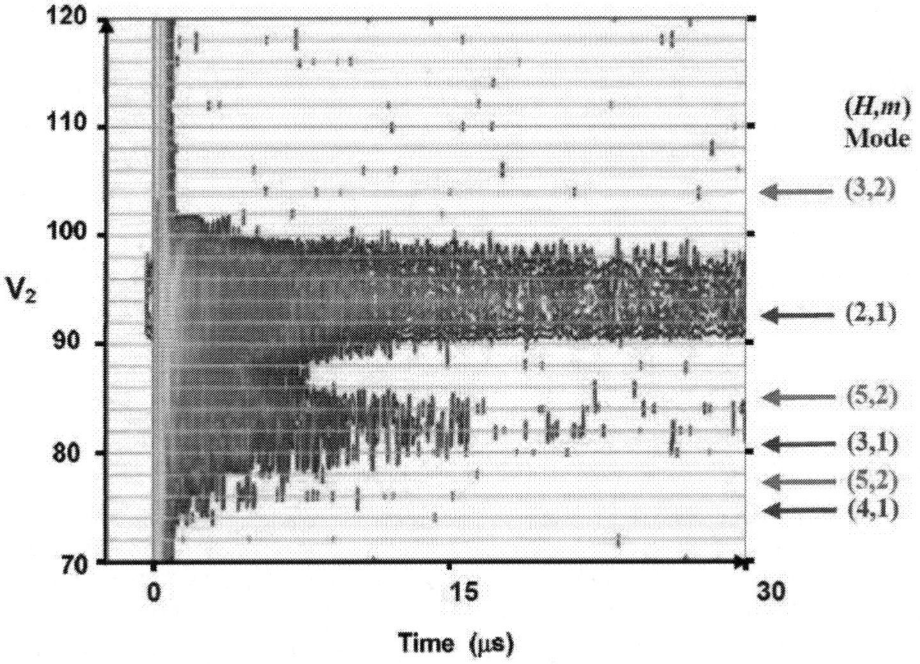

Figure 3. The yield of scattered electrons from the target region as a function of storage time and applied lens voltage V_2 for V_3/V_1 = 1. The yield scale is logarithmic and a maximum at zero μs. The deduced (*H,m*) mode voltages from equation 2 using the parameterized focal length data [4] are indicated.

5. Conclusions

A transfer optics model of a 4 lens electrostatic storage ring for low energy charged particles has been developed. Specific lens voltage ranges are predicted to achieve storage. Excellent agreement is achieved between experimentally observed electron storage and the model developed. The experimental results and theoretical modeling suggest HDA aberrations cause instability in specific regions of lens voltage.

Acknowledgements

Financial support from The Royal Society, The Australian Research Council and the Canadian Funds: CFI, OIT and NSERC are gratefully acknowledged.

References

[1] Andersen L H, Hebner O, Zajfman D 2004 *J. Phys. B.* **37** R57
[2] Tessier D R, Niu Y, Seccombe D P, Reddish T J, Alderman A J, Birdsey B G, Read F H, and Hammond P 2007 *Phys. Rev. Lett.* **99** 253201
[3] Hammond P, Reddish T J, Alderman A J, Tessier D R and Read F H *in preparation*
[4] Harting E and Read F H 1976 *Electrostatic Lenses* Elsevier, Amsterdam
[5] CPO Programs, available on the Web site http://www.electronoptics.com/

Journal of Physics: Conference Series **185** (2009) 012014 doi:10.1088/1742-6596/185/1/012014

Coherent rotational dynamics of molecules induced by intense ultrafast laser fields

Hirokazu Hasegawa and Yasuhiro Ohshima[1]
Institute for Molecular Science, Myodaiji, Okazaki 444-8585, Japan, and
The Graduate University for Advanced Studies, Myodaiji, Okazaki 444-8585, Japan

E-mail: ohshima@ims.ac.jp

Abstract. Coherent rotational dynamics of gas-phase molecules induced by the excitation with an intense nonresonant ultrafast laser field is discussed. In particular, utilities of quantum-state resolved spectroscopic probe after the nonadiabatic rotational excitation (NAREX) by the ultrafast pulse(s) are described with an example of double-pulse excitation experiments on NO, which provide detailed information on excitation pathways in NAREX. Future prospects of the present method are briefly discussed.

1. Introduction

When a gaseous molecular sample is irradiated by an intense nonresonant ultrafast laser pulse, the laser field exerts a torque that aligns the molecular axis along the laser polarization vector, due to the interaction with the molecular anisotropic polarizability. The process, now commonly termed as "nonadiabatic molecular alignment," has been attracting much attention because of interesting physics involved and possible broader applications [1]. Here the field–molecule interaction only remains in much shorter duration than the characteristic time for molecular rotation, and thus the nonadiabatic alignment is inherently accomplished by nonadiabatic rotational excitation (NAREX), where rotation of the molecules is coherently excited to create a rotational quantum wavepacket (WP), a non-stationary superposition of rotational eigenstates.

While most of the previous studies have been devoted to exploring the degree of molecular alignment and its time evolution by using various real-time measurements [2–6], the state distribution in NAREX has not been focused on despite of the following physical significances. First, details concerning the excitation process during exposure to laser fields are encoded in the state distribution. Second, molecules in highly anisotropic states exerted to NAREX will be particularly useful for studies on, e.g., quantum-state resolved collision or reaction dynamics. Third, NAREX can become a competent tool for changing and controlling the rotational state distribution of molecules, especially when recent elaborate methods, i.e., double-pulse pair [7–9] and shaped pulse [10,11], are invoked.

Recently, we have developed a method for measurements of the rotational-state distribution of a molecule irradiated by a nonresonant femtosecond (fs) laser pulse by utilizing a frequency-domain spectroscopic probe with almost full quantum-state resolution [12–14]. Two different classes of molecules have been examined so far; a diatomic molecule, nitric oxide (NO), and a symmetric-top molecule, benzene (C_6H_6). They have been selected primarily because of easy access for probing

[1] To whom any correspondence should be addressed.

distributions by resonant-enhanced multiphoton ionization (REMPI). The present manuscript describes some details of the experimental study on NO. In particular, utilities of double-pulse excitation are demonstrated to provide definite information on excitation pathways in NAREX.

2. Experimental

A pulsed molecular beam containing a trace amount of a NO diluted in a He or Ne buffer gas was produced in a differentially pumped vacuum chamber. The rotational temperature of a molecular ensemble in the beam was adiabatically cooled to < 2 K. The molecular beam was then irradiated with the fundamental output (~ 800 nm) from a fs Ti:Sapphire laser system (150 fs duration with 1 mJ/pulse) operated at 1 kHz. This ultrafast pulse is referred as a pump hereafter. In the case of double-pulse excitation experiments, the output from the fs laser systems is divided into a nearly iso-enegetic pulse pair with a variable time delay by using a Michelson interferometer. After the excitation with the pump pulse(s), molecules were ionized via (1 + 1) REMPI by the probe pulse (~ 226 nm) from the frequency-doubled output of a nanosecond XeCl excimer laser-pumped dye laser system. The bandwidth of the probe pulse was 0.4 cm^{-1}. The generated molecular ions are detected by a time-of-flight mass spectrometer (TOF-MS). The time delay between the pump and the probe pulses is fixed to 100 ns. Since the observed REMPI excitation spectra unchanged when the pump-probe delay time was varied down to 50 ns, collisional rotational relaxation is negligible. The polarization direction of the pump and the probe pulses were set to be parallel to each other, and along the ion-extraction field in the TOF-MS ionization region. In order to probe only a molecular ensemble exposed by a uniform pump pulse field, the probe pulse was kept as weak as possible (≤ 10 μJ/pulse) and tightly focused with a plano-convex lens with f = 170 mm, while the pump pulse was loosely focused by a plano-convex lens with f = 300 mm and its focal position was slightly shifted from the probe region.

3. Results and discussions

Figure 1 shows the (1+1) REMPI excitation spectra of A $^2\Sigma^+ - X\ ^2\Pi_{1/2}$ (0, 0) band of NO, recorded at various pump pulse energies, along with the corresponding rotational-state distributions in the X state.

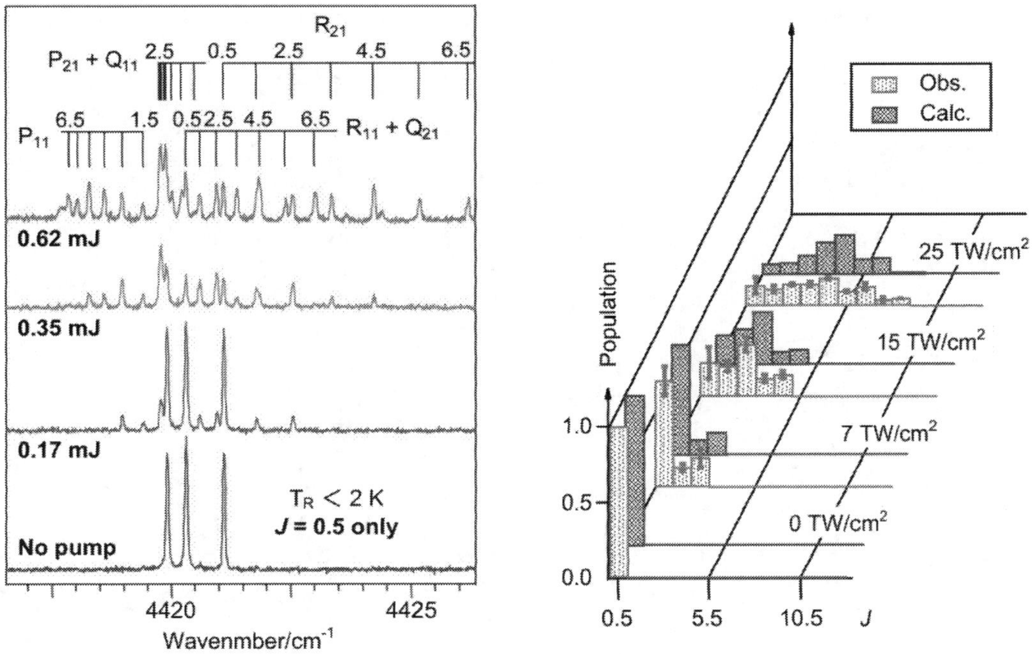

Figure 1. Left: Excitation spectra of NO, recorded after the irradiation of pump pulse with various energies. Right: Observed and calculated rotational-state distributions.

The delay-time dependence of the rotational-state distribution after the double-pulse excitation was examined as the REMPI signals for each transition with different J values in the $P_{11}(J)$ branch, as shown in Figure 2. Characteristic modulations in population are observed against the delay. The observed modulations result from the interference between the rotational eigenstates constructing the WP. Accordingly, information on the eigenstate components constituting the WP and the coupling strength between them can be extracted from the modulated signals. Figure 2 also displays the Fourier-transformed spectra of the observed beat signals for J = 1.5 – 6.5. Peaks appear at the positions corresponding to the rotational energy differences. The peak assignments are also shown in the top of Figure 2 (right). This figure shows that the rotational eigenstates constructing the WP can be classified into two groups. One group referred as group I, which is composed of J = 2.5, 4.5, and 6.5, has the common peaks such as 8B, 16B, 24B, and 40B. On the other hand, another group named as group II, which is made from J = 1.5, 3.5, and 5.5, has the common peaks such as 3B, 12B, 15B, 20B, and 32B.

The existence of two groups in the wave packet can be explained by the fact that the coupling matrix elements with $\Delta J = \pm 2$ are larger than those with $\Delta J = \pm 1$. Consequently, the initial J = 0.5 state is strongly coupled to J = 2.5. And then, the produced J = 2.5 state is also excited to J = 4.5 by the $\Delta J = \pm 2$ interaction. As the results, the excitation chain, $J = 0.5 \rightarrow 2.5 \rightarrow 4.5 \rightarrow 6.5 \rightarrow \ldots$, makes the group I. The group II is constructed by a similar interaction chain, $J = 0.5 \rightarrow 1.5 \rightarrow 3.5 \rightarrow 5.5 \rightarrow \ldots$, starting with relatively weak interaction between J = 0.5 and 1.5 through the $\Delta J = \pm 1$ interaction. The coupling between the two groups is very weak because of the substantially small $\Delta J = \pm 1$ interaction. The delay-time dependence of the rotational-state distribution was also calculated by numerically solving the time-dependent Schrödinger equation. The results are represented also in Figures 1 and 2. The agreement between the experiment and the calculation is remarkably good. The similar results have been reported separately by Meijer *et al.* [15].

The delay-time dependence of the rotational-state distribution exhibits a noteworthy feature. The state distribution changes periodically with the interval of 20 ps. This revival time is twice larger than well known value, $T_{rev} = 1/(2Bc)$ [1–7]. This difference originates from the energy level structure of NO. For a linear molecule in non-degenerate electronic states, such as N_2, O_2, and CO_2, the angular momentum J is an integer. Thus the rotational energy, E_J/hc = 0, 2B, 6B, 12B, …, $BJ(J+1)$, has the greatest common measure (GCM) of 2B. The rotational WP is invariant under the time evolution from

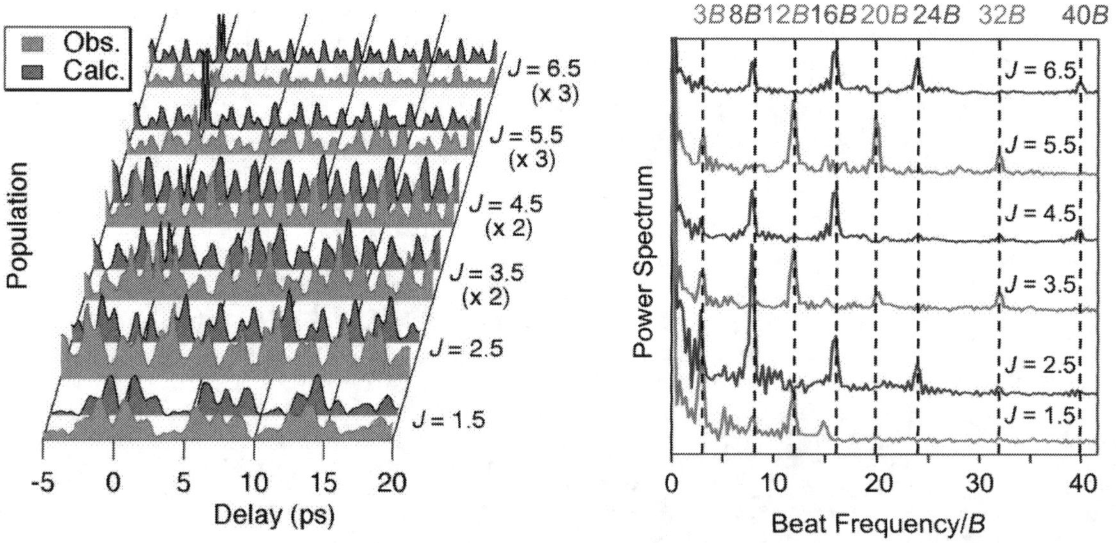

Figure 2. Left: Population dependence of each rotational state of NO on the time delay of double pump pulses. Right: Fourier-transformed signals of delay-dependent rotational population.

t to $t + T_{rev}$ because $\exp(-iE_J T_{rev}/\hbar)=1$. On the other hand, molecules in degenerate electronic states, such as NO and OH, have a half integer angular momentum J. The rotational energy is expressed as: $E_J/hc = B, 4B, 9B, \ldots, B(J + 1/2)^2$, with the GCM of B. The rotational WP of those molecules returns to the initial value by time evolution from t to $t + 1/(Bc)$. Whether the total angular momentum J is integer or half integer plays an essential role for the full revival time being $T_{rev} = 1/(2Bc)$ or $1/(Bc)$. The observed period of 20 ps for NO agrees well with $1/(Bc) = 19.7$ ps calculated with the precisely known value of $B = 1.696$ cm^{-1}.

4. Conclusion

As clearly demonstrated in the previous section, the quantum-state resolved probe after the NAREX by an intense nonresonant ultrafast laser field can afford an indispensable insight into the NAREX process, in particular, when the double-pulse excitation is applied. One of the big advantages of the present method is its state selectivity. We can follow the individual NAREX processes from different initial states by selecting appropriate states to be probed out from a mixed ensemble. By contrast, almost all the previously adopted methods [1–11] rely on the measurements of polar angular distribution averaged over the molecular ensemble, and thus meet difficulty in separating individual processes. In addition, the double-pulse excitation can be extended to reconstructing the rotational WP, which is created by the single pulse of the excitation pair. Such an experimental WP reconstruction has been successfully demonstrated quite recently by the authors' group [14].

Furthermore, nonadiabatic interaction with a nonresonant intense ultrafast laser field can coherently excite the other degree of freedom for molecular motions, i.e., vibration. Realization of such a nonadiabatic vibrational excitation (NAVEX) and the resultant creation of a vibrational WP is definitely the next step to be proceeded. We have extended the research into this direction.

Acknowledgments

This work was partly supported by Grants-in-Aid from MEXT Japan (18244120, 18750020, and 20050032), and the RIKEN-IMS joint program on "Extreme Photonics". An additional financial support from the Mitsubishi Foundation is also appreciated.

[1] Stapelfeldt H and Seideman T 2003 *Rev. Mod. Phys.* **75** 543
[2] Rosca-Pruna F and Vrakking, M J J 2002 *J. Chem. Phys.* **116** 6567, 6579
[3] Dooley P W, Litvinyuk I V, Lee K F, Rayner D M, Spanner M, Villeneuve D M, and Corkum P B 2003 *Phys. Rev.* A **68** 023406
[4] Péronne E, Poulsen M D, Bisgaard C Z, Stapelfeldt H and Seideman T 2003 *Phys. Rev. Lett.* **91** 043003
[5] Miyazaki K, Kaku M, Miyaji G, Abdurrouf A and Faisal F H M 2005 *Phys. Rev. Lett.* **95** 243903
[6] Stavros V G, Harel E and Leone S R 2005 *J. Chem. Phys.* **122** 064301
[7] Bisgaard C Z, Viftrup S S and Stapelfeldt H 2006 *Phys. Rev.* A **73** 053410
[8] Lee K F, Villeneuve D M, Corkum P B and Shapiro E A 2004 *Phys. Rev. Lett.* **93** 233601
[9] Lee K F, Shapiro E A, Villeneuve D M and Corkum P B 2006 *Phys. Rev.* A **73** 033403
[10] Renard M, Hertz E, Guérin S, Jauslin H R, Lavorel B and Faucher O 2005 *Phys. Rev.* A **72** 025401
[11] Horn C, Wollenhaupt M, Krug M, Baumert T, de Nalda R and Bañares L 2006 *Phys. Rev.* A **73** 031401
[12] Hasegawa H and Ohshima Y 2006 *Phys. Rev.* A **74** 061401
[13] Hasegawa H and Ohshima Y 2008 *Chem. Phys. Lett.* **454** 148
[14] Hasegawa H and Ohshima Y 2008 *Phys. Rev. Lett.* **101** 053002
[15] Meijer A S, Zhang Y, Parker D H, van der Zande W J, Gijsbertsen A and Vrakking M J J 2007 *Phys. Rev.* A **76** 023411

Journal of Physics: Conference Series **185** (2009) 012015 doi:10.1088/1742-6596/185/1/012015

Photoionization of Be, B^+, C^{2+}, N^{3+}, O^{4+}, and F^{5+} in the MCRRPA theory

Ju-Tang Hsiao[1], Hsiao-Ling Sun[1], Sheng-Fang Lin[1,2], and Keh-Ning Huang[1,2]

[1]Institute of Atomic and Molecular Sciences, Academia Sinica, P.O. Box 23-166, Taipei, Taiwan 106, Republic of China

[2]Department of Physics, National Taiwan University, Taipei, Taiwan 106, Republic of China

E-mail: jthsiao@ntu.edu.tw

Abstract. The multiconfiguration relativistic random-phase approximation theory is applied to the valence-shell photoionization along the Be isoelectronic sequence. Photoionization cross sections between $(1s^2 2s)^2S_{1/2}$ and $(1s^2 2p)^2P^o_{1/2}$ ionization thresholds are calculated. There are five Rydberg series of doubly excited states that manifest themselves as autoionization resonances in the photoionization cross section. A comparative study of Be, B^+, C^{2+}, N^{3+}, O^{4+}, and F^{5+} is presented.

1. Introduction

Photoionization spectra of atoms or ions have attracted much attention in recent years because of the interplay between correlation and relativistic effects plays an important role. Double-excitation resonances in the photoionization spectrum provide a stringent testing ground for the accuracy of theoretical models, while experimental studies have gained considerable momentum due to the the rapidly evolving experimental techniques.

Atomic systems in the Be isoelectronic sequence have been considered as suitable objects for systematic studies of the photoionization processes because of their relatively simple divalence structure. In these atomic systems, the photoionization spectrum dominated by the doubly excited states due to the interaction between two valence electrons. Nevertheless, many works have shown relativistic effects play an important role in the photoionization of small atoms, such as Be [1, 2], Ne [3], and Mg [4]. Therefore, it is worthwhile to investigate the interplay between correlation and relativistic effects on the photoionization of atoms or ions with a low nuclear charge.

Applications of the multiconfiguration relativistic random-phase approximation theory (MCRRPA) to the photoionization of Be [5, 6], Mg [7], Sr [8], and Zn [9] have been very successful. In the photon-energy region between $(1s^2 2s)^2S_{1/2}$ and $(1s^2 2p)^2P^o_{1/2}$ ionization thresholds of atomic systems along the Be isoelectronic sequence, double-excitation autoionization resonances are the most prominent feature in its photoionization spectrum. These double-excitation resonances may be classified into five Rydberg series: $(2pns)^1P^o_1$, $(2pns)^3P^o_1$, $(2pnd)^1P^o_1$, $(2pnd)^3P^o_1$ and $(2pnd)^3D^o_1$. In this work, photoionization of atomic systems along the Be isoelectronic sequence are studied by using the MCRRPA theory accounting for all relativistic excitation channels.

Journal of Physics: Conference Series **185** (2009) 012015 doi:10.1088/1742-6596/185/1/012015

2. MCRRPA Theory

A detailed formulation of the MCRRPA has been given in a previous paper [10] and only a summary of the essential features will be given here. The N-electron atomic system is described as a superposition of configuration wavefunctions with time-dependent weights. The time-dependent variational principle is employed to determine the response of the atomic system to a time-dependent external field. The resulting terms independent of the external field lead to the usual multiconfiguration Dirac-Fock (MCDF) description of an atomic state. Those terms proportional to the external field lead to the MCRRPA equations describing the linear response of the atomic state to the external field. If we start from a single-configuration reference state, the MCRRPA equations reduce to the RRPA equations. The multiconfiguration reference state is described by a linear combination of three configurations with two valence electrons coupled to zero angular momentum and even parity:

$$\Psi = C_1(1s^2_{1/2}2s^2_{1/2})_0 + C_2(1s^2_{1/2}2p^2_{1/2})_0 + C_3(1s^2_{1/2}2p^2_{3/2})_0, \tag{1}$$

where C_1, C_2, and C_3 are configuration-weight coefficients.

We consider seven excitation channels from the valence electrons in the electric-dipole approximation for photon energies between $(1s^22s)^2S_{1/2}$ and $(1s^22p)^2P^o_{1/2}$ ionization thresholds:

(i) Photoionization channels:
 (a) $(2s^2)^1S_0 \to (2s\epsilon p)^1P^o_1$,
 (b) $(2s^2)^1S_0 \to (2s\epsilon p)^3P^o_1$.

(ii) Photoexcitation channels:
 (c) $(2p^2)^1S_0 \to (2pns)^1P^o_1$, $n \geq 3$,
 (d) $(2p^2)^1S_0 \to (2pns)^3P^o_1$,
 (e) $(2p^2)^1S_0 \to (2pnd)^1P^o_1$,
 (f) $(2p^2)^1S_0 \to (2pnd)^3P^o_1$,
 (g) $(2p^2)^1S_0 \to (2pnd)^3D^o_1$.

The couplings between the *photoionization* and *photoexcitation channels* generate five autoionization resonance series in the photoionization spectrum.

3. Results and Discussions

We consider the incident photon with energy from the ionization threshold $(1s^22s)^2S_{1/2}$ up to the threshold $(1s^22p)^2P^o_{1/2}$, where only photoionization channels (a) and (b) are open. The couplings of the two open channels and remaining five photoexcitation channels producing five Rydberg series of resonances $(2pns)^1P^o_1$, $(2pns)^3P^o_1$, $(2pnd)^1P^o_1$, $(2pnd)^3P^o_1$, and $(2pnd)^3D^o_1$. The calculated cross sections for the Be isoelectronic sequence from Be to F^{5+} are presented in Figs. 1 and 2. For comparison, we use the dimensionless variable ε instead of the photon energy $h\nu$: $\varepsilon \equiv (h\nu - I_s)/\Delta I$, where $\Delta I \equiv I_p - I_s$, and I_s and I_p are ionization potentials for $(1s^22s)^2S_{1/2}$ and $(1s^22p)^2P^o_{1/2}$, respectively. By assuming the form for the ionization potential as $I_i = a(Z - b_i + c_i/Z)^2$, we obtain the fitting formula for the Be isoelectronic sequence as $\Delta I(Z) = AZ + B + C/Z$, where A=0.0713114, B=-0.1229696, C=-0.0666783, and Z is the nuclear charge.

In Fig. 1, two Rydberg series of double-excitation resonances, a broad series $(2pns)^1P^o_1$ and a narrow series $(2pnd)^1P^o_1$, can be clearly identified in the photoionization profile of the Be atom. Because the photoionization continuum $(2s\epsilon p)^1P^o_1$ and the double-excitation series $(2pns)^1P^o_1$ and $(2pnd)^1P^o_1$ all arise from the $n = 2$ atomic complex and with the same total spin and orbital angular-momenta, the couplings between them are very strong such that only these two Rydberg series are prominent in the spectra. The broad $(2pns)^1P^o_1$ and narrow $(2pnd)^1P^o_1$ resonance profiles reveal the shorter and longer lifetimes of the respective doubly excited states.

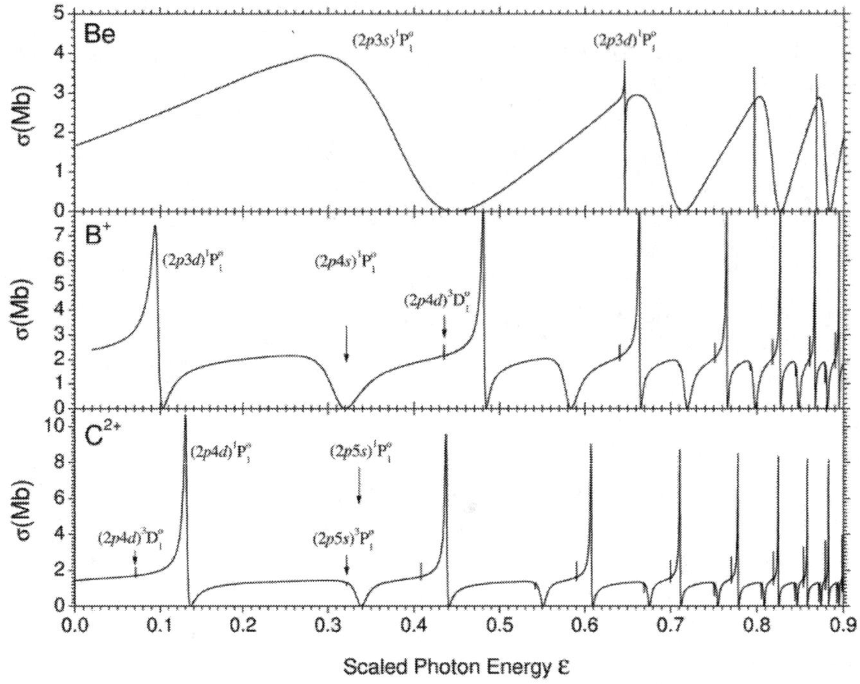

Figure 1. Photoionization cross sections for Be, B^+, and C^{2+} between the ionization thresholds $(1s^22s)^2S_{1/2}$ and $(1s^22p)^2P^o_{1/2}$. The broad-hump series $(2pns)^1P_1^o$ and the narrow-spike series $(2pnd)^1P_1^o$ sandwiched amid the series $(2pns)^1P_1^o$ for Be are shown in the upper panel. For B^+ in the middle panel and C^{2+} in the lower panel, the $(2pnd)^1P_1^o$ series consists of sharp peaks, while the $(2pns)^1P_1^o$ series appears as broad dips. Extremely sharp resonance structures are triplet resonances $(2pns)^3P_1^o$, $(2pnd)^3P_1^o$ and $(2pnd)^3D_1^o$.

As shown in Figs. 1 and 2, the characteristics of the photoionization profiles are however very similar among the Be-like ions. The $(2pnd)^1P_1^o$ series consists of sharp peaks, while the $(2pns)^1P_1^o$ series appears as broad dips. The photoionization cross section of each ion shows an average background with resonance structures, where the background decreases from roughly 2 Mb for the Be atom to about 0.5 Mb for the F^{5+} ion with increasing nuclear charge.

Because of low nuclear charge for Be, B^+, and C^{2+}, they behave rather nonrelativistically, and only two of five Rydberg series of resonances show up distinctly in the photoionization cross section. Experimental results [11, 12, 13] or the cross sections in nonrelativistic calculations [14, 15, 16] should be similar to the MCRRPA result but without the extremely sharp resonance structures of triplet resonances $(2pns)^3P_1^o$, $(2pnd)^3P_1^o$, and $(2pnd)^3D_1^o$. However, the gradual onset of relativistic effects with increasing nuclear charge, additional Rydberg series of resonances show up in the photoionization spectra of higher members of the Be isoelectronic sequence N^{3+}, O^{4+}, and F^{5+}.

Physically, the coupling strength between the bound and continuum components of the wave function of a doubly-excited state is measured by the resonance width, which also determines the nonradiative decay rate of an autoionization state. Because the orbital symmetries of the photoionization channels $(2s\epsilon p)^3P_1^o$ and $(2s\epsilon p)^1P_1^o$ are of P-type, the coupling of resonances $(2pnd)^3D_1^o$ with these photoionization channels are weaker such that their widths are much narrower. Our MCRRPA calculation includes relativistic and correlation effects in an *ab initio*

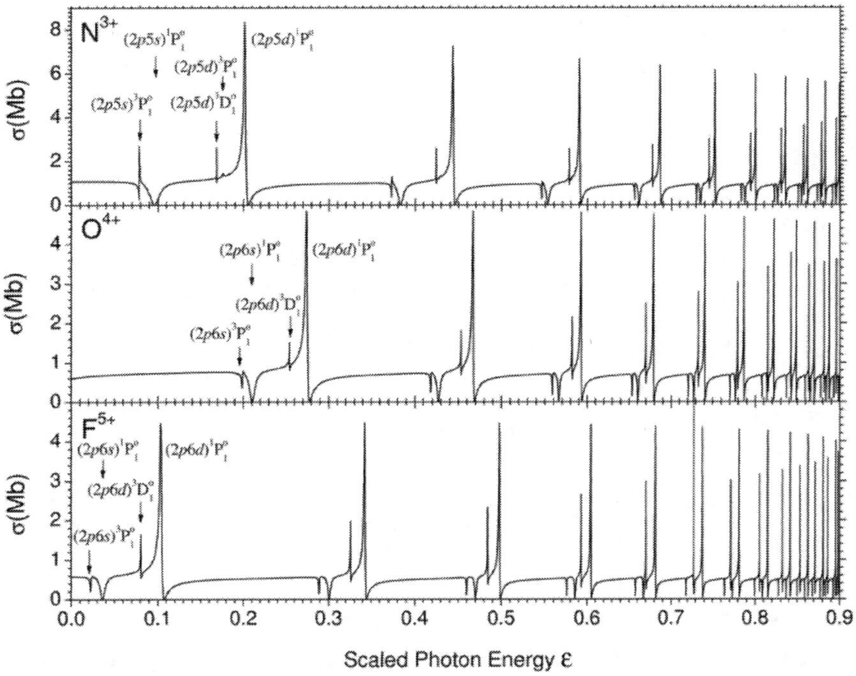

Figure 2. Photoionization cross sections for N^{3+}, O^{4+}, and F^{5+} between the ionization thresholds $(1s^22s)^2S_{1/2}$ and $(1s^22p)^2P^o_{1/2}$. The $(2pnd)^1P^o_1$ series consists of sharp peaks, while the $(2pns)^1P^o_1$ series appears as broad dips. Extremely sharp resonance structures are triplet resonances $(2pns)^3P^o_1$, $(2pnd)^3P^o_1$ and $(2pnd)^3D^o_1$.

manner and treats the initial and final states in a balanced fashion; therefore, it should in principle provide reliable positions and lifetimes of the doubly excited states.

Acknowledgment This research work was supported in part by the National Science Council of the Republic of China under Grant No. NSC93-2112-M001-036

References

[1] E. W. B. Dias, H. S. Chakraborty, P. C. Deshmukh, and S. T. Manson, J. Phys. B **32**, 3383 (1999).
[2] L.-R. Wang, J.-T. Hsiao, and K.-N. Huang, J. Phys. B **39**, L217 (2006).
[3] A. A. Wills, T. W. Gorczyca, N. Berrah, B. Langer, Z. Felfli, E. Kukk, J. D. Bozek, O. Nayandin and M. Alshehri, Phys. Rev. Lett. **80**, 5085 (1998).
[4] M. D. Lindsay, L. T. Cai, G. W. Schinn, C. J. Dai, and T. F. Gallagher, Phys. Rev. A **45**, 231 (1992).
[5] H.-C. Chi, K.-N. Huang, and K.-T. Cheng, Phys. Rev. A **43**, 2542 (1991).
[6] J.-T. Hsiao, L.-R. Wang, H.-L. Sun, S.-F. Lin, C.-L. Lu, and K.-N. Huang, Phys. Rev. A **78**, 013411 (2008).
[7] H.-C. Chi and K.-N. Huang, Phys. Rev. A **50**, 392 (1994).
[8] C.-M. Wu, H.-C. Chi, and K.-N. Huang, J. Phys. B **27**, 3927 (1994).
[9] L.-R. Wang, H.-C. Chi, and K.-N. Huang, Phys. Rev. Lett. **83**, 702 (1999).
[10] K.-N. Huang and W. R. Johnson, Phys. Rev. A **25**, 634 (1982).
[11] R. Wehlitz, D. Lukic, and B. Bluett, Rev. A **68**, 052708 (2003).
[12] S. Schippers, A. Müller, B. M. McLaughlin, A. Aguilar, C. Cisneros, E. D. Emmons, M. F. Gharaibeh, and R. A. Phaneuf, J. Phys. B **36**, 3371 (2003).
[13] A. Müller, R. A. Phaneuf, A. Aguilar, M. F. Gharaibeh, A. S. Schlachter, I. Alvarez, C. Cisneros, G. Hinojosa, and B. M. McLaughlin, J. Phys. B **35**, L137 (2002).
[14] D.-S. Kim, S. S. Tayal, H-L Zhou, and S. T. Manson, Phys. Rev. A **61**, 062701 (2000).
[15] D.-S. Kim and S. T. Manson, J. Phys. B **37**, 4013 (2004).
[16] D.-S. Kim and S. T. Manson, J. Phys. B **37**, 4707 (2004).

Journal of Physics: Conference Series **185** (2009) 012016 doi:10.1088/1742-6596/185/1/012016

Frequency shifts of the ^{88}Sr $^1S_0 - ^3P_1$ transition induced by buffer gas collisions

Tetsuya Ido[1,2], Nobuyasu Shiga[1], Ying Li[1], Hiroyuki Ito[1], and Shigeo Nagano[1]

[1] Space-Time Standards Group, National Institute of Information and Communications Technology (NICT), 4-2-1 Nukui-kitamachi, Koganei, Tokyo, 184-8795 Japan
[2] JST, PRESTO & CREST, 5 Sanbancho, Chiyoda-ku, Tokyo, 102-0075 Japan

E-mail: ido@nict.go.jp

Abstract. Buffer gas induced collision shift for $^{88}\mathrm{Sr}\,^1S_0 - ^3P_1$ transition is investigated by precision saturation spectroscopy of thermal gas in a heat cell. The cell was filled with rare gas of helium, neon, argon, and xenon as buffer gasees. Helium showed the largest fractional shift coefficient of $1.6 \times 10^{-9}\mathrm{Torr}^{-1}$. The result could be useful to evaluate the background gas collision shift of Sr lattice clocks.

1. Introduction

Latest progress of optical clocks realized an ionic clock with an accuracy of $\sim 10^{-17}$[1]. This rapid improvement urgently requires the physicists pursuing optical atomic clocks to consider other systematic shifts which have been so far ignored. Among those potential systematics, the collision shifts induced by residual gas in the vacuum chamber may ultimately limit the accuracy of optical clocks as the evaluation of the effects is not straightforward. Deliberate injection of the buffer gas into the chamber often causes the loss of trapped atoms, preventing a background gas pressure large enough to enable the finite measurement.

From the beginning of the laser spectroscopy, measurements of the frequency shifts and broadening induced by the buffer gas were performed by using a gas cell with buffer gas pressures of $1 - 10^3$Torr, where the shifts and broadening resulted in the order of $\sim$ 100MHz. In this work based on precision saturation absorption spectroscopy for the $^{88}\mathrm{Sr}(5s^2)^1S_0 - (5s5p)^3P_1$ spin-forbidden transition ($\lambda = 689.4$nm, linewidth : 7.4kHz), we detected the buffer gas induced shifts of a few kHz with low gas pressure of $\sim$ 10mTorr. Since this transition is quite similar to the lattice clock transition $^1S_0 -^3 P_0$ [2, 3], our experiment may give us valuable information to estimate the systematic shifts in real state-of-the-art optical clocks using neutral strontium atoms.

2. Experimental setup

Fig. 1 shows the experimental setup of the precision saturation absorption spectroscopy. The transition from the ground state $^1S_0$ to the $m_J = 0$ magnetic sublevel of $^3P_1$ was probed by FM sideband technique [4]. The pump and probe beam of the saturation spectroscopy is linearly polarized to excite the π transition with respect to a quantization axis taken to be parallel to a bias magnetic field of 3 Gauss. Only the probe beam is phase modulated by an electro-optic

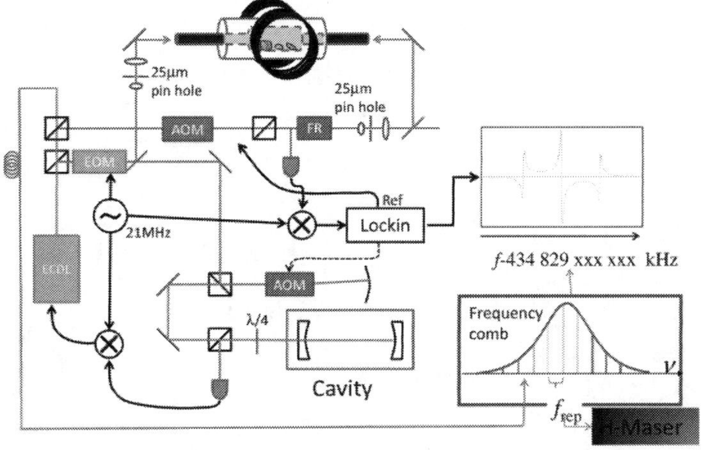

Figure 1. Configuration of the precision absorption spectroscopy. External cavity loaded diode laser (ECDL) is first locked to a high-finesse cavity, and then the beam is divided to a pump and probe beam for a saturation absorption spectroscopy. The absolute frequency of the stabilized laser is continuously calibrated by a frequency comb referenced to Atomic Time (TA). AOM: Acousto-optic modulator. EOM: Electro-optic modulator. FR: Faraday rotator

modulator (EOM). The pump beam free from the phase modulation is chopped by acousto-optic modulator and the signal induced by the saturation effects is lock-in detected from the demodulated signal [5]. Two pinholes of 25μm diameter are placed to guide the paths of pump and probe beam. The overlapping of the two beams are realized by the two pinholes. The $1/e^2$ beam radius of the pump and probe beam is 4.2mm and 3.3mm, respectively. The temperature of the heat pipe is typically 455 °C. A frequency comb referenced to Atomic Time (TA) always monitor the absolute frequency of the probe beam.

3. Results and discussion

The frequency shift induced by various buffer gas [6] is investigated. The coefficient of the frequency shift to the pressure of various buffer gases are summarized in Fig. 2. Only helium shows a positive frequency shift. The largest shift is obtained for helium with a shift coefficient of 680 kHz / Torr. This shift corresponds to the fractional shift of $1.6 \times 10^{-9}\text{Torr}^{-1}$. In our case, the broadening and shifting caused by collisions with perturbers can be described in the impact approximation, where the phase shift induced by the collision is the time integral of the differential interaction potential of $\mathrm{Sr}(^1S_0) - \mathrm{Rg}(^1S_0)$ and $\mathrm{Sr}(^3P_1) - \mathrm{Rg}(^1S_0)$ molecule, where Rg denotes rare gas. The positive shift in case of the He buffer gas indicates that repulsive part of the quasimolecular potential plays a crucial role. Based on the impact approximation with straight line trajectory, theoretical calculation of the shift and broadening coefficient is currently underway and will be described elsewhere.

The $^1S_0 -^3P_1$ transition of the strontium resembles the lattice clock transition $^1S_0 -^3P_0$. The ground state is shared and the excited states belong to the same multiplets. The quasimolecular potentials with the asymptote of $\mathrm{Sr}(^3P_1) - \mathrm{Rg}(^1S_0)$ consists of two potentials $^3\Sigma^+$ and $^3\Pi$, whereas only $^3\Pi$ contributes to the quasimolecular potential with $\mathrm{Sr}(^3P_0) - \mathrm{Rg}(^1S_0)$ asymptote . The effects of the $^3\Sigma^+$ and $^3\Pi$ molecular potentials for the $^1S_0 -^3 P_1$ case are in same orders of magnitude. Since the transition strength of the $^1S_0 -^3 P_0$ clock transition is too small to

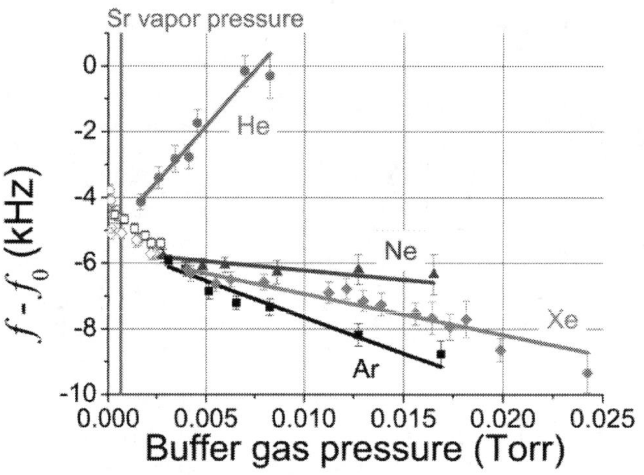

Figure 2. Frequency shift of the resonance frequency induced by various kinds of buffer gas. f_0 is the laser frequency of true atomic resonance which was obtained from balistically expanding ultracold atoms [7]. The points in low buffer gas pressures comparable to the Sr vapor pressure are excluded in the linear fitting because the area under spectroscopy no longer has same background gas pressure as the value provided by vacuum gauges.

perform saturation absorption spectroscopy, the information obtained here works for the coarse evaluation of the background gas collision shifts in real lattice clocks.

4. Acknowledgements

The authors thank theoretical help and comments from R.Ciuryło and M. Kajita. Experimental help from H. Ishijima, A. Yamaguchi, and M. Koide are also acknowledged.

References

[1] Rosenband T, *et al.* 2008 *Science* **319** 1808
[2] Akatsuka T, Takamoto M, and Katori H 2008 *Nature Physics* **4** 954
[3] Ludlow, A D, *et al.* 2008 *Science* **319** 1805
[4] Hall J L, Hollberg L, Baer T, and Robinson H G 1981 *Appl. Phys. Lett.* **39** 680
[5] Drever R W P, Hall J L, Kowalski F V, Hough J, Ford G M, Munley A J and Ward H 1983 *Appl. Phys. B* **31** 97
[6] Allard N, and Kielkopf J 1982 *Rev. Mod. Phys.* **54** 1103
[7] Ido T, Loftus T H, Boyd M M, Ludlow A D, Holman K W, and Ye J 2005 *Phys. Rev. Lett.* **94** 153001

Journal of Physics: Conference Series **185** (2009) 012017 doi:10.1088/1742-6596/185/1/012017

Generalisation of scattering theory to charged particles

A. S. Kadyrov[1], I. Bray[1], A. M. Mukhamedzhanov[2] and A. T. Stelbovics[1]

[1] ARC Centre for Antimatter-Matter Studies, Curtin University of Technology, GPO Box U1987, Perth, WA 6845, Australia

[2] Cyclotron Institute, Texas A&M University, College Station, Texas 77843, USA

E-mail: a.kadyrov@curtin.edu.au

Abstract. Generalisation of scattering theory to charged particles is presented. It is based on a surface-integral approach. New general definitions for the breakup amplitude in a three-body system valid for both short-range and long-range Coulombic interactions are given.

1. Introduction

In scattering of charged particles the incident and scattered waves are distorted right out to infinity due to the long-range nature of the Coulomb interaction. Therefore the Coulomb potential often receives a special treatment known in scattering theory as screening and renormalization [1]. The *ad-hoc* renormalization procedure leads to the correct cross sections for the two-body problem, however, it is based on prior knowledge of the exact answer and has no *ab initio* theoretical justification. A renormalization method was implemented successfully for the three-body problem when only two particles are charged [2]. However, no practical renormalization method exists that is valid for a system of three charged particles above the breakup threshold.

Another issue is how to extract the breakup amplitude from the wavefunction when the latter is available. The conventional theory fails to provide a formal post-form definition of the breakup amplitude for three charged particles in terms of the total wavefunction with outgoing scattered waves. Therefore, the Coulomb interaction is screened and the formula for the short-range case is used. However, it is well known that the short-range definition of the breakup amplitude diverges when the screening radius is taken to infinity. Thus we have a situation where we cannot use the theory unless we screen the Coulomb interaction, and when we do, we end up with quantities which diverge as screening is removed. This leaves no choice but to invoke renormalization to fix unphysical results. As mentioned above this is not possible in all cases of interest. Therefore, a new approach to Coulomb few-body problems that does not need renormalization is required. Below we present a different approach to the same problem based on surface integrals. We use the recently derived analytic forms of the total scattering wavefunctions in asymptotic domains [3, 4] in order to develop well-defined prior and post forms of the breakup amplitude valid for short-range and Coulombic potentials.

2. Surface-integral approach to scattering theory

We shall now briefly explain the derivation of the new definitions for the breakup amplitude in order to outline the surface-integral approach. Consider scattering of particle α with incident momentum $\vec{q}_{\alpha n}$ off a bound pair (β,γ) in initial state $\phi_{\alpha n}$ of energy $E_{\alpha n}$ above the breakup threshold. The total wavefunction describing this process satisfies the Schrödinger equation

$$(E-H)\Psi^{+}_{\alpha n}(\vec{r}_\alpha,\vec{\rho}_\alpha) \;=\; 0, \tag{1}$$

with outgoing-spherical wave boundary conditions. Here $H = H_0 + V$, $H_0 = -\Delta_{\vec{r}_\alpha}/2\mu_\alpha - \Delta_{\vec{\rho}_\alpha}/2M_\alpha$ is the free Hamiltonian, $V = v + V_\alpha + V_\beta + V_\gamma$ is the full interaction, v is a three-body interaction and V_α is a Coulombic interaction between particles β and γ, $E = E_{\alpha n} + q^2_{\alpha n}/2M_\alpha = k^2_\alpha/2\mu_\alpha + q^2_\alpha/2M_\alpha$ is the total energy of the system. We use the Jacobi coordinates where $\vec{r}_\alpha$ and $\vec{k}_\alpha$ are respectively the relative coordinate and momentum between particles β and γ. The coordinate of particle α relative to the center of mass of the pair (β,γ) is $\vec{\rho}_\alpha$, with $\vec{q}_\alpha$ being the canonically conjugate relative momentum. The corresponding reduced masses are denoted by μ_α and M_α. Another process which may take place within the same system at the same energy E is called $3 \to 3$ scattering. The wavefunction Ψ^-_0 describing this process is also an eigenstate of the same Hamiltonian H, however we impose on it incoming-wave boundary conditions. It develops to the final state where all three particles are in the continuum.

In scattering theory we deal with functions which are not square-integrable (L^2). Non-L^2 functions make certain integrals emerging in the theory divergent. In case of integrals containing the interaction potential a standard procedure which ensures existence of the integrals is limiting the range of the potential which irreversibly distorts the nature of the problem. Alternatively, we first formulate the scattering problem in a finite region of coordinate space and then extend it to the full space. To this end we introduce a partial inner product of two arbitrary functions Ψ_i and Ψ_f in the space of functions describing various states and arrangements in a three-body system according to $\langle\Psi_f|\Psi_i\rangle_{R_0} = \int_{R\le R_0} d\vec{r}_\alpha d\vec{\rho}_\alpha \Psi^*_f\Psi_i$. Here R is a hyperradius in the 6-dimensional configurations space $R = \left(\mu_\alpha\, r^2_\alpha + M_\alpha\, \rho^2_\alpha\right)^{1/2}$. The integration is limited to the volume of a 6-dimensional hypersphere of radius R_0.

We want to use $\Psi^+_{\alpha n}$ and Ψ^-_0 as starting points to derive prior and post forms of the breakup amplitude . If we split $\Psi^+_{\alpha n}$ into the initial-channel wave $\Phi^+_{\alpha n}$ and scattered wave $\Psi^{sc+}_{\alpha n}$ then Eq. (1) transforms to

$$(E-H)\Psi^{sc+}_{\alpha n}(\vec{r}_\alpha,\vec{\rho}_\alpha) \;=\; (H-E)\Phi^{+}_{\alpha n}(\vec{r}_\alpha,\vec{\rho}_\alpha). \tag{2}$$

We multiply Eq. (2) by Ψ^{-*}_0 from the left and integrate the result over a volume of a hypersphere of radius R_0:

$$\langle\Psi^-_0|(E-H)\Psi^{sc+}_{\alpha n}\rangle_{R_0} \;=\; \langle\Psi^-_0|(H-E)\Phi^+_{\alpha n}\rangle_{R_0}. \tag{3}$$

For simplicity we assume V to be real. Then we have $\langle(E-H)\Psi^-_0|\Psi^{sc+}_{\alpha n}\rangle_{R_0} = 0$, which is true for any R_0 simply due to $(E-H)\Psi^-_0 = 0$. Now we subtract this from Eq. (3). Despite the fact that both Ψ^-_0 and $\Psi^{sc+}_{\alpha n}$ are non-L^2 functions, terms of the form $\langle\Psi^-_0|(E-V)\Psi^{sc+}_{\alpha n}\rangle_{R_0}$ are finite due to the limited space (regardless of the nature of the potential). Therefore, canceling them we get

$$-\langle\Psi^-_0|H_0\Psi^{sc+}_{\alpha n}\rangle_{R_0} + \langle H_0\Psi^-_0|\Psi^{sc+}_{\alpha n}\rangle_{R_0} = \langle\Psi^-_0|(H-E)\Phi^+_{\alpha n}\rangle_{R_0}. \tag{4}$$

Now we calculate the limit of this equation as $R_0 \to \infty$. An essential feature of the term on the left-hand side (LHS) of Eq. (4) is that it is easily transformed into an integral over the hypersurface of radius R_0 so that the result depends only on the behavior of the wavefunctions on this surface. Therefore, the knowledge of the wavefunctions anywhere inside the surface is not required. Then it can be evaluated using the asymptotic forms of the wavefunctions. If $R_0 \to \infty$ with all interparticle distances being large then for the limit of the LHS of Eq. (4) we have

$$\frac{1}{2(\mu_\alpha M_\alpha)^{3/2}} \lim_{R_0\to\infty} R_0^5 \int d\hat{\vec{r}}_\alpha d\hat{\vec{\rho}}_\alpha \int_0^{\pi/2} d\varphi_\alpha \sin^2\varphi_\alpha \cos^2\varphi_\alpha \left[\Psi_0^{-*}\frac{\partial\Psi_{\alpha n}^{\mathrm{sc}+}}{\partial R} - \Psi_{\alpha n}^{\mathrm{sc}+}\frac{\partial\Psi_0^{-*}}{\partial R}\right]_{R=R_0}. \quad (5)$$

Here we first transformed H_0 into (R,ω)-variables, where ω is a 5-dimensional hyperangle $(\widehat{\vec{r}}_\alpha, \widehat{\vec{\rho}}_\alpha, \varphi_\alpha)$ with $\varphi_\alpha = \arctan\left[(\mu_\alpha/M_\alpha)^{1/2} r_\alpha/\rho_\alpha\right]$. Then we made use of Green's theorem to transform the volume integral into the surface integral. The asymptotic forms of Ψ_0^- and $\Psi_{\alpha n}^{\mathrm{sc}+}$ were given in Ref. [3] and [4], respectively. The asymptotic form of scattered wave $\Psi_{\alpha n}^{\mathrm{sc}+}$ contains the breakup amplitude as a function of the coordinates. The above integral is infinitely oscillatory as $R_0 \to \infty$. Calculating it by means of the stationary-phase method we find that it reduces to $T(\vec{k}_\alpha, \vec{q}_\alpha)$ which is the breakup amplitude. Therefore, the limit of Eq. (4) is written as

$$T(\vec{k}_\alpha, \vec{q}_\alpha) = \lim_{R_0\to\infty} \langle \Psi_0^- | (H-E)\Phi_{\alpha n}^+ \rangle_{R_0}. \quad (6)$$

We can also use Ψ_0^- as a starting point. If we separate Ψ_0^- into Coulomb-distorted three-body plane wave Φ_0^- and scattered wave $\Psi_0^{\mathrm{sc}-}$ then the Schrödinger equation for Ψ_0^- can be written as

$$(E-H)\Psi_0^{\mathrm{sc}-}(\vec{r}_\alpha, \vec{\rho}_\alpha) = (H-E)\Phi_0^-(\vec{r}_\alpha, \vec{\rho}_\alpha). \quad (7)$$

Take complex conjugate of Eq. (7) and multiply it by $\Psi_{\alpha n}^+$ from the right. Then integrating the result over the volume of a hypersphere of radius R_0 we get

$$\langle (E-H)\Psi_0^{\mathrm{sc}-} | \Psi_{\alpha n}^+ \rangle_{R_0} = \langle (H-E)\Phi_0^- | \Psi_{\alpha n}^+ \rangle_{R_0}. \quad (8)$$

From Eq. (1) we have $\langle \Psi_0^{\mathrm{sc}-} | (E-H)\Psi_{\alpha n}^+ \rangle_{R_0} = 0$. Subtracting this from (8) we get

$$-\langle H_0\Psi_0^{\mathrm{sc}-} | \Psi_{\alpha n}^+ \rangle_{R_0} + \langle \Psi_0^{\mathrm{sc}-} | H_0\Psi_{\alpha n}^+ \rangle_{R_0} = \langle (H-E)\Phi_0^- | \Psi_{\alpha n}^+ \rangle_{R_0}. \quad (9)$$

We again investigate the limit of this equation as $R_0 \to \infty$. However, this time we have the situation of three free particles in the initial state. In the final state we may have three free particles or two of the particles may form a bound state. In the latter case the limit of LHS of Eq. (9) is written as

$$\frac{1}{2M_\alpha} \lim_{R_0\to\infty} R_0^2 \int d\hat{\vec{r}}_\alpha d\hat{\vec{\rho}}_\alpha \left[\Psi_{\alpha n}^+ \frac{\partial}{\partial\rho_\alpha}\Psi_0^{\mathrm{sc}-*} - \Psi_0^{\mathrm{sc}-*}\frac{\partial}{\partial\rho_\alpha}\Psi_{\alpha n}^+\right]_{\rho_\alpha=R_0}. \quad (10)$$

An essential difference from Eq. (5) here is that only one of the three-dimensional volume integrals is transformed into the surface integral. The remaining volume integral is limited due to presence of a bound state in channel α. Calculating the limit of this equation as $R_0 \to \infty$ we obtain

$$T(\vec{k}_\alpha, \vec{q}_\alpha) = \lim_{R_0\to\infty} \langle (H-E)\Phi_0^- | \Psi_{\alpha n}^+ \rangle_{R_0}. \quad (11)$$

The same analysis can be performed on the right-hand side of Eqs. (6) and (11) to yield $T(\vec{k}_\alpha, \vec{q}_\alpha)$. So the general prior and post forms of the breakup amplitude, valid for both short-range and Coulombic potentials, are written as

$$T^{\mathrm{prior}}(\vec{k}_\alpha, \vec{q}_\alpha) = \langle \Psi_0^- | (H-E)\Phi_{\alpha n}^+ \rangle, \quad (12)$$
$$T^{\mathrm{post}}(\vec{k}_\alpha, \vec{q}_\alpha) = \langle (H-E)\Phi_0^- | \Psi_{\alpha n}^+ \rangle. \quad (13)$$

3. Consistency of the results

The new definitions are consistent with the well-known ones for short-range interactions. To see this we note that when the interactions are short-ranged we have $\Phi^+_{\alpha n}(\vec{r}_\alpha, \vec{\rho}_\alpha) \to e^{i\vec{q}_{\alpha n}\vec{\rho}_\alpha}\phi_{\alpha n}(\vec{r}_\alpha)$. This leads to $(H - E)\Phi^+_{\alpha n}(\vec{r}_\alpha, \vec{\rho}_\alpha) = (V - V_\alpha)e^{i\vec{q}_{\alpha n}\vec{\rho}_\alpha}\phi_{\alpha n}(\vec{r}_\alpha)$. At the same time if we have three particles in the final channel then $\Phi^-_0(\vec{r}_\alpha, \vec{\rho}_\alpha) \to e^{i\vec{k}_\alpha\vec{r}_\alpha + i\vec{q}_\alpha\vec{\rho}_\alpha}$. Then we have $(H - E)\Phi^-_0(\vec{r}_\alpha, \vec{\rho}_\alpha) = Ve^{i\vec{k}_\alpha\vec{r}_\alpha + i\vec{q}_\alpha\vec{\rho}_\alpha}$. Therefore, the general forms of the amplitudes (12) and (13) reduce to the standard definitions for short-range interactions

$$T^{\text{prior}}(\vec{k}_\alpha, \vec{q}_\alpha) = \langle \Psi^-_0 | V - V_\alpha | \phi_{\alpha n}, \vec{q}_{\alpha n} \rangle, \tag{14}$$

$$T^{\text{post}}(\vec{k}_\alpha, \vec{q}_\alpha) = \langle \vec{q}_\alpha, \vec{k}_\alpha | V | \Psi^+_{\alpha n} \rangle. \tag{15}$$

4. Summary

Summarizing, we have given a brief account of a formalism of scattering theory which is applicable to three-body systems with both short-range and Coulombic potentials. The new formulation is based on a surface-integral approach. It was made possible by the recently obtained analytical forms of the three-body asymptotic wavefunctions. New representations for the breakup amplitude in a three-body system have been given. Further details of the present formalism can be found in Ref. [5].

Acknowledgments

This work was supported by the Australian Research Council and the U.S. Department of Energy under Grant No. DE-FG02-93ER40773.

References

[1] Gorshkov V G 1961 *Zh. Eksp. Teor. Fiz.* **40** 1481–1490
[2] Alt E O, Sandhas W and Ziegelmann H 1978 *Phys. Rev. C* **17** 1981–2005
[3] Alt E O and Mukhamedzhanov A M 1993 *Phys. Rev. A* **47** 2004–2022
[4] Kadyrov A S, Mukhamedzhanov A M, Stelbovics A T, Bray I and Pirlepesov F 2003 *Phys. Rev. A* **68** 022703
[5] Kadyrov A S, Bray I, Mukhamedzhanov A M and Stelbovics A T 2008 *Phys. Rev. Lett.* **101** 230405

Positronium formation for e^+-Na scattering at low and intermediate energies

M Z M Kamali[1], T P P Hiew[2] and Kuru Ratnavelu[2]

[1]Centre for Foundation Studies in Science, University of Malaya, Kuala Lumpur, MALAYSIA.
[2]Institute of Mathematical Sciences, University of Malaya, Kuala Lumpur, MALAYSIA .

E-mail: mzmk2000@yahoo.com

Abstract. There have been various theoretical calculations for positron-sodium scattering that attempt to calculate accurate positronium(Ps) formation cross sections at low and intermediate energies. The recent hyperspherical close-coupling (HSCC) calculations (Le *et al.* 2005 Phys. Rev. A 71, 032713) has handled the contribution from the positronic bound states to a great extent. Their calculations have further strengthened the argument that the theoretical Ps cross sections at low energies deviate from the predictions of the experimental measurements (Surdutovich *et al.* 2002 Phys. Rev. A 65, 032713). Here we examine whether the inclusion or the neglect of the positronic bound states in the calculation have any significant effects on the low energy Ps formation cross sections. Thus, we attempt to investigate if large-scale CC calculations are deficient in this respect.

1. Introduction

There is no question that the positronium(Ps) channel gives a very significant effect on the positron-sodium(Na) scattering system [1-4] at low and intermediate energies. In the earlier calculations [5-8], the close-coupling equations have been used to treat this system by excluding the possibility of Ps formation. Abdel Raouf [9], performed the first close coupling calculation which included only one atomic state as well as one Ps channel. This motivated the realistic close-coupling calculations which explicitly included the Ps channels [10-12]. Walters and co-workers [4,13] as well as Ryzhikh and Mitroy [3], working independently, predicted that the Ps formation cross section will rapidly diminish as the incident energy decreases. The prediction was experimentally [2] verified for Li but not for Na. At incident energies about or less than 1 eV, the Ps formation cross section for e^+-Na seems to increase tremendously. Questions have been raised about the inadequacy of CC calculations for Li and Na as these atoms can form positronic states [14-17]. The recent theoretical work by Bromley's group[1], tries to incorporate the positronic bound state in the HSCC calculations by implementing two model potentials which are the $FCLX_{pol}$ (frozen core+local exchange+polarization) and the SFG (Schweizer-Faßbinder-Gonzalez-Ferez). The present work has been largely motivated by the disagreement between the theoretical and experimental measurements especially at these low incident energies ($<=$ 1eV). In this paper, the CC calculations [18] are extended to 15 (CC(9,6)) and 12 states (CC(9,3)). The atomic states used are taken from Zhou's private communication [19]. Our aim is to investigate the effects of including the larger n-states in Ps formation cross sections.

2. Theory

The present close-coupling method is based on the work of Mitroy and Ratnavelu [11]. Briefly, we present the pertinent details. The Schrödinger equation, for a (N+1)-body system is given by:

$$(H - E)\Psi(r_i, r_0) = 0 \tag{1}$$

where E is the total energy of the $(N+1)$ body system. The H can be divided into H_e and H_{atom} where H_e , is the Hamiltonian of the incident particle with respect to the target atom and H_{atom} represents the N-electron Hamiltonian of the target atom,

$$H_e = -\frac{1}{2}\nabla_0^2 + \frac{Z}{r_0} - \sum_i \frac{1}{r_{i0}} \tag{2}$$

$$H_{atom} = \sum_i \left(-\frac{1}{2}\nabla_i^2 - \frac{Z}{r_i} \right) + \frac{1}{2}\sum_{j \neq i} \frac{1}{r_{ij}} \tag{3}$$

Thus, incorporating the Ps channels, an alternative partitioning of the Hamiltonian can be used:

$$H = H_{Ps}(m) + H_{ion}(m) + H_{int}(m) \; , \qquad \forall m \in \{1, \ldots\ldots, N\} \tag{4}$$

where

$$H_{Ps}(m) = -\frac{1}{4}\nabla_R^2 + \frac{Z}{r_0} - \frac{Z}{r_m} - \sum_{i \neq m} \left(\frac{1}{r_{i0}} - \frac{1}{r_{im}} \right) \tag{4a}$$

$H_{Ps}(m)$ is the interaction of the positronium atom containing the mth electron with the residual ion,

$$H_{ion}(m) = \sum_{i, i \neq m} \left(-\frac{1}{2}\nabla_1^2 - \frac{Z}{r_i} \right) + \frac{1}{2} \sum_{\substack{i,j \\ i \neq j, i \neq m, j \neq m}} \frac{1}{r_{ij}} \tag{4b}$$

$H_{ion}(m)$ is the Hamiltonian of the residual ion with the mth electron removed

$$H_{int}(m) = -\nabla_{\rho m}^2 - \frac{1}{r_{m0}} \tag{4c}$$

and $H_{int}(m)$ is the internal Hamiltonian for a positronium atom containing the mth electron. Thus, the Schrodinger equation can be written as

$$(E - H)\left[\sum_\alpha \Psi_\alpha(r_i) F_\alpha(r_0) + \sum_\beta \phi_\beta(\rho) \Omega(r_{i'}) G_\beta(R) \right] = 0 \tag{5}$$

By using standard procedures, the Schrodinger equation can be reduced to the momentum-space Lippmann-Schwinger equations for a positron with momentum $\mathbf{k}$ on a sodium atom which can be written as

$$\langle \mathbf{k}'\Psi_{\alpha'} | T | \mathbf{k}\Psi_\alpha \rangle = \langle \mathbf{k}'\Psi_{\alpha'} | V | \mathbf{k}\Psi_\alpha \rangle + \sum_{\alpha''} \int d^3k'' \frac{\langle \mathbf{k}'\Psi_{\alpha'} | V | \mathbf{k}''\Psi_{\alpha''} \rangle \langle \mathbf{k}''\Psi_{\alpha''} | T | \mathbf{k}\Psi_\alpha \rangle}{E^{(+)} - \varepsilon_{\alpha''} - \varepsilon_{core} - \frac{1}{2}k''^2}$$
$$+ \sum_{\beta''} \int d^3k'' \frac{\langle \mathbf{k}'\Psi_{\alpha'} | V | \mathbf{k}''\Omega\phi_{\beta''} \rangle \langle \mathbf{k}''\Omega\phi_{\beta''} | T | \mathbf{k}\Psi_\alpha \rangle}{E^{(+)} - \varepsilon_{\beta''} - \varepsilon_{core} - \frac{1}{4}k''^2} \tag{6a}$$

$$\langle \mathbf{k}'\Omega\phi_{\beta'} | T | \mathbf{k}\Psi_\alpha \rangle = \langle \mathbf{k}'\Omega\phi_{\beta'} | V | \mathbf{k}\Psi_\alpha \rangle + \sum_{\alpha''} \int d^3k'' \frac{\langle \mathbf{k}'\Omega\phi_{\beta'} | V | \mathbf{k}''\Psi_{\alpha''} \rangle \langle k''\Psi_{\alpha''} | T | \mathbf{k}\Psi_\alpha \rangle}{E^{(+)} - \varepsilon_{\alpha''} - \varepsilon_{core} - \frac{1}{2}k''^2}$$
$$+ \sum_{\beta''} \int d^3k'' \frac{\langle \mathbf{k}'\Omega\phi_{\beta'} | V | \mathbf{k}''\Omega\phi_{\beta''} \rangle \langle \mathbf{k}''\Omega\phi_{\beta''} | T | \mathbf{k}\Psi_\alpha \rangle}{E^{(+)} - \varepsilon_{\beta''} - \varepsilon_{core} - \frac{1}{4}k''^2} \tag{6b}$$

Journal of Physics: Conference Series **185** (2009) 012018 doi:10.1088/1742-6596/185/1/012018

3. Results and Discussion

The present work uses two sets of calculations:

(a) CC(9,6): consists of nine atomic states (Na(3s, 3p, 4s, 3d, 4p, 5s, 4d, 5p, 5d)) + six Ps (Ps(1s, 2s, 2p, 3s, 3p, 3d)).

(b) CC(9,3): similar to the CC(9,6) except for the Ps states we reduced it to only three Ps states (Ps(1s, 2s, 2p)).

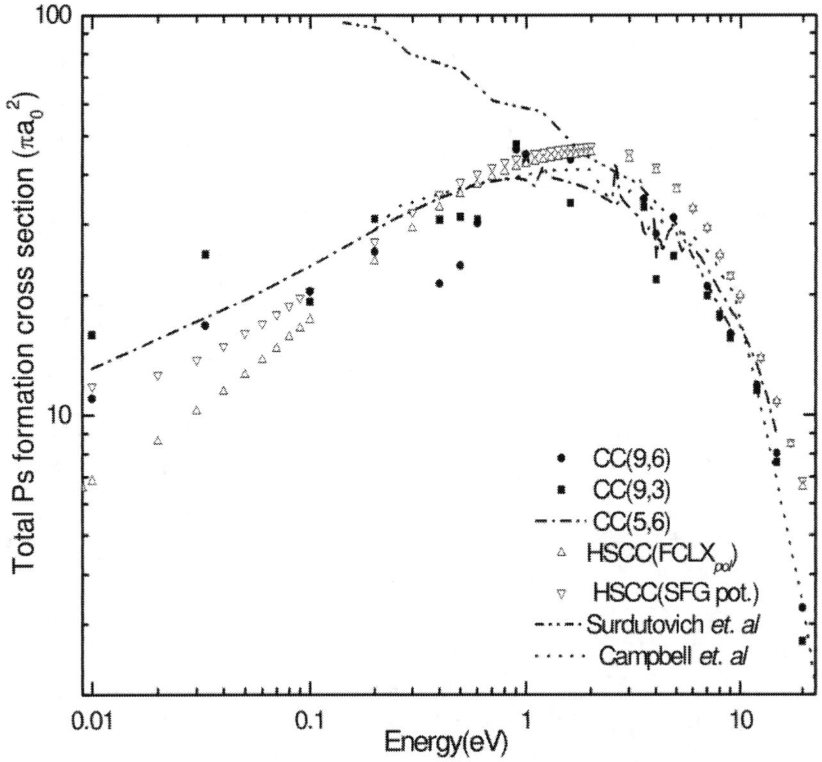

Figure 1. Total Ps formation cross section for e^+-Na scattering

The present results (CC(9,6) and CC(9,3)) for the total Ps formation cross section are displayed in Figure 1. The present cross sections are shown with the HSCC cross section[1] and the experimental measurements [2]. The previous CC calculations [3, 4] are also shown. The present results seems to be in good agreement with the measured data at energy above 1 eV. At energies ranging from 3 to 5 eV, a structure is seen except in the HSCC data. At energies <= 1 eV, the present results seems to support previous theoretical works. Further, the present results displays some structures in this low energy region. This could plausibly be due to the limitation of the present CC calculations.

In Figure 2, the total Ps(1s) formation cross section at energies within 0.01- 2 eV are studied. The inclusion of (n = 3) Ps states in the CC(9,6) calculations reduces the cross sections nearer to the CC(5,6) and the HSCC (SFG pot) cross sections for energies up to 0.2 eV. The present CC(9,6) calculations shows obvious numerical artefacts at 0.4 and 0.5 eV. Our various convergence studies do not improve the cross section at these two energies.

4. Conclusion

The present calculations show similar trends with the HSCC and CC(5,6) in the low energy region. However, at energies below 1 eV, the present CC(9,6) and CC(9,3) calculations are in reasonable agreement with the qualitative trends of the HSCC except at 0.4 and 0.5 eV.

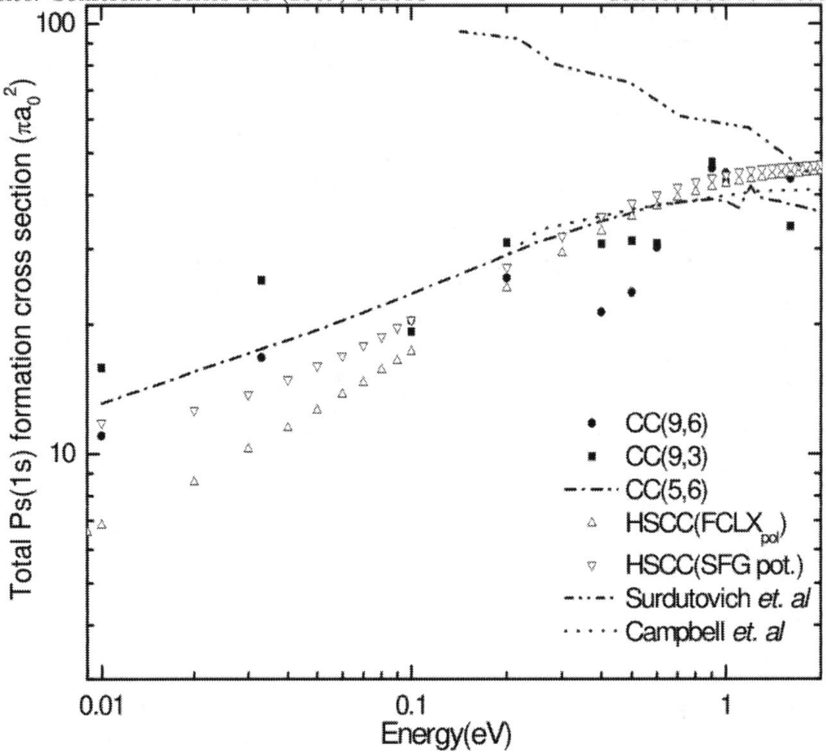

Figure 2. Total Ps(1s) formation cross section for e^+-Na scattering from 0.01 eV - 2 eV

Acknowledgements:

The funding of this work by the SAGA grant (Academy of Sciences Malaysia:Account No: 66-02-03-0077) under KR and MZMK is gratefully acknowledged.

References

[1] Anh Thu Le , Bromley M W J and Lin C D 2005 *Phys. Rev.* A **71**, 032713.
[2] Surdutovich E, Johnson J M, Kaupilla W E, Kwan C K and Stein T S 2002 *Phys. Rev.* A **65**, 032713.
[3] Ryzhikh G G and Mitroy J, 1997 *J. Phys.* B **30**, 5545
[4] Campbell C P, McAlinden M T, Kernoghan A A and Walters H R J 1998 *Nucl. Instrum. Methods Phys. Res.* B **143**, 41.
[5] Ward S J, Horbatsch M, McEachran R P and Stauffer A D 1988 *J. Phys.* B **21** L611.
[6] Ward S J, Horbatsch M, McEachran R P and Stauffer A D 1989a *J. Phys.* B **22** 1845.
[7] Ward S J, Horbatsch M, McEachran R P and Stauffer A D 1989b *J. Phys.* B **22** 3763.
[8] Sarkar K P, Basu M and Ghosh A S 1988 *J. Phys.* B **21** 1649.
[9] Abdel-Raouf M A 1988 *J. Phys.* B **21** 2331.
[10] Hewitt R N, Noble C J and Bransden B H 1993 *J. Phys.* B **26** 3661.
[11] Mitroy J and Ratnavelu K 1994 *Aust. J. Phys.* **47** 721.
[12] McAlinden M T, Kernoghan A A and Walters H R J 1994 *Hyperfine Int.* **89** 161.
[13] McAlinden M T, Kernoghan A A, and Walters H R J 1997 *J.Phys.* B **30** 1543.
[14] Ryzhikh G G and Mitroy J 1997 *Phys. Rev. Lett.* **79** 4124.
[15] Strasburger K and Chojnacki H 1998 *J. Chem. Phys.* **108** 3218.
[16] Ryzhikh G G, Mitroy J and Varga K 1998 *J. Phys. B* **31**, L265.
[17] Yuan J, Esry B D, Morishita T and Lin C D 1998 *Phys. Rev.* A **58** R4.
[18] Natchimuthu N and Ratnavelu K 2001 *Phys. Rev.* A **63** 052707.
[19] Zhou Y 1998 *private communication.*

Cluster state entanglement of charge qubits

Matthew Lubelski Katz and **Jingbo Wang**

School of Physics, The University of Western Australia, Australia

E-mail: wang@physics.uwa.edu.au

Abstract. The cluster state is a special, highly entangled quantum state that provides a universal resource on which to perform measurement-based quantum computation. In this paper, we propose a scheme for the generation of such entanglement in a solid-state medium by the suppression of resonant tunneling of a ballistic electron by a nearby single-electron charge qubit. To investigate the viability of this scheme for the creation of cluster states, we performed numerical simulations and studied the entanglement interaction in detail. We show that high levels of entanglement can be achieved, with an average concurrence of 0.968.

1. Introduction

The cluster state model of quantum computation is a promising new approach to the design and formulation of quantum computers. In the "conventional" circuit model of quantum computation a register of qubits is acted upon by a series of successive logic gates in a manner analogous to that of its electronic counterpart. The cluster state model differs from this approach by instead using a system of highly entangled qubits as the universal computational resource [1]. Computation is then performed through a series of parallel single-qubit measurements that propagate quantum correlations through the system. While the results of these measurements are random, they can be classically combined to form a deterministic output and so universal computation achieved.

Through the choice of measurements, cluster state computation offers a unique way of designing quantum algorithms. More than this, it provides a robust scheme for implementing a practical computing device. Such a device does not suffer from many of the disadvantages associated with the quantum circuit implementation as it uses only single-qubit measurements to perform computation. Furthermore, it has been shown that a cluster state quantum computer would be more resilient to decoherence effects [2] and provide the ability for efficient error correction [3, 4]. Before cluster state computation can be practically performed, the controlled generation of the entanglement necessary to form the cluster state must be realized and this problem is the focus of this work.

2. Theoretical model

In our model the charge qubit is based on the single-electron lateral coupled quantum dots experimentally implemented by Petta *et al* [5]. These are constructed from a GaAs/AlGaAs heterostructure containing a two-dimensional electron gas (2DEG) 100nm below the surface and

Journal of Physics: Conference Series **185** (2009) 012019 doi:10.1088/1742-6596/185/1/012019

controlled by Ti/Au top gates. The ballistic electron is produced by a single-electron source with an energy uncertainty of 1% and is confined to a GaAs quantum wire. This wire contains double symmetric resonant tunneling barriers positioned so that they lie over the center of one of the quantum dots. A schematic diagram of the proposed system is shown in figure 1. Currently, quantum wires have been produced down to diameters of a few nm [6] and so the ballistic electron is strongly radially confined and can be treated as one-dimensional. The barriers are formed of two InAs slices embedded in the quantum wire heterostructure. Such slices have been experimentally produced with widths of less then 50 nm and have a maximum barrier height of approximately 1.4 meV [7]. While the exact geometry of these potential barriers has yet to be established, similar experiments [8] with InP doped InAs wires have shown a good match to rectangular potential functions. Resonant tunneling occurs when the ballistic electron has an energy that closely matches one of the virtual energy eigenstates of the potential well formed by these barriers. Around these specific energies the transmission probability profile of the electron is sharply peaked and the barriers are effectively "transparent".

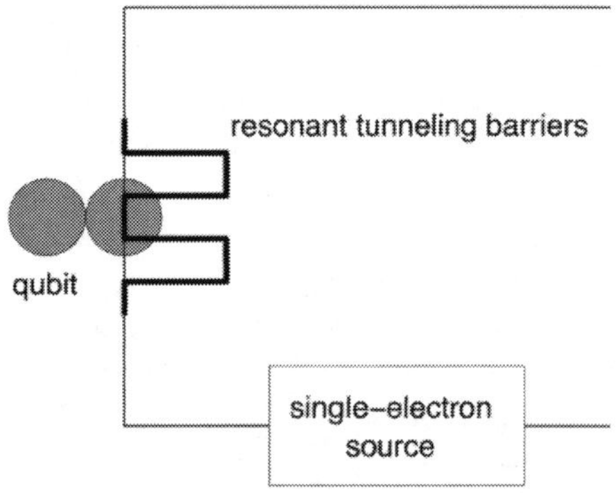

Figure 1. A schematic diagram of the proposed entanglement system. The ballistic electron is pumped towards the barriers in the quantum wire by a single-electron source, entanglement occurs with the charge qubit through the Coulombic suppression of resonant tunneling through the double barriers.

To simulate this system, the electrons in the qubit and quantum wire are described by the three-dimensional joint wavefunction $\psi(x_1,y_1,x_2,t)$, where x_1 and y_1 represent the position of the electron in the qubit, x_2 represents the position of the electron in the quantum wire, and t the time. The total Hamiltonian of the system is given by $\hat{H}=\hat{H}_q+\hat{H}_w+\hat{H}_{\text{int}}$, where $\hat{H}_q$ is the Hamiltonian of the isolated qubit, $\hat{H}_w$ is the Hamiltonian of the isolated ballistic electron, and $\hat{H}_{\text{int}}$ describes the Coulombic interaction that acts on the combined Hilbert space of the two electrons. The dynamics of the system are governed by the time-dependent Schrödinger equation $i\partial\psi/\partial t=\hat{H}\psi$, which is solved using the numerical scheme described by Wang and Midgley [9] and Hines *et al* [10].

Entanglement occurs because the Coulombic interaction of the qubit distorts the potential experienced by the ballistic electron and so alters its resonant tunneling probability profile. By varying the separation and width of the barriers, as well as the ballistic electron's incident energy, this transmission probability profile can be finely controlled. Through optimization the system is tuned so that resonant tunnelling will occur when the qubit electron is localized to the dot furthest from the wire and is suppressed when the qubit electron is localized to the dot closest to the wire, labelled by the $|0\rangle$ and $|1\rangle$ computational basis states respectively. The transmission coefficients of the ballistic electron with the charge qubit in initial states $|0\rangle$ and $|1\rangle$ states are plotted in figure 2 for a range of initial ballistic electron energies.

It has been shown that, while the particular mechanism of entanglement is unimportant, for a quantum system to be used as a cluster state then the localized entanglement between pairs of nearest neighbor qubits must be at a maximum [11]. If this is not the case, measurements on the individual qubits will not result in the complete transfer of quantum information and computational errors will occur.

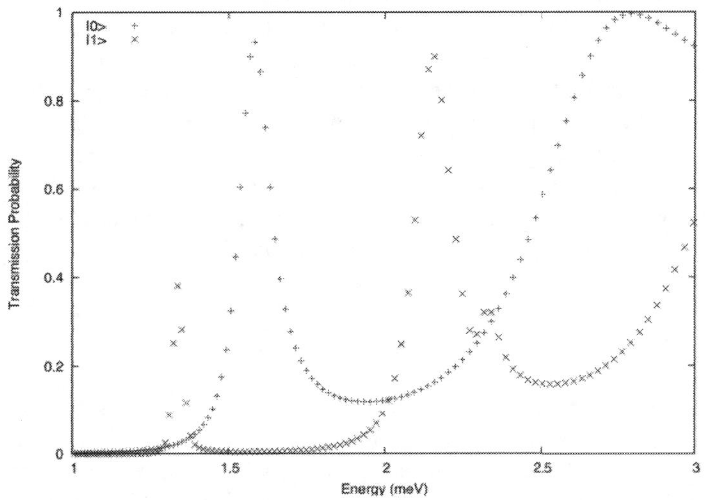

Figure 2. Transmission probabilities of the ballistic electron with 1% energy uncertainty for initial qubit states |0⟩ and |1⟩ over a range of initial ballistic electron energies using optimized barrier parameters.

To quantify the degree of bipartite entanglement present in the system we use the concurrence or "entanglement of formation" [12]. For an arbitrary N qubit system, the concurrence is given by $C(\psi) = \left|\langle \psi | \sigma_y^1 \otimes \sigma_y^2 \otimes ... \sigma_y^N | \psi^* \rangle\right|$, where ψ^* is the complex conjugate of ψ in a fixed basis, and $\sigma_y^i = \{\{0,-i\},\{i,0\}\}$ is the Pauli matrix acting on the Hilbert space of ith qubit in this basis. The concurrence of the system is then given in the range 0 to 1, where 1 means a maximally entangled state and 0 means a separable state. Shown in figure 3 is the average concurrence between the qubit and ballistic electrons (N=2) for a range of initial energies. As might be expected, the maximum average concurrence of 0.968 occurs when the difference in the transmission probability between the |0⟩ and |1⟩ states is at its maximum, namely at the ballistic electron energy of 1.585 meV.

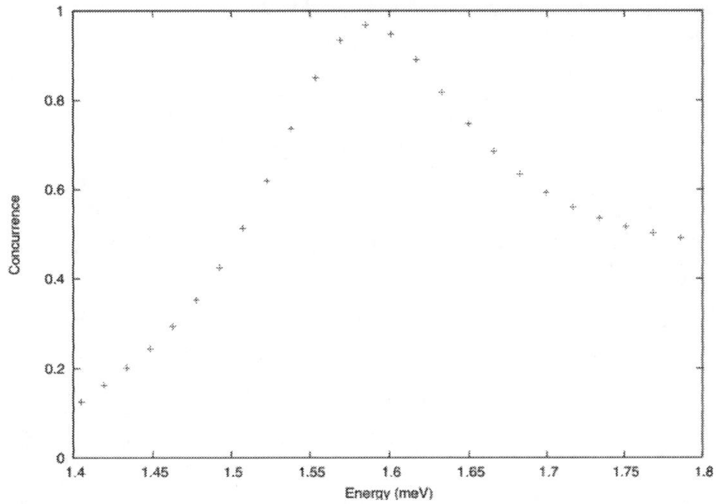

Figure 3. The average concurrence for a range of initial ballistic electron energies with optimized barrier parameters.

3. Conclusions

In this work we have carried out a detailed simulation of a solid-state system for generating controlled quantum entanglement between a single-electron charge qubit and a ballistic electron by utilizing resonant tunneling in a quantum wire. The simulation was closely modeled on experimentally realizable components and treated the ballistic and qubit electrons and their interactions in detail. It was shown that the tunneling dynamics are highly sensitive to the system parameters. For symmetric rectangular potential barriers with an optimized height, width and separation of 1.4 meV, 50 nm and 90 nm respectively, the system produced entanglement with an average concurrence value of 0.968. This entanglement occurred due to the change in kinetic energy of the ballistic electron caused by the Columbic repulsion from the qubit electron and is effectively elastic leaving the probability density of the qubit largely unchanged.

There are several key avenues of research that still need to be explored if this resonant tunneling scheme is to be shown to be suitable for cluster state generation. Further simulation needs to be performed to establish the exact nature of the phase shifts of the ballistic and qubit electrons to see whether it will be possible to compensate for these shifts during computation. It would also be desirable to try different potential barrier configuration in order to see if a more consistent concurrence could be produced across the whole range of energies contained within the ballistic electrons wave packet. If this can be achieved then it would facilitate entanglement pumping and so, with multiple tunneling events, the entanglement could be increased to an arbitrarily high value.

References

[1] R. Raussendorf and H. J. Briegel. Phys. Rev. Lett. 86, 5188 (2001)
[2] W. Dür and H. J. Briegel. Phys. Rev. Lett. 92, 180403 (2004)
[3] S. Y. Looi, L. Yu, V. Gheorghiu, and R. B. Gri!ths. Phys. Rev. A 78, 042303 (2008)
[4] M. A. Nielsen and C. M. Dawson. Phys. Rev. A 71, 042323 (2005)
[5] J. R. Petta, A.C. Johnson, C. M. Marcus, M. P. Hanson, A. C. Gossard. Phys. Rev. Lett. 93, 186802 (2004)
[6] W. E. Buhro and H. Yu. Advanced Materials, 15, 416 (2003)
[7] N. Panev, A. I. Persson, N. Sköld, and L. Samuelson. App. Phys. Lett. 83, 2238 (2003)
[8] M. T. Björk, B. J. Ohlsson, C. Thelander, A. I. Persson, K. Deppert *et al.* App. Phys. Lett. 81, 4458 (2002)
[9] J. B. Wang and S. Midgley, Phys. Rev. B 60, 13668 (1999)
[10] C. Hines, K. Jacobs, and J.B. Wang, J. Phys. A 40, 609 (2007)
[11] M. Van den Nest, A. Miyake, W. Dür, and H. J. Briegel. Phys. Rev. Lett. 97, 150504 (2006)
[12] W. K. Wootters, Phys. Rev. Lett. 80, 2245 (1998)

Acknowledgment

The authors would like to thank Chris Hines for valuable discussions. This work was supported by iVEC - 'The hub of advanced computing in Western Australia', The University of Western Australia, and the Australian Research Council.

Journal of Physics: Conference Series **185** (2009) 012020 doi:10.1088/1742-6596/185/1/012020

Anisotropic Interactions and Valence Ionic States of Dibenzenechromium Observed by Collision-Energy-Resolved Penning Ionization Electron Spectroscopy with He*(2^3S) Metastable Atoms

Naoki Kishimoto and Koichi Ohno

Department of Chemistry, Graduate School of Science, Tohoku University, Aramaki, Aoba-ku, Sendai 980-8578, Japan

E-mail: n-kishi@mail.tains.tohoku.ac.jp

Abstract. Penning ionization of dibenzenechromium with He*(2^3S) metastable atoms was observed by collision-energy-resolved two-dimensional electron spectroscopy. Collision energy dependence of partial ionization cross sections (CEDPICS) indicates that attractive interaction around the benzene rings (Bz) results in negative slope of CEDPICS for ionization from valence π_{Bz} orbtials. Negative CEDPICS of bands 1 and 2 was ascribed to strong electron correlation effects in mixing configurations of direct ionization from metal d orbitals and shake-up excitation as shown by CI calculation. Repulsive interaction was also found around the σ_{CH} bonds and the Cr atom.

1. Introduction

Electronic properties of organometallic compounds have attracted considerable interest. The electronic structure of organometallic compounds, however, has not been investigated well mainly due to strong electron correlation effects and high spin states. In order to investigate the electronic structure of dibenzenechromium, electron spectroscopy methods [1-3] and theoretical calculations of ionization energies [4-6] have been utilized. Recently, we have published [7] an article on the electronic structure of ferrocene by collisional Penning ionization electron spectroscopy (PIES). By this method, orbital reactivity can be connected with spatial extending of molecular orbtials (MOs) outside of the boundary surface of a colliding excited rare gas atom A* to a target molecule M [8] ($A^* + M \rightarrow A + M^+ + e^-$), since the transition probability for the electron transfer from an occupied MO to the inner singly occupied orbital of A* depends mainly on the spatial overlap between the related orbtials. The change of orbital reactivity as a function of collision energy between metastable He*(2^3S) atoms and target molecules, which is collision energy dependence of partial ionization cross sections (CEDPICS) [9], can be observed by the two-dimensional (collision-energy/electron-energy-resolved) Penning ionization electron spectroscopy (2D-PIES) [10,11]. In the case of ferrocene [7], negative CEDPICS by mixing configurations were obtained for the first two ionic states.

2. Methods

In our experimental set up [10-12], beams of electronically excited metastable He atoms were produced by a discharge nozzle source with a tantalum hollow cathode. Sample molecules were put in a sample cell set under the collision cell and vaporized at ca. 325 K. In the two-dimensional Penning ionization measurement, the metastable atom beam was modulated by a pseudorandom chopper [12], and then introduced into a reaction cell. The velocity distribution of the metastable He* beam was monitored by secondary electrons emitted from a stainless steel plate inserted in the reaction cell.

We performed ab initio molecular orbital calculations with the Gaussian 03 package of programs [13]. The geometries of the neutral target molecules were optimized with density functional theory (DFT) with B3LYP correlation functional [14] and 6-311G** basis functions. In the electron density contour maps, thick solid curves indicate the repulsive molecular surface approximated by atomic spheres of van der Waals radii [15]. For the ionization potential energies, P3 propagator [16] and SAC/SAC-CI [17] calculations were performed with 6-311G** basis for Cr and cc-pVDZ basis for C and H atoms. Based on the well-known resemblance between He*(2^3S) and Li(2^2S) [18], a Li(2^2S) atom in the ground state was utilized in place of a He*(2^3S) atom for potential surface calculations.

3. Results and Discussion

3.1. Ionization from ligand π and σ orbitals

Ionization bands 3 and 5 in Figure 1(a) can be assigned to benzene π orbitals (π_{Bz}) based on post SCF calculations listed in Table 1. As described in the previous PIES studies for sandwich compounds [7,19], arene π bands show relatively large intensity rather than σ_{CH} band in PIES, which can be ascribed to the extending spatial distribution of arene π orbtials and electron exchange process [20] of Penning ionization in which an electron of a target molecule transfers into the 1s orbital of He*(2^3S) and the excited electron is ejected to the continuum state simultaneously.

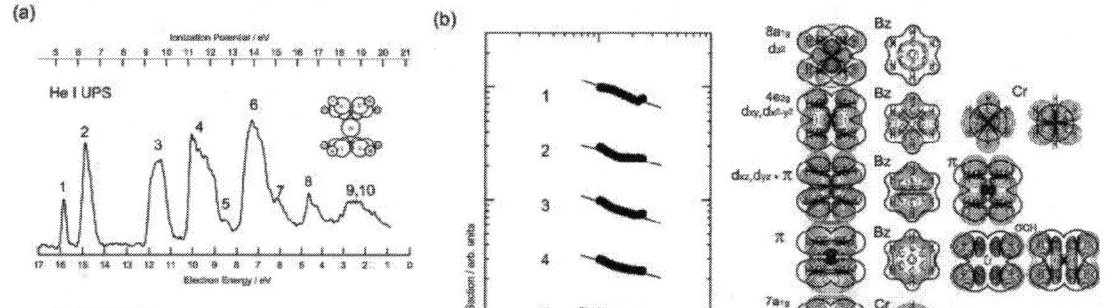

Figure 1. (a) He I UPS and He*(2^3S) PIES of $(Bz)_2Cr$. (b) CEDPICS of $(Bz)_2Cr$ by collision with He*(2^3S) atoms. Cr and Bz indicates electron density maps on the planes at the Cr atom and 0.85 Å above the Bz ring, respectively.

Negative slope of CEDPICS (Figure 1(b) and Table 1) for band 3 (π_{Bz}, $m = -0.35$) can be ascribed to attractive interaction between the target molecule and He*(2^3S) atoms around the π_{Bz} orbital region. It is known that slower He* atoms can penetrate into the inner reactive region than the faster He* atoms because of the lower centrifugal barrier, which can lead to negative CEDPICS.[2,6-8] When the attractive interaction of R^{-s} is dominant, ionization cross section $\sigma(E_c)$ can be expressed[17] by

$$\sigma(E_c) \propto E_c^{-2/s} \quad (1)$$

Journal of Physics: Conference Series **185** (2009) 012020 doi:10.1088/1742-6596/185/1/012020

This formula was derived from the impact parameter for which orbiting occurs for atomic targets, and the slope of log $\sigma(E_c)$ vs. log E_c plot can be connected to the function form ($m = -2/s$). Attractive potential well of ca. 80 meV was obtained for the model potential calculation for the access of Li to the π_{Bz} orbital region (Figure 2). On the other hand, repulsive interaction was obtained around the σ_{CH} orbital region, which is consistent with the smaller negative slope of CEDPICS for band 6 (σ_{CH}, $m = -0.13$) and band 7 (σ_{CH}, $m = -0.17$) than that for band 3 (π_{Bz}, $m = -0.35$). In addition, the gentle negative slope of CEDPICS for band 4 ($m = -0.27$) can be explained by the overlapping π_{Bz} and σ_{CH} ionization bands in PIES. Band 5 was assigned to π_{Bz} ($7a_{1g}$) orbital based on the post SCF calculations and the large negative CEDPICS ($m = -0.33$). The small intensity of band 5 in UPS and PIES can be ascribed to the mixing of single (small intensity for ionization from $7a_{1g}$ (0.87)) and shake-up ionization configurations calculated by the SAC/SAC-CI method (Table 1).

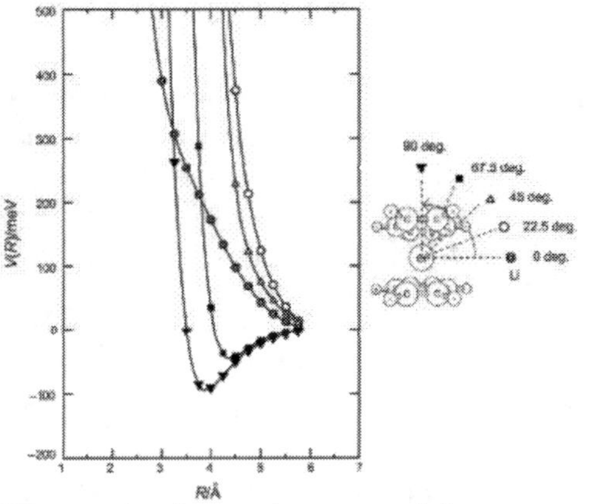

Figure 2. Interaction potential energy curves $V(R)$ as a function of distance R between Cr and Li.

Band 8 shows enhanced intensity in PIES, which was assigned mainly to two valence MOs. Although the corresponding $5a_{2u}$ and $6a_{1g}$ MOs have σ_{CH} character, they have in-phase extending distributions outside the repulsive surface within or between the two rings. This in-phase extending of MOs results in relatively larger Penning ionization probability than other MOs. The small negative CEDPICS ($m = -0.07$) is due to repulsive interaction around the σ_{CH} orbital region.

Table 1. Ionization potentials (eV) by He I UPS and theoretical calculations, band assignments, and slope parameter (*m*) of CEDPICS for $(C_6H_6)_2Cr$

Band	IP	Koopmans	SAC-CI[a]	OP2[b]	P3[c]	*m*
1	5.36	10.01 ($8a_{1g}$, d_{z^2})	4.54 (0.80)[d]	7.14	-	-0.34
2	6.32	6.53 ($4e_{2g}$, $d_{xy}, d_{x^2-y^2}$)	5.28 (0.89)	8.58	-	-0.25
3	9.72	9.63 ($4e_{1g}$, $d_{xz}, d_{yz}+\pi_{Bz}$)	8.84 (0.93)	10.77	9.96 (0.90)	-0.35
		10.56 ($6e_{1u}$, π_{Bz})	9.29 (0.93)	10.86	10.20 (0.89)	
4	11.16	12.79 ($6a_{2u}$, π_{Bz})	11.07 (0.94)	11.41	11.86 (0.87)	-0.27
	11.50	13.32 ($3e_{2u}$)	11.17 (0.92)	12.04	11.83 (0.89)	
	11.80	13.53 ($3e_{2g}$)	11.23 (0.94)	12.29	12.08 (0.90)	
5	12.66	14.85 ($7a_{1g}$, π_{Bz})	12.73 (0.87)[e]		13.42 (0.85)	-0.33
6	13.94	16.10 ($5e_{1u}$)	13.76 (0.93)	13.96	14.34 (0.88)	-0.13
		16.33 ($1b_{1g}$)	13.86 (0.92)		14.44 (0.86)	
		16.22 ($3e_{1g}$)			14.54 (0.89)	
		16.43 ($1b_{2u}$)			14.32 (0.86)	
7	15.06	17.44 ($2b_{2g}$)			15.32 (0.86)	-0.17
		17.66 ($2b_{1u}$)			15.51 (0.86)	
8	16.56	19.04 ($5a_{2u}$)			16.96 (0.87)	-0.07
	16.94	19.61 ($6a_{1g}$)				
9	18.5	22.16 ($2e_{2u}$)				
10	18.9	22.51 ($2e_{2g}$)				

[a] Intensities are shown in the parentheses. Coefficient threshold for the ground state configuration was lowered from 0.005 (default value) to 0.001 in calculations.
[b] Second-order quasi-particle approximation Green's function method. Reference [6].
[c] Pole strengths are shown in the parentheses.
[d] Mixing of direct and shake-up configurations (main: 0.89 ($8a_{1g}$) + 0.22 ($8a_{1g}$,$4e_{1g} \rightarrow 6e_{1g}$) – 0.19 ($8a_{1g}$,$4e_{2g} \rightarrow 5e_{2g}$)) was obtained for the 1^2A_{1g} state.
[e] Mixing of direct and shake-up configurations (main: 0.93 ($7a_{1g}$) + 0.13 ($6a_{2u}$,$4e_{2g} \rightarrow 4e_{2u}$)) was obtained for the 2^2A_{1g} state.

3.2. Ionization from metal *d* orbitals

The ionization energy order for the first two ionic states obtained by post SCF calculations is $8a_{1g}$ (d_z^2) < $4e_{2g}$ ($d_{xy}, d_{x^2-y^2}$), which is different from Koopmans approximation due to strong electron correlation effects on the first ionic state. The P3 calculation failed for ionization from both $8a_{1g}$ and $4e_{2g}$ MOs probably due to the strong electron correlation effects. The intensity order of the first two bands of dibenzenechromium is band 1($8a_{1g}$(d_z^2)) < band 2($4e_{2g}$ ($d_{xy}, d_{x^2-y^2}$)) in UPS, which is reversed in the PIES spectrum. The intensity of band 2 is larger (nearly 1.5 times) than that of band 1 in the PIES spectrum. This result in the PIES spectrum was ascribed to larger spatial extending of $8a_{1g}$(d_z^2) MO than $4e_{2g}$ ($d_{xy}, d_{x^2-y^2}$) MO in a previous paper [19]. When one pays attention to the mixing configuration for band 1 (Table 1), however, this enhancement may be caused by the large contribution of ionization from spatially extending $4e_{1g}$ ($d_{xz}, d_{yz} + \pi_{Bz}$) MO.

The slope value for band 1 ($m = -0.34$) is similar to π_{Bz} ionization (band 3, $m = -0.35$), which is consistent with attractive interaction around the π_{Bz} MO region where the d_z^2 orbital extends. Interaction potential energy curve for the access of Li to the Cr atom is repulsive. Although mixing of configurations for the first ionic state includes ionization from the $4e_{2g}$ ($d_{xy}, d_{x^2-y^2}$) MO that extends around the Cr atom, configurations for ionization from z-components ($8a_{1g}$(d_z^2), $4e_{1g}$($d_{xz}, d_{yz} + \pi_{Bz}$)) are dominant in this study. The observed negative slope value of CEDPICS for band 2 ($4e_{2g}$, $m = -0.25$) is smaller than that for band 1 reflecting the repulsive interaction around the metal atom. It should be noted that the mixing of configurations for the second ionic state includes shake-up configurations from the $4e_{2g}$ ($d_{xy}, d_{x^2-y^2}$) MO.

Acknowledgements

This research was supported by a Grant for Scientific Research from the Ministry of Education, Culture, Sport, Science, and Technology of Japan. The authors are grateful to Prof. Masahiro Ehara in IMS (Japan) for advising details in theoretical calculations.

References

[1] Evans S., Green J.C., and Jackson S.E., 1972 J. Chem. Soc. Faraday Trans. II **68**, 249.
[2] Cabelli D.E., Cowley A.H., and Lagowski J. J., 1982 Inorg. Chimi. Acta. **57**, 195.
[3] Brennan J.G., Cooper G., Green J.C., Kaltsoyannis N., MacDonald M.A., Payne M.P., Redfern C.M., and Sze K.H., 1992 Chem. Phys. **164**, 271.
[4] Guest M. F., Hillier I.H., Higginston B.R., and Lloyd D.R., 1975 Mol. Phys. **29**, 113.
[5] Weber J., Geoffroy M., Goursot A., and Penigault E., 1978 J. Am. Chem. Soc. **100**, 3995.
[6] Böhm M.C., 1983 J. Chem. Phys. **78**, 7044.
[7] Kishimoto N., and Ohno K., 2009 J. Phys. Chem. A **113**, 521.
[8] Ohno K., Mutoh H., and Harada Y., 1983 J. Am. Chem. Soc. **105,** 4555.
[9] Ohno K., Takami T., Mitsuke K., and Ishida T., 1991 J. Chem. Phys. **94**, 2675.
[10] Ohno K., Yamakado H., Ogawa T., and Yamata T., 1996 J. Chem. Phys. **105,** 7536.
[11] Ohno K., 2004 Bull. Chem. Soc. Jpn. **77**, 887.
[12] Kishimoto N., and Ohno K., 2007 Int. Rev. Phys. Chem. **26**, 93.
[13] Frisch M. J., *et al.* Gaussian 03 (Gaussian, Inc., Pittsburgh, PA, 2003).
[14] Becke A. D., 1993 J. Chem. Phys. **7**, 5648.
[15] Pauling L., The Nature of the Chemical Bond; Cornell University, Ithaca, NY, 1960.
[16] Ortiz J. V., 1996 J. Chem. Phys. **104**, 7599.
[17] Nakatsuji H., in Computational Chemistry - Reviews of Current Trends, Vol. 2, ed. J. Leszczynski, World Scientific, 1997, pp.62-124.
[18] Niehaus A., 1981 Adv. Chem. Phys. **45**, 399.
[19] Masuda S., Aoyama M., and Harada Y., 1990 J. Am. Chem. Soc. **112**, 6445.
[20] Hotop H., Niehaus A., 1969 Z. Phys. **228**, 68.

Journal of Physics: Conference Series **185** (2009) 012021 doi:10.1088/1742-6596/185/1/012021

Magnetic resonance of atoms passing through a magnetic lattice

Y Kobayashi and A Hatakeyama

Tokyo University of Agriculture and Technology, Koganei, Tokyo 184-8588, Japan

E-mail: `hatakeya@cc.tuat.ac.jp`

Abstract. We have observed magnetic resonance of Rb atoms passing through a periodic magnetic field produced by a stack of planar arrays of parallel current-carrying wires ("magnetic lattice"). The magnetic resonance occurs between the Zeeman-split sublevels of Rb atoms when the frequency of the field oscillation that the atoms experience equals the transition frequency. A pump laser polarizes Rb atoms with a specific velocity selected by the Doppler effect and a probe laser detects the magnetic resonance. The obtained narrowest resonance lines have widths mainly determined by the transit-time broadening.

1. Introduction

We study resonance transitions in the internal states of atoms passing through a static periodic field. This motion-induced resonance occurs when the frequency of the field oscillation that the atoms experience equals the transition frequency. The principle of the resonance is the same as that of "Okorokov effect" [1] or "resonant coherent excitation" [2], which has been studied extensively using fast ion beams passing through crystals for resonances at high frequencies even in the x-ray regime. Its principle is, however, quite general, and we are interested in its extension to very low energy experiments, such as resonances in the rf or microwave region. If one can precisely control both the atomic motion and field periodicity, it is possible to control the atomic internal states with this resonance technique in the same way as using electromagnetic radiation. The motion-induced resonance is strongly velocity-dependent in nature, and it is particularly worth noting that the internal excitation should occur at the expense of the atomic kinetic energy. These features are not attainable with the standard rf or microwave resonance technique, and we expect that motion-induced resonance will find useful applications as a new type of atom control method. We have already reported motion-induced magnetic resonance using a thin cell of Rb vapor, to which a periodic field is applied with a pair of arrays of parallel current-carrying wires [3, 4]. Its simple experimental setup was satisfactory for the demonstration of motion-induced resonance in the rf region. However, the atoms in the thin cell collided with the surfaces of the cell quite often, which degraded the sharpness of the resonance. For further investigations, we require a longer coherent interaction of atoms with a periodic field, together with an improvement in the signal-to-noise ratio of resonance spectra. We have therefore built a new experimental system, in which an effusive atomic beam of Rb passes through a stack of planar arrays of parallel current-carrying wires ("magnetic lattice") placed in a vacuum chamber. The spatial period of the magnetic field produced by the magnetic lattice is $a = 1$ mm. The atomic velocity v is about 500 m/s, selected with a collinear laser

Journal of Physics: Conference Series **185** (2009) 012021 doi:10.1088/1742-6596/185/1/012021

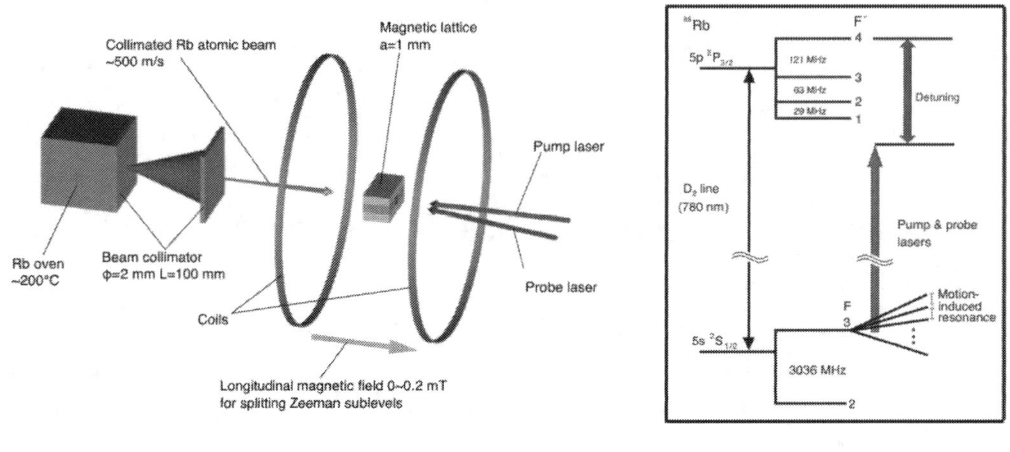

Figure 1. Schematic of experimental setup. Inset: Relevant energy levels of ^{85}Rb. The laser detuning is also shown.

beam using the Doppler effect. The frequency of the field oscillation the atoms see is then $f = v/a \sim 500$ kHz. The laser beam polarizes the atoms by optical pumping, and another probe laser detects magnetic resonance transitions between the Zeeman sublevels of the ground state. As scanning a longitudinal magnetic field, we have observed resonance peaks much sharper and clearer than the previous cell experiments. The narrowest resonance profiles obtained in the measurement have widths almost only determined by the transit time through the periodic field.

2. Experiment

A schematic of the experimental setup is shown in Fig. 1. Rb atoms emerge from an oven held at about 200°C with a broad Maxwell-Boltzmann velocity distribution, from which we selectively use atoms in a narrow velocity range, as described later. The effused atoms are collimated by two holes 2 mm in diameter separated by $L = 100$ mm. The resultant beam divergence is hence 2/100.

A magnetic lattice consists of a stack of printed circuit boards (PCBs), each of which has a periodic current carrying trace. A central part of the magnetic lattice viewed in the beam direction is shown in Fig. 2(a). Each board has a rectangular shape 60 mm in width and 30 mm in length along the beam direction. Totally 11 PCBs are stacked with gaps of 0.8 mm between each other, through which a large fraction of Rb atoms in the beam pass. The angles of the magnetic lattice can be adjusted to accomplish fine alignment between the lattice and the beam.

A schematic diagram of the circuit is shown in Fig. 2(b). A current-carrying copper trace (0.4 mm in width) goes back and forth with a spatial period of $a = 1$ mm. The current at the edges of the board is halved to compensate for the distortion of the periodic field produced by an array of a finite number of current-carrying wires [5]. Figure 2(c) schematically illustrates how this compensation works. As a whole 28 spatial periods are generated. Also refer to Ref. [3] for the details of the periodic magnetic field produced by arrays of parallel current-carrying wires.

A energy level diagram of ^{85}Rb relevant to the measurement is given in the inset of Fig. 1. A circularly polarized pump laser (wavelength: 780 nm) polarizes the atoms of ^{85}Rb in the $F = 3$ ground state by optical pumping. Magnetic resonance transitions between the Zeeman-split sublevels of this $F = 3$ state are detected with a probe laser (wavelength: 780 nm) through its absorption. The polarization of the probe laser is modulated between left- and right-circularly

Journal of Physics: Conference Series **185** (2009) 012021 doi:10.1088/1742-6596/185/1/012021

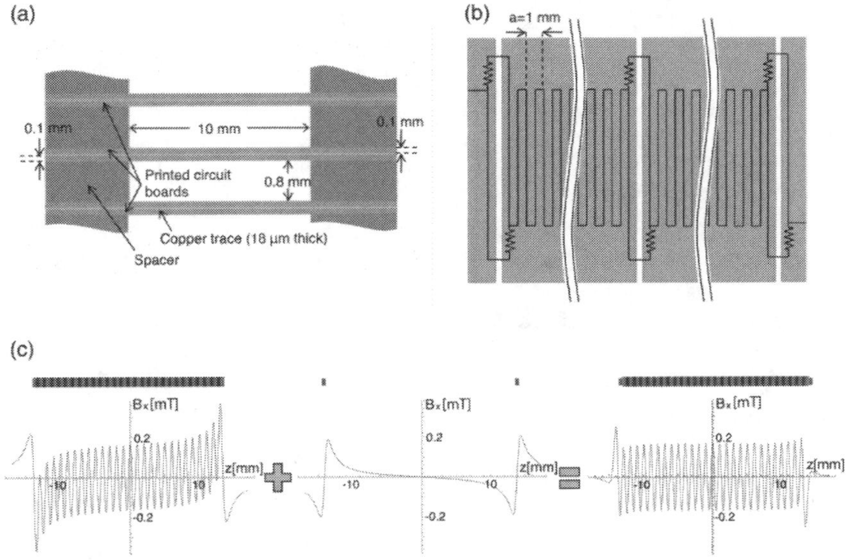

Figure 2. (a) Central part of the magnetic lattice viewed in the atomic beam direction. (b) Schematic diagram of the circuit on printed circuit boards (PCBs). One PCB corresponds to a green shaded region. PCBs are shown side by side, although they are actually piled up. (c) Compensation for a distorted periodic field by use of two additional correction wires. The calculated magnetic field is a component perpendicular to PCBs at the center between two adjacent PCBs for a current of 1 A.

polarizations at 42 kHz by a photoelastic modulator for the lock-in detection. Resonance signals are observed as scanning a longitudinal magnetic field applied by a pair of coils along the pump beam and thus changing the Zeeman splitting. Other two pairs of coils (not shown in the figure) are used to cancel stray magnetic fields.

Another important role of the laser beams is to select atoms within a narrow velocity range from a broad Maxwell-Boltzmann distribution using the Doppler effect. The laser frequency is stabilized to one of the resonance lines of saturation absorption spectrum of ^{85}Rb and is further shifted using an acousto-optic modulator. The resultant detuning of the laser frequency from the $F = 3 \rightarrow F' = 4$ cyclic transition determines a selected velocity. The width of the selected velocity distribution is estimated to be about $5 - 10$ m/s, which is mainly determined by the natural linewidth (6 MHz) of the $F = 3 \rightarrow F' = 4$ transition .

3. Results

Figures 3(a) and (b) demonstrate motion-induced resonance for atoms with a velocity of 512 m/s selected by the laser detuning of 656 MHz. The x- and y-axes are the scanning longitudinal magnetic field and the lock-in signal of the transmitted probe laser intensity, respectively. A current of 5 mA generated a periodic magnetic field with an amplitude of 0.8 μT at the center between two adjacent PCBs. The periodic magnetic field and the lasers were weak enough to obtain the narrowest signals in the present experimental setup. The traces were averaged 1000 times over 500 s. As clearly seen in the figures, a motion-induced resonance signal appears only when the periodic magnetic field is applied. The signal is peaked at 510 kHz, in a good agreement with the expected value of 512 kHz. Note that the large peaks seen in both figures at the zero longitudinal magnetic field are so-called "Hanle resonances", which were produced

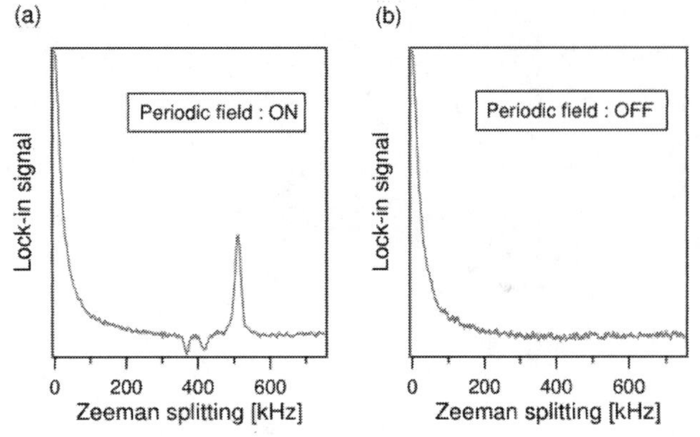

Figure 3. Observed spectra with a periodic magnetic field (a) on and (b) off.

by a residual transverse magnetic field.

One may notice that there are two small dips in Fig. 3(a). They are also signals of motion-induced resonance for atoms with different velocities. We intended to observe atoms with a velocity of 512 m/s through the $F = 3 \rightarrow F' = 4$ cyclic transition, but actually, although smaller in number, atoms with a velocity of 368 m/s or 417 m/s were additionally observed through the transition of $F = 3$ to $F' = 2$ or 3, respectively. Note that the lasers had different detunings with respect to these transitions and hence selected different velocities as well. The two small signals change in the opposite direction compared to the main resonance peak, because the laser transmission decreases for these transitions when motion-induced resonance occurs, while it increases for the $F = 3 \rightarrow F' = 4$ transition.

The line broadening due to the time of flight through the magnetic lattice, the fundamental limit of the linewidth in our setup, is estimated to be 16.3 kHz (full width at half maximum) [6], which is the main contribution to the obtained linewidth of 20 kHz. This implies that other factors contributing to the linewidth, such as dephasing caused by the collisions of atoms with the printed circuit boards, are not critical. Among them, however, the finite width of the velocity selection certainly causes an additional broadening, estimated to be about 10% increase from the transit-time broadening. The inhomogeneity of the longitudinal magnetic field is another probable reason for the broadening.

Acknowledgments
This work was supported by KAKENHI (No. 20684017) and the “Improvement of Research Environment for Young Researchers” program of The Ministry of Education, Culture, Sports, Science, and Technology, Japan.

References
[1] Okorokov V V 1965 *J. Nucl. Phys. (Moscow)* **2** 1009 [*Sov. J. Nucl. Phys.* **2** 719]; Okorokov V V 1965 *Zh. Eksp. Teor. Fiz. Pis'ma Red.* **2** 175 [*JETP Lett.* **2** 111]
[2] Datz S, Moak C D, Crawford O H, Krause H F, Dittner P F, Gomez del Campo J, Biggerstaff J A, Miller P D, Hvelplund P and Knudsen H 1978 *Phys. Rev. Lett.* **40** 843
[3] Hatakeyama A, Enomoto Y, Komaki K and Yamazaki Y 2005 *Phys. Rev. Lett.* **95** 253003
[4] Hatakeyama A 2008 *Appl. Phys.* B **92** 615
[5] Sidorov A I, McLean R J, Rowlands W J, Lau D C, Murphy J E, Walkiewicz M, Opat G I and Hannaford P 1996 *Quantum Semiclass. Opt.* **8** 713
[6] Demtröder W 1996 *Laser Spectroscopy: Basic Concepts and Instrumentation, 2nd ed* (Berlin, Springer)

Journal of Physics: Conference Series **185** (2009) 012022 doi:10.1088/1742-6596/185/1/012022

Electron controlled chemistry

E. Krishnakumar
Tata Institute of Fundamental Research, Homi Bhabha Road, Mumbai 400005.

E-mail: ekkumar@tifr.res.in

Abstract. The role of dissociative electron attachment in controlling electron induced chemical processes is discussed in terms of the unique characteristics of this process. The selectivity of molecular dissociation that exists in the gas phase using low energy electrons alone or in combination with laser excitation and in condensed phase are described and recent efforts in chemical control using electrons are summarized.

1. Introduction

It is by now established that free electrons are the facilitators of all chemical reactions that take place in plasmas. The interaction of free electrons with atoms and molecules lead to the formation of excited species, ions and radicals [1, 2] which then reacts among themselves or with atoms and molecules in their ground or excited states to produce new species. In many of these reactions, the role of electrons is not only in providing the necessary energy to overcome the barrier for a reaction, but also in creating a pathway which may not have been possible otherwise. A wide variety of processing plasmas used for various industrial purposes like lighting, semiconductor etching, plasma assisted chemical vapour deposition and high temperature resistant coatings and paints, gas lasers, pollution control and nanolithography are examples where the electron induced chemical reactions play the decisive role. It is also realized that the electron induced processes are the inevitable links in the creation of molecules, including biological molecules in interstellar medium and radiation damage in biological systems. In most of these situations, it is the process of dissociative electron attachment (DEA) that plays a crucial role. The DEA process occurs through a resonant capture of the electron by the molecule and the unique characteristics of the formation and decay of the resonance give rise to possible control of the dissociation of the molecule. There is a fast growing body literature which deals with dynamics of this process at the individual molecule level in gas phase and in condensed phase, in the form of thin films, adsorbed on surfaces as single species or mixed, and in the form of mixed clusters. All these are headed towards better understanding of electron induced processes and ways of controlling it for specific applications.

2. Chemical control through DEA

The negative ion resonance formed in electron-molecule collision can decay either through electron ejection or through dissociation. The decay of these negative ion resonances through autodetachment generally leads to vibrationally or electronically excited states of molecules, many of which (like high vibrational level excitation and spin forbidden transitions) may not be possible through the direct scattering process. If the life time against autodetachment is long enough and the potential energy surface is repulsive, the resonance may decay through dissociation yielding a stable negative ion and a

Journal of Physics: Conference Series **185** (2009) 012022 doi:10.1088/1742-6596/185/1/012022

neutral fragment. The entire process starting from the formation of the resonance and its subsequent dissociation to form a stable negative ion is called dissociative electron attachment (DEA). The process of DEA has several unique properties which allow selective chemical transformations. These are the kinematics, the decay dynamics and the selection rules.

The most striking feature of electron collision on molecules which lead to DEA is that even near zero energy electrons can dissociate a large number of molecules. This has to be clearly contrasted with photon processes, where the photons need a minimum energy equal to the bond dissociation energy to break a molecule. This unique feature of electron collisions arises from the positive electron affinity of many atomic and molecular species which allows the dissociation limit of the negative ion resonance to be lower than the ground state energy of the neutral molecule. The minimum energy requirement for the DEA process is obtained by subtracting the relevant electron affinity (EA) from the bond dissociation energy (D_0). Depending on D_0 and EA, the minimum energy could even be negative. Since DEA deals with free electrons, the minimum energy in this case would be zero energy. However, the negative sign of the minimum energy signifies that the molecule could even capture a bound electron from another species, e.g. a Rydberg state, and undergo dissociation [3]. For a given molecule, depending on D_0 of the particular bond and the electron affinity of the fragment, there will be distinct threshold energies for various fragmentation channels. The distinct thresholds for different channels give a handle on breaking the molecule selectively using the electron energy as a control parameter.

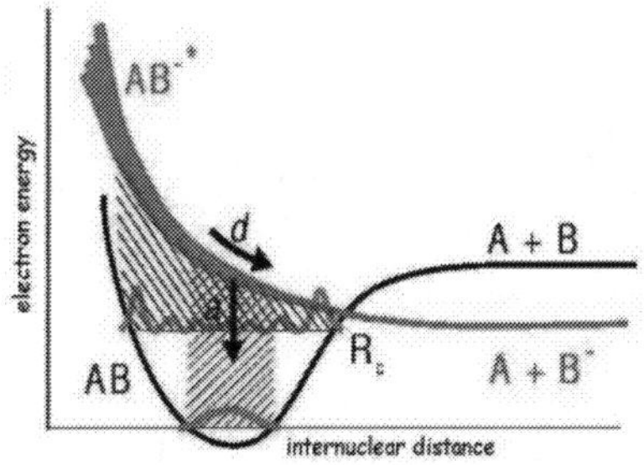

Figure 1. Schematic of DEA from a molecule in vibrational excited state as compared to the ground vibrational level.

After its formation, the negative ion resonance could decay either through autodetachment or through dissociation. Both the dissociation time and the electron detachment time are functions of the inter-nuclear co-ordinates. As shown by O'Malley [4], the DEA cross section is an exponential function of the ratio of the two time scales – the autodetachment time and the dissociation time. The exponential dependence on the timescales allows dramatic changes in the DEA cross section by altering the Franck-Condon overlap region between the neutral molecule and the resonance. This is shown in figure 1 for the case of a hypothetical molecule AB. When the molecule is in the vibrational ground state at the time of capture, the resonance takes a time τ_d to cross the internuclear distance R_c, beyond which it cannot decay by electron ejection and thus ahs to necessarily dissociate. When the molecule is vibrationally excited, the Franck-Condon region extends over larger internuclear distance. The electron capture occurring at the right turning point of the vibration leaves the resonance very close to R_c, wher τ_d is very small τ_a is larger. This could lead to several orders of magnitude increase in the DEA cross section as observed in the case of heated gases [1, 2] and Na_2 molecules selectively prepared in high vibrational levels [5]. This behaviour provides a way for mode selective chemistry. What is needed is the selective excitation of the vibrational mode in a molecule using a laser followed by collision of electrons of appropriate energy. However, very few experiments have been reported in this respect, mainly due to the problems in preparing the molecules in excited vibrational levels with sufficient target densities.

While pure vibrational excitation provides a way for controlling the DEA process, electronic excitation also provide ways of controlling the DEA process. The electronic excitation leads to change in quantal state, shape, size and symmetry of the molecules. The change in shape and size generally leads to increased extent of the electron cloud as well changes in bond distances. Electronic excitation also increases the polarizability of the molecule which is the dominant term in the electron –molecule interaction potential. All these factors lead to changes in the capture cross section. In addition, structural changes as well as vibrational excitation that accompany electronic excitation lead to change in the Franck-Condon overlap regions considerably. The effect due to this would be similar to or far more dramatic than what is seen with pure vibrational excitation. The change in symmetry of a molecule may also lead to extensive changes in the DEA process due to the symmetry based selection rules in the electron capture process [6]. Hence change in symmetry through electronic excitation may lead to opening up a new channel or closing an existing one [7]. In short, electronic excitation followed by electron collision offer a means of controlling molecular dissociation.

So far we discussed the case of isolated molecules in a gaseous medium. What about the possibility of controlling electron induced processes when they are in aggregation, like in a cluster or in condensed phase? It is seen that the negative ion resonances in molecules are formed even when they are condensed on a surface [8]. However, the DEA process is considerably modified by the environment starting with the change in symmetry of the molecule due to the presence of the neighbouring atoms or molecules [9, 10]. In addition to this, changes in the capture probability, lifetime, energy and decay channels of the resonance could be modified in the condensed phase. For molecules adsorbed on a well ordered substrate, the preferential orientation could play a decisive role in the capture probability due to the symmetry rules. Charge transfer between the negative ion resonance and various elements of their local environment are also seen to affect the resonance properties and DEA cross sections. Considerable body of literature exist on these effects [11-13]. Apart from the changes in the DEA process, there are other features inherent to the condensed phase like the inelastic scattering of the electron with other species before it reaches the "target molecule". More importantly the presence of other species in the immediate neighbourhood may provide a direct channel for secondary chemical reactions of the radicals and ions produced by the electrons. Chemical control may be achieved by controlling the DEA process through changes in the lifetime, energy and the decay channels by appropriate choice of substrate and/or co-adsorbed molecules with the products of the DEA then undergoing further reactions with the neighbouring molecules.

3. Gas Phase Studies

3.1. Control through threshold energies

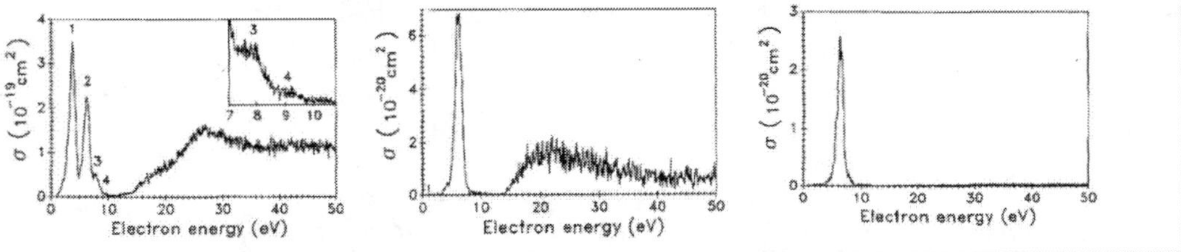

Figure 2. The dissociative attachment cross sections for dominant ions from CS_2. The left extreme shows S^-, the middle one for CS^- and rightmost one for S_2^- formation. As discussed in the text, below 4 eV only S^- is produced, but at the higher energy resonance at 6 eV all the three types of ions produced. The continuous signal at higher energies is due to the polar dissociation process of the excited neutral molecules (from Ref. 14).

Dissociative attachment process is known to show state selectivity from early days of work in this area. As discussed above, this selectivity is due to different bond dissociation energies and the electron affinity of the fragment which eventually carries away the extra electron. Figure 2 illustrates this for the case of CS_2 [14]. It is found that all possible fragment ions are formed from CS_2 through the DEA process. However, only S^- is formed at the lowest resonance, whereas all the fragment ions are found to be formed at higher energy resonance. In this case, by keeping the electron energy below 4 eV the dissociation is directed to only the S^- + CS channel. There are many other examples of similar behaviour, particularly those involving halogen containing molecules like chloro-flouro methanes. Since the C-F bond dissociation energy is larger that of C-Cl bond dissociation and the electron affinity of Cl is slightly larger than F, it is found that Cl^- is formed at near zero electron energies, while $F^-$is formed at electron energies above 2 eV [2, 15, 16]. Such a property has recently been cleverly utilized in complete chemical transformation of a molecular film of 1,2-$C_2F_4Cl_2$ (see below). Another example of this is the site selective dissociation of DNA bases by slow electrons (of energy less than 3 eV) in the H abstraction channel of the DEA process [17].

3.2 Control by initial excitation of the molecule

As discussed above vibrational excitation followed by electron attachment has been found to enhance the DEA cross sections by several orders of magnitude [1, 2, 5]. Most of the results indicated an increase of cross sections in the existing channels. However, the DEA to vibrationally excited C_6H_4BrCl produced by heating, the two competing DEA channels leading to the formation of Cl^- and Br^- were found to have rather anomalous temperature dependence [18] pointing to the possibility of controlling or enhancing specific dissociation channel in a given molecule using a combination of mode selective laser excitation followed by electron collision. There are very few results in this direction, where molecules are prepared in selected states before electron attachment. DEA to electronically excited O_2 ($^1\Delta_g$) molecules produced in a microwave discharge showed the presence of new resonances that could not be formed from the ground state molecules due to the selection rules [7]. The measurements on electronically excited SO_2 molecules formed by UV laser excitation showed that while one of the resonances seen in the ground state attachment gets enhanced in cross section by a factor of six, another resonance gets completely suppressed [19]. Initially the absence of the resonance was explained as due to symmetry based selection rules. However, another set of measurements in which the exciting laser could be tuned through a series of vibronic levels (which are expected to lead to different electronic states through vibronic couplings) showed that the different Franck-Condon region from which electron attachment occurs may be responsible for the absence of the second resonance [20]. These measurements also showed that excitation of vibrational levels above R_c (Fig. 1) may not lead to an increase in the DEA cross sections. As seen in figure 2, the DEA to CS_2 molecules in their ground state produces S_2^- in addition to other fragment negative ions. S_2^- can be formed only through a complex dynamics of the negative ion resonance, which breaks two C-S bonds and forms an S-S bond. This entails strong bending mode vibrations in the resonance. Considering this, one expected that the DEA cross section for S_2^- formation from an electronic state of CS_2 with bent geometry would show an enhancement. This could be done by exciting the ground state molecules which are in the linear state to the bent excited state using a 308 nm laser. However, the DEA data from such a measurement showed that while the cross sections for S^- and CS^- increased several times, S_2^- was absent beyond detection limits [21]. These results indicate that while electronic excitation dramatically changes the DEA process, detailed measurements are needed for understanding and exploiting this.

3.3 Control through functional group dependence

The chemical properties of all molecules are defined by the functional groups existing in them. Molecules with the same functional groups show very similar optical absorption bands [22]. In the case of electron-molecule collisions, electron transmission spectra indicated the presence of similarity in the formation of resonant attachment in water and its alkyl derivatives [23]. They had also observed that the electron transmission spectra of benzene derivatives are largely uniform and resemble that of benzene, despite large differences in symmetry [24]. It is also seen that the compilation of data on resonances in aromatic molecules and heterocyclic molecules shows distinctly similar resonance positions in them [25]. More recent electron transmission measurements on selected amino acids showed the presence of almost identical shape resonances arising from the electron attachment to the π^* orbital of the COOH group [26], indicating the functional group dependence in electron attachment. However, the presence of functional group dependence in DEA was brought out only very recently through a set of systematic measurements in various organic molecules [27, 28]. These measurements showed that electron attachment to alcohols, carboxylic acids and amines lead to formation of core excited resonances which fragment to give mostly H^- ions and these resonances could be identified with the resonances seen from H_2O, CH_4 and NH_3 depending on the presence of the O-H, C-H or N-H bonds in the molecules. Measurements on molecules which have H atoms substituted with D atoms at selected sites showed that the resonant behaviour is localized at specific sites for a given electron energy depending on the functional group and the dissociation of the negative ion resonance occurs only at that particular functional group. This was explained as due to both energy and charge localization at a given site following electron capture. That each of the functional group has characteristic electron attachment energy allows using electron energy as a control parameter to selectively break C-H, N-H and O-H bonds in molecules. The distinguishing feature of the functional group dependence is that the site/bond selectivity is not related to the minimum energy requirement for breaking a given bond as discussed in the case of CS_2 and the halogen containing molecules. The selectivity occurs in an energy range above 4 eV which is above the threshold energy needed in breaking any of the three bonds under discussion. The site/bond selectivity arising from the functional group dependence has also been observed in bigger molecules like Thymine [29] and higher alcohols [30]. It has also been shown that one can identify the contribution from the methyl group and the aromatic ring in the molecules using methyl substitution at specific sites in the ring [29].

In order to channel the site/bond selective dissociation of molecules into practically useful situations, it is necessary to characterize the state of the radical that is created in the DEA process. One would also like to know if it is a two particle fragmentation or a three particle fragmentation since the kinetic energy of the electron giving rise to the observed site selective fragmentation is fairly high. If it is a two-particle fragmentation, how is the excess energy distributed in terms of the kinetic energy and internal energy of the fragments. These questions were answered by measuring the kinetic energy and angular distribution of the H^- ions using the recently developed Velocity Map Imaging technique [31]. The measurements indicated that the functional group dependence exists even in the angular distribution of the fragments. It showed that the orientation of the particular bond with respect to the incoming electron determines the capture process, rather than the whole molecule [32]. The kinetic energy of H^- shows that for the case of N-H and O-H bonds the dissociation is a two-body process, while that for C-H it is a three-particle break up.

4. Clusters

There are several experiments reporting dissociative electron attachment to clusters. But the one relevant to the present discussion is the well known nucleophilic displacement (S_N2) reaction achieved using DEA to molecules in a mixed cluster. The cross section for the production of Cl^- from CH_3Cl is seen to be at undetectable limits ($< 10^{-23}$ cm^2). However, when this molecule was made into a cluster with NF_3 and the cluster was subjected to low energy electron collision, copious signal of Cl^- and $Cl^-.CH_3Cl$ could be obtained. The DEA to NF_3 molecules is known to prodce F^- with very large cross section as a broad resonance centred at the electron energy of 2 eV [33]. These ions after being formed in the cluster react with the CH_3Cl molecules within it forming the Cl^- and $Cl^-.CH_3Cl$ ions. The electron energy at which these ions are produced is identical to that forming F^- from NF_3, confirming the nucleophilic displacement reaction [34]. This study in a cluster indicates that such reactions could be made to take place in condensed phase, probably with far more efficiency providing increased potential for practical applications.

5. Control in condensed phase

While gas phase studies provide information at the single molecule level, it is the condensed phase that is most likely to lead to practical applications of the chemical control by electrons, since the condensed phase provides larger density of one or more species in close proximity which allows the chemical reactions to proceed efficiently. However, it must be said that practical application of electron controlled chemistry already exists in gas phase in the semiconductor industry, though this was empirically arrived at. For example, the optimisation of electron temperature in a flourocarbon plasmas to produce CF_x (x = 1 to 3) radicals most efficiently in order to etch the SiO_2 substrates. In the context of condensed phase, efforts are on towards transforming one chemical entity into another, using low energy electrons and towards nanolithography. For example, ozone was synthesized in the condensed phase by electron bombardment of multilayer films of molecular oxygen condensed at temperatures below 30 K on metal surfaces [35] and CO was formed by low energy electron collision on thin films of acetone condensed at low temperature on a solid Ar substrate [36]. Electron induced reactions of cyclopropane in the condensed phase producing propene [37], bond selective dissociation of alkanethiol based self-assembled monolayers adsorbed on gold substrate [38] and oxidation of InP(110) surface through DEA to O_2 condensed on it [39] are further examples. Complete chemical transformation of a film of $C_2F_4Cl_2$ into Cl_2 using subexcitation electrons through bond selective DEA is an example of chemical control in bulk using electron energy as a control parameter [40]. Partial fluorination of hydrogenated amorphous Si(111) surface by low energy electron irradiation of adsorbed CF_4 [41] and electron induced functionalization of diamond by small organic groups [42] point to possible applications of electron controlled chemistry in nanolithography. As pointed out earlier [41], by choosing adsorbates having appropriate DEA resonances, it may be possible to induce reactions on semiconductor surfaces efficiently in a spatially selective way in nanometre scale by low energy electron irradiation using a Scanning Tunnelling Microscope (STM).

The subject of control in electron induced chemistry will not be complete without referring to the recent and exciting developments at physical and chemical manipulation of isolated single molecules on surfaces using STM. The inelastic collision of tunnelling electrons from the STM tip has been used to break individual bonds [43-45] in molecules. The role of resonant attachment in dissociating the molecules using STM was identified in the case of O_2 [46]. STM was used to induce and control molecular motion surfaces [47-50]. Conversion of a *trans*-2-butene molecule into a 1,3-butadiene molecule on the Pd(110) surface [51] and dissociation of iodobenzene into iodine atom and phenyl radical followed by joining of two phenyl radicals to form a biphenyl molecule [52, 53] have demonstrated enormous possibilities of chemical control at individual molecule level using STM.

6. Outlook

From the success so far at efforts in using free electrons to control chemical reactions it appears that it is likely to provide practical applications ahead of the more glamorous activity taking place using coherent control and femtosecond lasers. However, there is much to be done both in terms of generating basic data on electron interaction with molecules in gas phase, in clusters and in condensed phase, which could be used for fine tuning various processes leading to practical applications. There is very limited efforts so far in applying the electron based control for a practical application in gas phase, unless one considers plasma as an extension of the gas phase work. Most of the work in plasmas has been empirical in nature. In order to apply the bond selective chemistry in plasmas, it may be necessary to control the electron energy distribution in plasmas to finer ranges than what has been achieved [54-58]. The basic information generated on the interaction of low energy electrons in the gas phase and in the condensed phase along with the use of STM is very likely to bring out new applications of electron controlled chemistry in the near future.

7. References

[1] L.G. Christophorou 1984 *Electron-molecule Interactions and Their Applications – Vol I,*

Academic Press, Orlando, Florida
[2] Illenberger E and Momigny Illenberger E and Momigny J 1992, *Gaseous Molecular Ions: An Introduction to Elementary Processes Induced by Ionization,* Springer-Verlag
[3] Dunning F B 1995 *J. Phys. B: At. Mol. Opt. Phys.* **28** 1645
[4] O'Malley T F 1966 *Phys. Rev.* **150** 14
[5] Ku¨lz M, Keil M, Kortyna A, Schellhas B, Hauck J, Bergmann K, Meyer W, and Weyh D 1996 *Phys. Rev A* **53** 3324
[6] Dunn G H 1962 *Phys. Rev. Lett.* **8** 62
[7] Belic D S and Hall R I 1981 *J. Phys. B: At. Mol. Opt. Phys.* **14** 365
[8] Sanche L 1984, *Phys. Rev. Lett.* **53** 1638
[9] Azria R, Parenteau L and Sanche L 1987, *Phys. Rev. Lett.* **59** 638
[10] Hashemi R and Illenberger E 1991 *Chem. Phys. Lett.* **187** 623
[11] Bass A D and Sanche L 2003, *Radiat. Phys. Chem.* **68** 3
[12] Huels M A, Parenteau L and Sanche L 1994, *J. Chem. Phys.* **100** 3940
[13] Sanche L 1990, *J. Phys. B: At. Mol. Opt. Phys.* **23** 1597
[14] Krishnakumar E and Nagesha K 1993, *J. Phys. B: At. Mol. Opt. Phys.* **25** 1645
[15] Illenberger E 1992, *Chem. Rev.***92** 1589
[16] Aflatooni K and Burrow P D 2001 *Int. J. Mass Spectrom.* **205** 149
[17] Abdul–Carime H, Gohlke S and Illenberger E 2004, *Phys. Rev. Lett.* **92** 168103
[18] Rosa A, Barszczewska W, Nandi D, Ashok V, Kumar S V K, Krishnakumar E, Bruning F and Illenberger E 2001, *Chem. Phys. Lett.* **342** 536
[19] Krishnakumar E, Kumar S V K, Rangwala S A, and Mitra S K 1997, *Phys. Rev. A* **56** 1945
[20] Kumar S V K, Ashoka V S and Krishnakumar E 2004, *Phys. Rev. A* **70** 052715
[21] Rangwala S A, Kumar S V K, and Krishnakumar E 2001, *Phys. Rev. A* **64** 012707
[22] Robin M B 1975, *Higher excited states of polyatomic molecules, Vol. I and II* (Academic Press, New York).
[23] Mathur D and Hasted J B 1975 *Chem. Phys. Lett.* **34** 90
[24] Mathur D and Hasted J B 1975 *J. Phys. B: At. Mol. Opt. Phys.* **9** L31
[25] Hasted J B and Mathur D 1984 in *Electron Molecule Interactions and Their Applications, vol. 1,* Ed L G Christophorou (Academic Press) p 136
[26] Aflatooni K, Hitt B, Gallup G A and Burrow P D 2001, *J. Chem. Phys.* **115** 6489
[27] Prabhudesai V S, Kelkar A H, Nandi D and Krishnakumar E 2005, *Phys. Rev. Lett.* **95** 143202
[28] Prabhudesai V S, Nandi D, Kelkar A H and Krishnakumar E 2008, *J. Chem. Phys.* **128** 154309
[29] Ptasinska S, Deniff S, Grill V, Mark T D, Illenberger E and Scheier P 2005, *Phys. Rev. Lett.* **95** 093201
[30] Ibanescu B C, May O, Monney A and Allan M 2007, *Phys. Chem. Chem. Phys.* **9** 3163
[31] Nandi D, Prabhudesai V S, Krishnakumar E and Chatterjee A, 2005 *Rev. Sci. Instr.* **76** 053107
[32] Krishnakumar E, Prabhudesai V S and Bhargava Ram N 2007 *J. Phys.:Conf. Ser.* **88** 012073
[33] Nandi D, Rangwala S A, Kumar S V K and Krishnakumar E 2001 *Int. J. Mass Spectrom.* **205** 111
[34] Langer J, Matejcik S and Illenberger E 2000 *Phys. Chem. Chem. Phys.* **2** 1001
[35] Lacombe S, Cemic F, Jacobi K, Hedhili M N, Le Coat Y, Azria R and Tronc M 1997 *Phys. Rev. Lett.* **79** 1146
[36] Lepage M, Michaud M and Sanche L 2000 J. Chem. Phys. **113** 3602
[37] Winterling H, Haberkern H and Swiderek P 2001 *Phys. Chem. Chem. Phys.* **3** 4592
[38] Olsen C and Rowntree P A 1998 *J. Chem. Phys.* **108** 3750
[39] Chen Y, Luo Y –S, Seo J M and Weaver J H 1991 *Phys. Rev. B* **43** 4527
[40] Balog R and Illenberger E 2003 *Phys. Rev. Lett.* **91** 213201
[41] Di W, Rowntree P and Sanche L 1995 *Phys. Rev. B* **52** 16618
[42] Lafosse A, Bertin M, C´aceres D, J¨aggle C, Swiderek P, Pliszka D, and Azria R 2005 *Eur. Phys. J. D* **35** 363

[43] de Vivie-Riedle R, Kurtz L and Hofman A 2001 *Pure Appl. Chem.* **73** 525
[44] Stipe B C, Rezaei M A, Ho W, Gao S, Persson M and B. I. Lundqvist B I 1997 *Phys. Rev. Lett.* **78** 4410
[45] Hla S W, Bartels L, Meyer G and Rieder K H 2000 *Phys. Rev. Lett.* **85** 2777
[46] Martel R, Avouris Ph and Lyo I -W 1996 *Science* **272** 385
[47] Eigler D M, Lutz C P and Rudge W E 1991 *Nature* **352** 600
[48] Stipe B C, Rezaei M A, and Ho W 1998 *Phys. Rev. Lett.* **81** 1263
[49] Komeda T, KimY, Kawal M, Persson B N J and Ueba H 2002 *Science* **295** 2055
[50] Pascual J I, Lorente N, Song Z, Conrad H and Rust H –P 2003 *Nature* **423** 525
[51] Kim Y, Komeda T and Kawai M 2002 *Phys. Rev. Lett.* **89** 126104
[52] Hla S –W, Bartels L, Meyer G and Rieder K -H 2000 *Phys. Rev. Lett.* **85** 2778
[53] Sloan P and Palmer R E 2005 *Nature* **434** 367
[54] Bai K H, Choi C K and Chang H Y 2004 *Plasma Sources Sci. Technol.* **13** 662
[55] Itagaki N Y. Kawai Y and S. Kawakami S 2002 *J. Vac. Sci.& Tech. A: Vacuum, Surfaces, and Films* **20**1969
[56] Hong J I, Seo S H, Kim S S, Yoon N S, Chang C S, and Chang H Y 1999 *Phys. Plasmas* **6** 1017
[57] Shimizu T, Iizuka S, Kato K and Sato N 2003 *Plasma Sources Sci. Technol.* **12** S21
[58] Kato K, Emi J and Iizuka S 2007 *Thin Solid Films* **515** 1494

Some features in multi-component Bose-Einstein condensates

Li-Hua Lu and You-Quan Li

Zhejiang Institute of Modern Physics and Department of Physics,
Zhejiang University, Hangzhou 310027, P. R. China

E-mail: yqli@hbar.zju.edu.cn

Abstract. We give a brief description of theoretical studies on multi-component systems and detailed discussions on systems with atom-to-molecule conversion. The conversion of two species of atoms into stable molecules through the Feshbach resonance assisted by stimulated Raman adiabatic passage in photoassociation is studied with the help of mean-field Langrange density. The conversion efficiency is large for the fully bosonic system as well as the boson-fermion mixture as long as the general two-photon resonance condition is satisfied.

1. Background introduction

The research interests on multi-component Bose system were earlier initiated from the study on spinor bosons [1] with two discrete internal degrees of freedom that come from the nearly degenerate hyperfine levels or other sources, for instance, trapped ^{87}Rb with two specific hyperfine levels being active [2]. Li and coworkers [3] studied one-dimensional SU(2) bosons with δ-function interactions. They used Bethe ansatz to obtain the ground and low-lying state spectra with definite quantum numbers, and showed that the ground state is fully isospin polarized and there is a quadratic spin wave mode. Yang and Li [4] proved rigorously that for a large class of real Hamiltonians, the ground state is a fully isospin polarized state. It was demonstrated [5, 6, 7] that the multi-component bosonic systems have much richer physics than their single-component counterpart due to the additional degrees of freedom. The Bethe-ansatz method was further applied to three-component systems in one dimension, such as three-component bosons and Bose-Fermi mixtures [8]. The ground state of the latter system is shown to be the state with merely bosons.

The Freshbach resonance which couples the close channel and open channel makes the conversion of ultracold atoms into molecules possible. It is difficult to cool molecular gas directly, so such a technique plays an important role in the realization of molecular BECs. Additionally, the stimulated Raman adiabatic passage (STIRAP) technique is also considered to convert atoms into molecules [9, 10]. Although the STIRAP technique can create stable molecules rather than the bare Feshbach resonance, the two techniques can both provide cubic power terms in the Lagrangian, which makes the multi-component system present more fascinating pictures.

Here, we consider a four-component system with the atom-to-molecule conversion via the technique of STIRAP aided by Feshbach resonance. In next section, we give the general model. In Sec. 3, we discuss some concrete systems. First, taking ^{40}K-^{87}Rb mixture as an example and assuming the existence of a coherent population trapping (CPT) state, we numerically study the

Journal of Physics: Conference Series **185** (2009) 012023 doi:10.1088/1742-6596/185/1/012023

atom-to-molecule conversion efficiency of the system. The fidelity is also used to characterize the property of the system and to reflect the relation between the atom-to-molecule conversion efficiency and the adiabatic property of the system. Second, we discuss the system consisting of fermionic atoms in different hyperfine states and their compounded molecules. Our main conclusions are summarized in the last section.

2. Model for four-component system

We use $|\,a,b\rangle$ denote for the ground state of free atom pair in the open channel, $|\,m\rangle$ and $|\,g\rangle$ for the quasibound and ground molecular states in the closed channel, respectively. As shown in Fig. 1 (left panel), states $|\,a,b\rangle$ are coupled with $|\,m\rangle$ through a Feshbach resonance with coupling strength α' and detuning $\mathcal{E}'$. Meanwhile, a laser field drives transitions between states $|m\rangle$ and $|g\rangle$, where the coupling strength is denoted by Ω' and the detuning by Δ'. In general, the system may contain fermions and bosons, it will be convenient to study the system by means of the Euler-Lagrange equation with the help of mean-field Lagrange density [11]

$$\begin{aligned} L &= \sum_i \Big[\frac{i\hbar}{2}(\psi_i^*\frac{\partial\psi_i}{\partial t} - \psi_i\frac{\partial\psi_i^*}{\partial t})\Big] - \frac{\alpha'}{2}(\psi_m^*\psi_a\psi_b + \text{H.c.}) + \frac{\Omega'}{2}(\psi_m^*\psi_g + \text{H.c.}) \\ &\quad -\mathcal{E}'\psi_m^*\psi_m - \Delta'\psi_g^*\psi_g - \frac{1}{2}\sum_{i\neq j}\lambda'_{ij}|\psi_i|^2|\psi_j|^2 - \sum_i H'_i, \end{aligned} \tag{1}$$

where ψ_i $(i=a,b,m,g)$ represents the complex probability amplitude of the ith componet and H'_i denotes the effective intraspecies interaction, *i.e.*, $H'_i = \frac{1}{2}\lambda'_{ii}|\psi_i|^4$ for bosonic component but $H'_i = \frac{3}{5}A'_i|\psi_i|^{10/3}$ for fermionic component with $A'_i = \hbar^2(6\pi^2)^{2/3}/2m_i$. Here the coefficients $\lambda'_{ij} = \lambda'_{ji} = 2\pi\hbar^2 a_{ij}/m_{ij}$ are the interaction strengths between particles with a_{ij} being the s-wave scattering length and $m_{ij} = m_i m_j/(m_i+m_j)$ being the reduced mass.

3. Concrete examples

3.1. Fermi-Bose atomic mixture and molecules

For the case of conversion of Fermi-Bose mixture into molecules, the system consists of one bosonic component and three fermionic components, then the dynamical equations is written as

$$\begin{aligned} i\dot{\phi}_a &= \sum_{i\neq a}\lambda_{ai}|\phi_i|^2\phi_a + A_a(\phi_a^*\phi_a)^{2/3}\phi_a + \frac{\alpha}{2}\phi_b^*\phi_m, \\ i\dot{\phi}_b &= \sum_{i}\lambda_{bi}|\phi_i|^2\phi_b + \frac{\alpha}{2}\phi_a^*\phi_m, \\ i\dot{\phi}_g &= \sum_{i\neq g}\lambda_{gi}|\phi_i|^2\phi_g + A_g(\phi_g^*\phi_g)^{2/3}\phi_g - \frac{\Omega}{2}\phi_m + \Delta\phi_g. \\ i\dot{\phi}_m &= \sum_{i\neq m}\lambda_{mi}|\phi_i|^2\phi_m + A_m(\phi_m^*\phi_m)^{2/3}\phi_m + \frac{\alpha}{2}\phi_a\phi_b - \frac{\Omega}{2}\phi_g + \mathcal{E}\phi_m - i\gamma\phi_m, \end{aligned} \tag{2}$$

where an overhead dot refers to the derivation with respect to time and γ is a phenomenological parameter introduced to characterize the decay of quasibound molecules. Here $\phi_i = \psi_i/\sqrt{n_a+n_b}$, $\lambda_{ij} = \lambda'_{ij}(n_a+n_b)/\hbar$, $\alpha = \alpha'\sqrt{n_a+n_b}/\hbar$, $A_i = A'_i(n_a+n_b)^{2/3}/\hbar$, $\Omega = \Omega'/\hbar$, $\mathcal{E} = \mathcal{E}'/\hbar$, and $\Delta = \Delta'/\hbar$ with n_a and n_b denoting the initial atom densities of the corresponding species as long as there are no molecules in the system at the initial time. From Eqs. 2, we can find that the dynamical equations for bosonic and fermionic components are different because their effective self-interactions are in different powers of $|\psi|$. Taking a mixture consisting of ^{40}K

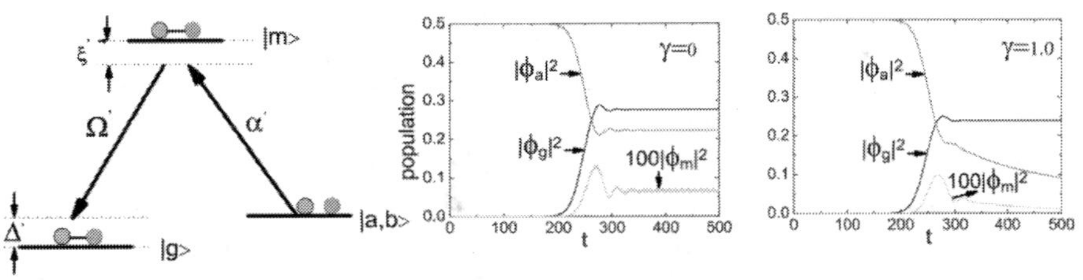

Figure 1. Left panel: scheme of the typical energy diagram for the four component system. Center and right panel: time evolutions of the population of particles for different decays γ of the quasibound molecules. The parameters are $\delta = 0$, $\lambda_{ab} = 0.23$, $\lambda_{bb} = 0.0056$, $\lambda_{am} = \lambda_{ag} = \lambda_{bm} = \lambda_{bg} = \lambda_{mg} = 0$, $A_a = 0.3$, $A_m = A_g = 0.09$ and $\mathcal{E} = -4.1$. Time is in unit of $1/\alpha$ and all other coefficients are in units of α.

and ^{87}Rb atoms as well as their compounded molecules as an example, we solve numerically Eqs.(2) and show the main result in Fig. 1. In the numerical calculation, the time-dependent Rabi frequency is chosen as $\Omega(t) = \Omega_{max}\left[1-\tanh\left(\frac{t-t_0}{\tau}\right)\right]$, where $\Omega_{\max}$, t_0 and τ are constants to be determined by the laser coupling of the two molecular states. Additionally, we assume there are no molecules in the system at the initial time and Δ evolves according to the corresponding resonance condition given in Ref. [11]. The magnitudes of the parameters (such as λ_{ij}, A_i, $\mathcal{E}$ and α) affecting the atom-to-molecule conversion efficiency are determined by means of the method in Ref. [11]. From Fig. 1, we can find that about 60% atoms can be converted into molecules if there is no decay of the quasibound molecules. Although such a conversion efficiency is higher than that of the bare Feshbach resonance, one can improve the conversion efficiency of STIRAP technique through optimizing the parameters of the system.

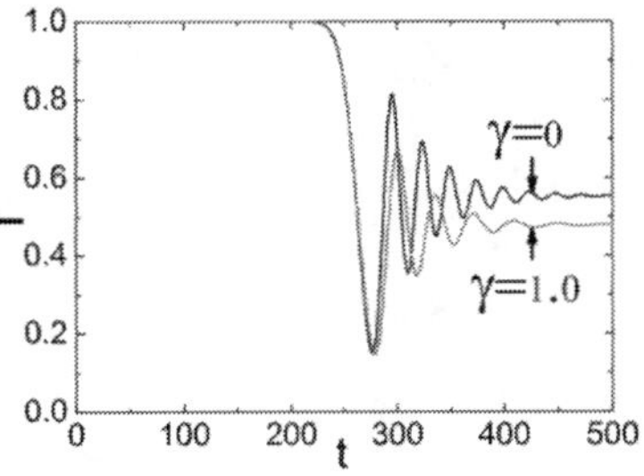

Figure 2. Time dependence of fidelity of the CPT state for different γ. The parameter choice is the same as in Fig.1

As we know [12, 13, 14] that the success of the STIRAP technique requires that there exists a CPT state that can be adiabatically followed. In Ref. [13], we gave the definition of the fidelity for multi-component systems with atom-to-molecule conversion and showed that the adiabatic property relating to the atom-to-molecule conversion efficiency can be characterized by the adiabatic fidelity which is defined as $f(\phi,\phi') =| \langle\phi \mid F^*(\phi)F(\phi') \mid \phi'\rangle \mid$ with $F(\phi) = \mathrm{diag}\left(\phi_b/|\phi_b|, \phi_a/|\phi_a|, \sqrt{2}, \sqrt{2}\right)$. Based on this expression, we numerically calculate the adiabatic fidelity of the CPT state. The result is plotted in Fig. 2. From this figure, one can see that the magnitude of fidelity remains about 1 until the time $t = 275/\alpha$, which implies that during this time the CPT state can be followed adiabatically. However, after $t = 275/\alpha$ the value of fidelity becomes much smaller than 1, which implies that the system deviates away from the CPT state at that time. So one can improve the conversion efficiency by optimizing the parameters to achieve a higher adiabatic fidelity.

3.2. Two-species fermionic atoms and molecules

Now let us discuss the case of conversion of two-species fermionic atoms into bosonic molecules. Although the fermionic atoms ^{6}Li and ^{40}K have been successfully converted into bosonic molecules via the bare Feshbach resonance in experiments, respectively, the technique of STIRAP aided by Feshbach resonance is an alternative approach due to the compounded molecules via such a technique being in ground state with longer life-time. Since there are two fermionic and two bosonic components in such a system, the dynamical equations for the probability amplitudes ψ_i can be also obtained according to the Euler-Lagrange equation. Using the same method as in Sec. 3.1, one can study the problem of the atom-to-molecule conversion via the technique of STIRAP aided by Feshbach resonance. We have shown (see [13] for details) that the technique of STIRAP aided by Feshbach resonance can efficiently convert two-species fermionic atoms into stable molecules. Comparing this technique with the bare Feshbach resonance, one can find that for ^{6}Li atoms, the atom-to-molecule conversion efficiency via the STIRAP technique aided by Feshbach resonance is higher than that via the bare Feshbach resonance, which is in contrast to the case for ^{40}K atoms.

4. Summary and acknowledge

After a brief description of theoretical studies on multi-component systems, we focused on the four-component systems in the presence of atom-to-molecule conversion. We studied the atom-to-molecule conversion for two-species atoms via the technique of STIRAP aided by Feshbach resonance. We showed that the fermionic and bosonic components satisfy different dynamical equations due to their effective self-interaction being in different powers of $|\psi|$. Taking the system consisting of ^{40}K-^{87}Rb mixtures and their compounded molecules as an example, we showed that the STIRAP technique aided by Feshbach resonance can efficiently convert Fermi-Bose mixture into molecules. We also studied the relation between the atom-to-molecule conversion efficiency and the adiabatic property of the system with the help of the adiabatic fidelity of CPT state. To achieve the higher atom-to-molecule conversion efficiency, one can improve the value of fidelity by optimizing the parameters of the system. For the case of converting fermionic atoms in different hyperfine states into molecules, we showed that the STRIAP technique aided by Feshbach resonance is more effective than the bare Feshbach resonance for ^{6}Li atoms rather than ^{40}K atoms.

The work is supported by NSFC Grant Nos. 10674117 and 10874149.

Reference

[1] T.-L. Ho, Phys. Rev. Lett. **81**, 742 (1998);
[2] C. J. Myatt, E. A. Burt, R. W. Ghrist, E. A. Cornell and C. E. Wieman, Phys. Rev. Lett. **78**, 586 (1997).
[3] Y. Q. Li, S. J. Gu, Z. J. Ying, and U. Eckern, cond-mat/0107578; Europhys. Lett. 61(3), 368(2003).
[4] K. Yang and Y. Q. Li, Inter. J. Mod. Phys. **17**, 1027 (2003).
[5] H. Pu, W. Zhang, and P. Meytre, Phys. Rev. Lett. **87**, 140405 (2001).
[6] C. P. Search, A. G. Rojo, and P. R. Berman, Phys. Rev. A **64**, 013615 (2002).
[7] R. A. Battye, N. R. Cooper, and P. M. Sutcliffe, Phys. Rev. Lett. **88**, 080401 (2002).
[8] Y. Q. Li, S. J. Gu, and Z. J. Ying, J. Phys. A: Math. Gen. 36, 2821(2003);
Z. X. Hu, Q. L. Zhang, and Y. Q. Li, J. Phys. A: Math. Gen. 39, 351 (2006).
[9] M. Mackie, R. Kowalski and J. Javanainen, Phys. Rev. Lett. **84**, 3803 (2000).
[10] K. K. Ni, S. Ospelkaus, M. H. G. de Miranda, A. Pe'er, B. Neyenhuis, J. J. Zirbel, S. Kotochigova, P. S. Julienne, D. S. Jin, J. Ye, Science **322**, 231-235 (2008).
[11] L. H. Lu and Y. Q. Li, Phys. Rev. A **76**, 053608 (2007).
[12] H. Y. Ling, H. Pu, and B. Seaman, Phys. Rev. Lett. **93**, 250403 (2004).
[13] L. H. Lu and Y. Q. Li, Phys. Rev. A **77**, 053611 (2008).
[14] S. Y. Meng, L. B. Fu, and J. Liu, Phys. Rev. A **78** 053410(2008).

Journal of Physics: Conference Series **185** (2009) 012024 doi:10.1088/1742-6596/185/1/012024

Recent progresses on atomic physics with highly charged ions in Lanzhou

X. Ma[1], X. L. Zhu[1], S. F. Zhang[1,2], B. Li[1,2], S. Y. Xu[1,2], L. J. Meng[1,2], W. T. Feng[1,2], D. C. Zhang[1], D. B. Qian[1], H. P. Liu[1], S. C. Yan[1,2], L. T. Sun[1], J. Y. Li[1], X. L. Tu[1,2], M. Wang[1], Z. G. Hu[1], H. S. Xu[1], G. Q. Xiao[1], H. W. Zhao[1], J. W. Xia[1], Y. J. Yuan[1], W. L. Zhan[1]

[1]Institute of Modern Physics, Chinese Academy of Sciences, Lanzhou, 730000, China

[2]The Graduate University of the Chinese Academy of Sciences, Beijing,100049, China

E_mail: x.ma@impcas.ac.cn

Abstract. Storage rings, dedicated small accelerators and experimental setups have been established during past few years in Lanzhou China. The commissioning of the heavy ion Cooler Storage Rings gained great success. These advances give good opportunities for atomic physics researches using highly charged ions. The current status of the accelerators, some experimental setups will be introduced. Examples of some research results will be selected and reported. The future aspects of atomic physics related to ion–matter interactions will be outlined.

1. Introduction
The interactions between highly charged ions and gas phase atoms/molecules, electrons, surfaces, solid materials, as well as bio-materials attract more and more interests in recent years, many new developments are reported. Highly charged ions can be produced in electron cyclotron resonance (ECR) ion source, electron beam ion trap/source (EBIT/S), and high energy accelerators. With the advent of heavy ion storage rings, new possibilities are open for atomic and nuclear physics researches with highly charged ions [1,2]. Here, we report the recent progresses on the cooler storage rings, dedicated small accelerators and experimental setups established for atomic physics researches during past few years in Lanzhou, China.

2. A platform for multi-discipline researches with highly charged ions
With the increasing interest in the application of low- and intermediate energy highly charged ions, a dedicated platform for multi-disciplinary researches with highly charged ions has been constructed at the Institute of Modern Physics (IMP) in Lanzhou, where the projectile energy is extended to 320q keV (where q is the projectile charge state). The new facility includes a 320kV high voltage platform, transportation beam-lines, and five experimental terminals. To reduce the power load to the platform, an all permanent ECR ion source was designed, which has a dimension of magnetic body of Ø650 mm ×562 mm. The overall magnetic field has been measured and compared to the theoretical results, good agreement was obtained. The injection magnetic field of the source is 1.28 T and the extraction magnetic field is 1.07 T, the details of the magnetic design can be found in reference [3].

The beam lines are equipped with five experimental terminals: (1) ultra-high vacuum chamber for surface studies, (2) reaction microscope for atomic physics collision dynamics as well as spectroscopy with atomic or molecular target beams [4], (3) multi-purpose experimental terminal for material science, the design is flexible and allows the external users to bring their own set-ups for experiments, (4) atomic physics terminal for atomic collision and micro-capillary experiments, and (5) a terminal for biophysics searches, respectively.

The commissioning of the platform was successful. Ions such as Ar^{q+} (8<q<16), Xe^{q+} (10<q<30), $Ne^{6,7,8+}$, $C^{4,5+}$, and $O^{3,5,6+}$, etc were provided for the experimental studies of atomic collisions, ion-surface, ion-solid, and ion-capillary array interactions. Recently the extraction of highly charged metallic ions was tested for lead. Figure 2 shows the intensity distributions of different charge states for Pb ions for microwave power at 350W. The Pb^{31+} ions were observed and further optimization is being under progress.

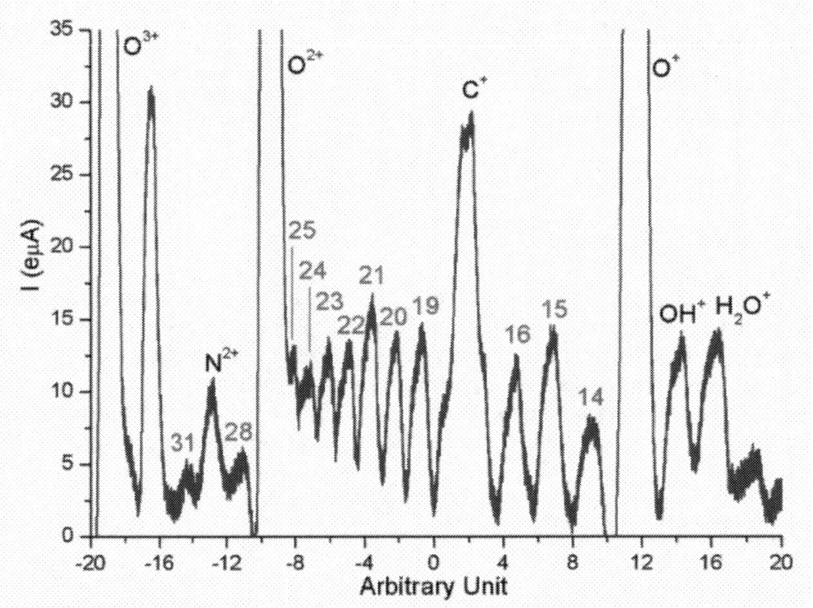

Figure 1. Typical charge state distribution spectrum for highly charged Pb ions.

3. Physics programs at the cooler storage ring

The project of cooler storage rings at Heavy Ion Research Facility in Lanzhou (HIRFL-CSR) have been completed at the Institute of Modern Physics (IMP) [5], the facility will provide heavy and highly charged ions up to bare uranium with high quality. The HIRFL-CSR is a multipurpose Cooling Storage Ring system, which consists of a main ring (CSRm) and an experimental ring (CSRe), as shown in figure 2. The two existing cyclotrons SFC (K=69) and SSC (K=450) of the HIRFL will be used as its injector system. The heavy ion beams with the energy range of 10~50 MeV/u from the HIRFL-SSC will be accumulated, cooled and accelerated to the high energy range of 100~600 MeV/u for heavy ions and the energy up to GeV/u for light ions in the CSRm. The ions then can be extracted to produce radioactive ion beams (RIBLL) or after striping, injected into the CSRe. The secondary beams can be injected into the CSRe for internal-target experiments or extracted for external experiments. High precision spectroscopy can be performed in the CSRe at the internal target, and the density of internal target reached several 10^{12} atoms/cm^2. Both rings are equipped with new generation electron coolers [6]. The 7 MeV/u C^{4+} and 22 MeV/u Ar^{8+} after stripping to C^{6+} and Ar^{18+} respectively, have been injected into the CSRm, cooled, and accelerated up to 1000MeV/u, up to $2x10^9$ ions are stored The 2.9 MeV/u Xe^{27+} ions have also been injected into the CSRm cooled and accelerated up to 235 MeV/u, and $1x10^8$ ions have been stored.

The researches of highly charged ions are making progresses in last few years, see e.g. [7,8]. The planned programs closely related to atomic physics are the following:

- High precision X-ray spectroscopy for few-electron heavy ions
- QED effects in strong Coulomb fields
- Polarization measurements of hard X-rays
- Collision dynamics at relativistic velocities
- Laser assisted recombination processes (ion-e, ion-atoms)
- Dielectronic recombination processes (DR)
- Nuclear properties investigated by atomic spectroscopy
- Mass measurements for radioactive ions far from stability
- Laser cooling of heavy ion beams
- Molecular ions related processes

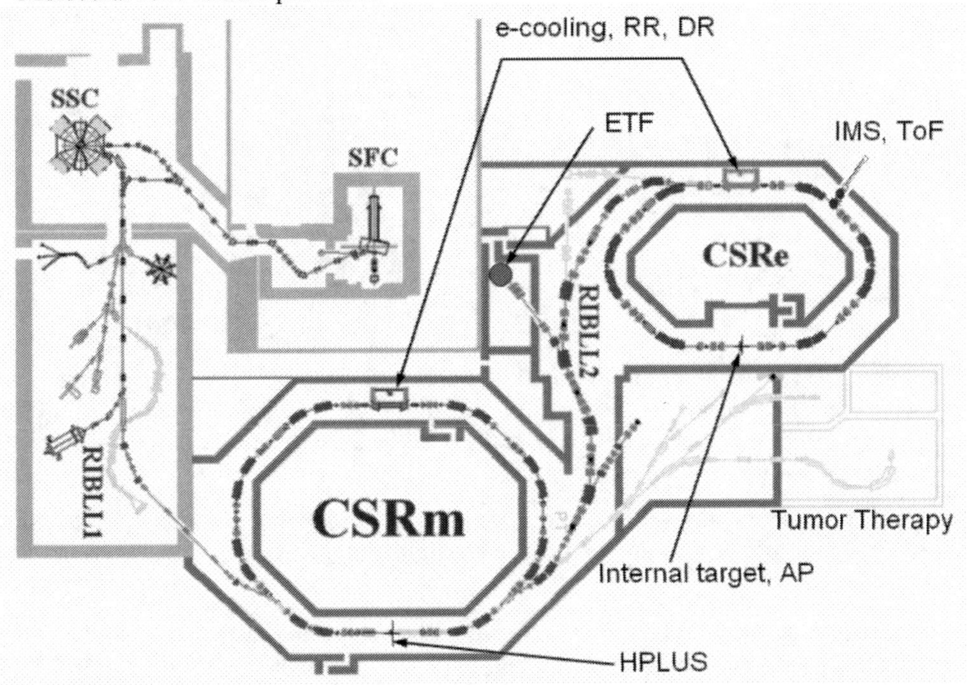

Figure 2. Schematic overview of HIRFL-CSR and some experimental areas.

4. Examples of some recent results

One experiment performed at the reaction microscope at 320kV platform was the state-elective studies for electron capture. The population to different quantum states for electron capture with the different impact parameter for He^{2+} on helium collision is shown in Figure 3, the strong dependence on impact parameters was demonstrated. The uppermost panel corresponds to a large transverse momentum of $8 < p_{\perp R} \leq 10$ a.u., and the lowermost panel corresponds to a small transverse momentum of $0 < p_{\perp R} \leq 2$ a.u. (small scattering angle, larger impact parameters *b*). It clearly shows that electron is mainly captured into the excited state for small impact parameter collisions, and that electron prefers the ground state at large impact parameters. The highly differential results tested stringently theoretical models [7,8].

The CSRe can be used as high-resolution mass analyzer, and the masses are determined from the precise measurement of their revolution frequencies. The CSRe was set to isochronous mode for commissioning run of mass measurement [9]. The fragments of $A/q = 2$ from fragmentation of 368MeV ^{36}Ar projectile were measured by their revolution frequencies. The mass spectrum is shown in figure 4. A mass resolution $\Delta m/m$ was estimated to be 10^{-6}.

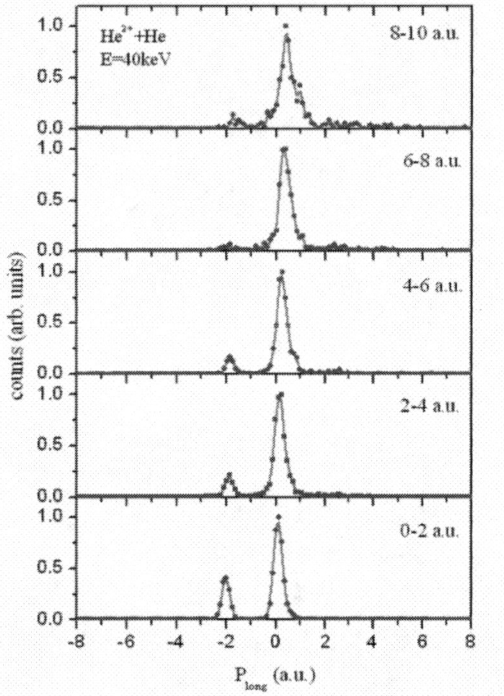

Figure 3. State-selective population for single electron capture versus transverse momentum (impact parameters b), upper panel for small b, lower panel for large b.

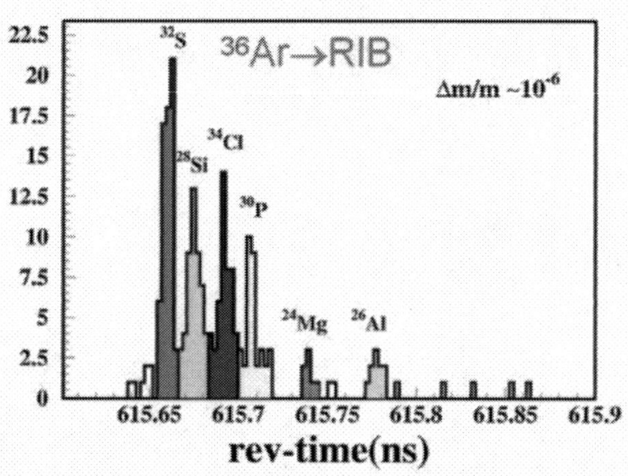

Figure 4. Fragment mass spectrum obtained at CSRe under isochronous mode for 368MeV ^{36}Ar projectile fragmentation.

In summary, the developments in accelerators and experimental equipments provide good opportunities for atomic physics related to ion –matter interactions. The facilities are open to outside users.

This work was supported by the key projects of Pilot Project of the Knowledge Innovation Program launched by the Chinese Academy of Sciences and partly by the National Nature Science Foundation of China under grant No. 10434100

References

[1] R. Schuch and S. Boehm, J. Phys. Conf. Series 88(2007)012002
[2] T. stöhlker, et. al., Nucl. Instr. Meth. B235(2005)494
[3] L. T. Sun, H. W. Zhao, et al., Rev. Sci. Instr. 77 (2006) 03A319
[4] X. Ma, X. L. Zhu, et al, Sciences in China (Series G) 51 (2008)755
[5] J. W. Xia, W.L. Zhan, B.W. Wei, et. al., Nucl. Instr. Meth. A 488 (2002) 11
[6] X. D. Yang, V.V. Parkhomchuk, W.L. Zhan, J.W. Xia, et al., Proceedings of COOL2007, Sept. 9-14, 2007, Bad Kreuznach, Germany, pp59
[7] G. W. F. Drake, W. Nörtershäuser, Z.-C. Yan, Canadian Journal of Physics 83(2005)311-325
[8] J. Eichler, T. Stöhlker, Physics Reports 439(2007)1-99
[9] J. Anton, B. Fricke, X. Ma, et. al., Phys. Lett. A 369(2007)85
[10] X. Ma, X. L. Zhu, B. Li, et. al., J. Phys. Conf. Series 88(2007)012019
[11] Tu Xiao-Lin, Xu Hu-Shan, et al., Chinese Physics C 33(2009), to be published.

Journal of Physics: Conference Series **185** (2009) 012025 doi:10.1088/1742-6596/185/1/012025

Super-elastic electron scattering from calcium over all angles.

William MacGillivray*, Andrew Murray, Martyn Hussey, George King, Alex Knight-Percival & Sher Satti.

*Southern Cross University, PO Box 157, Lismore, NSW 2480 Australia.
School of Physics & Astronomy, University of Manchester, Manchester, UK

william.macgillivray@scu.edu.au, Andrew.Murray@manchester.ac.uk .

Abstract. Experimental results are reported for the Atomic Collision Parameters of calcium over all scattering angles as derived from super-elastic scattering experiments performed using a new Magnetic Angle Changing (MAC) spectrometer. Results at 55eV incident energy are compared to theoretical calculations, and new data at 65eV are reported here for the first time.

1. Introduction.

The super-elastic scattering technique is a powerful method which provides highly accurate collision cross section data for testing state of the art theories. In comparison to more traditional coincidence techniques, super-elastic scattering data accumulates many thousands of time faster, and so can be used over a wide range of incident energies and scattering geometries [1 & references therein]. The method uses a high power continuous wave (CW) laser beam to prepare a target atom in a well defined excited state, the polarization, power and direction of the laser controlling the excited sub-state populations and coherences. An incident electron of well defined energy then de-excites the laser-prepared target, and so carries energy from the interaction in a super-elastic collision process. Since the 'shape' of the electron charge cloud in the laser excited target is well controlled, the probability of de-excitation in a given scattering direction depends on this shape. The interaction can be considered as the time-reversal of electron-photon coincidence methods, and so allows the Atomic Collision Parameters (ACPs) to be deduced. In the Natural frame of reference, these parameters are the angular momentum transferred to the target by the electron collision $L_\perp$, the alignment of the atom given by P_{lin}, γ and the spin-flip probability ρ_{00}. The experiments described here determine $L_\perp, P_{lin}$ and γ.

In contrast to all previous super-elastic measurements, the spectrometer used in these studies incorporates a MAC device that allows *all* scattering angles to be accessed [2]. This device uses a well controlled magnetic **B**-field to steer electrons to and from the interaction region while ensuring no field reaches the electron gun and scattered electron momentum analyzer. Correct adjustment of the MAC currents ensures that all electron scattering angles from 0° to beyond 180° can be accessed, allowing the most robust test of collision theories which can be made.

A complexity that arises when using a MAC device for super-elastic studies is that the **B**-field influences the shape of the laser excited target due to Zeeman shifting of the target states. These effects have been modeled using a Quantum Electro-dynamic (QED) theory [3], which considers the dynamics of the laser interactions assuming that the **B**-field produces a small perturbation. In the case of the 4^1P_1 state of calcium as studied here, the QED model indicates that $L_\perp$ can be directly

Journal of Physics: Conference Series **185** (2009) 012025 doi:10.1088/1742-6596/185/1/012025

determined by ensuring the laser remains on resonance with the appropriate $|m_J\rangle = |\pm 1\rangle$ sub-state excited by circularly polarized light, whereas the alignment parameters P_{lin}, γ require the effects of the Larmor precession to be considered during excitation by linearly polarized light. Under steady state conditions it has been shown that the alignment angle γ is directly related through a constant offset phase angle to the peak of the super-elastic data obtained by rotating the direction of linearly polarized light, whereas P_{lin} is obtained through a simple linear scaling to the data. These offset and scaling factors are due solely to the laser interaction in the **B**-field, and so are independent of electron scattering angle. As such, they were determined from experiment by comparing data with and without the MAC operating at lower scattering angles, and then using these experimentally measured factors to determine the ACP's for higher angles, where only the MAC could be used.

2. The Experimental Apparatus.

Figure 1a shows the apparatus used in the experiments. Calcium vapour is delivered from a well collimated effusive oven which produces an atomic beam of diameter ~2mm in the interaction region. Laser light at ~423nm from a Coherent MBR-110 Ti:Sapphire laser pumping an MBD-200 frequency doubler enters the vacuum chamber through a window on the top flange. The light polarization is varied using a combination of Glan-laser polarizer, $\lambda/2$ & $\lambda/4$ plates. The radiation passes axially through the MAC coils and interacts with the atomic beam to prepare atoms in a well defined state. Scattered electrons from the gun that de-excite the laser-prepared atoms are measured by the electron analyzer. The analyzer can rotate from ~10° to ~140°, and is prevented from taking data at lower angles due to scattering of the primary electron beam from the analyzer body (thereby producing unacceptably high background counts). Measurements at higher angles are prevented by the physical size of the electron gun and analyzer, as shown in figure 1b. The MAC device is used for scattering angles higher than 140°, and effectively steers electrons from this region back into the analyzer.

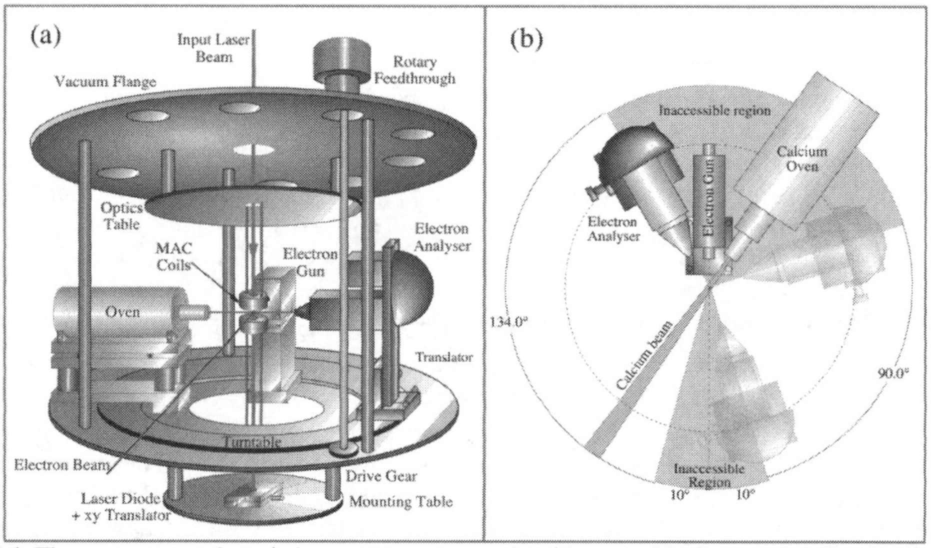

FIG 1. The apparatus, and restrictions on scattering angle without the MAC operating. For details, see text.

To ensure the input laser beam is accurately aligned, a visible laser diode (VLD) is located inside the apparatus to provide an external tracer beam once the chamber is evacuated. The VLD is adjusted using a set of non-magnetic oil-free vacuum XY translators [4] as shown. A liquid nitrogen cold trap [5] opposite the oven ensures calcium vapour does not deposit onto sensitive components inside the chamber. The direct electron beam from the gun ($I_{beam} \sim 5\mu A$) is dumped into an extended Faraday cup which allows both un-deflected and MAC-deflected electrons to be caught.

Journal of Physics: Conference Series **185** (2009) 012025 doi:10.1088/1742-6596/185/1/012025

For the experiments described here the incident electron energy ranged from 55eV to 65eV, and so the pass energy of the analyzer was set 2.93eV higher so as to measure super-elastically scattered electrons. Results at 55eV were collected using the MAC device in operation, whereas the very recent results at 65eV were taken without the MAC operating. As such, the data at 55eV covers the angular range up to and beyond 180°, whereas the results at 65eV are confined to angles from 15° to 125°.

3. Experimental Results.

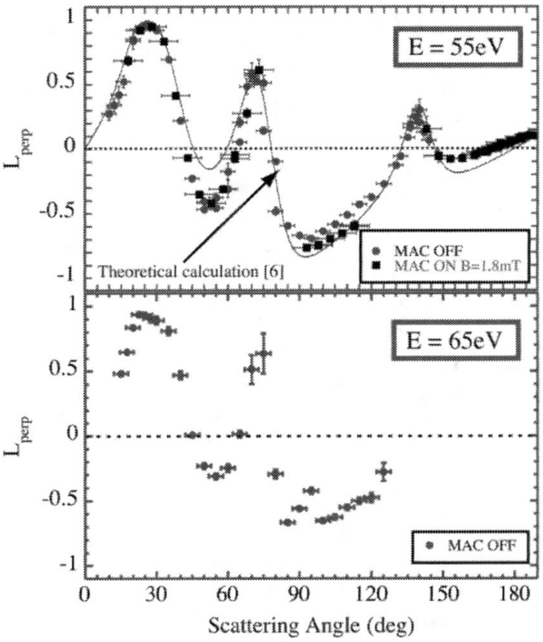

FIG 2. The angular momentum transferred to the atom during the collision, as measured by $L_{\perp}$.

Figure 2 shows the results at 55eV and 65eV for the angular momentum transferred to the atom during the collision, as measured by $L_{\perp}$. The results at 55eV are compared to the relativistic distorted wave calculations of Stauffer and colleagues [6]. A difference is seen between theory and experiment around 50° where the experimental data is more pronounced than suggested by theory, but overall the agreement is very good for this parameter. Experimental results with and without the MAC operating are also shown. The data at 65eV is similar in form to that at 55eV as might be expected. There is no theory at present with which this data can be compared.

Figure 3 shows the results for γ & P_{lin} at these energies. Once again there is impressive agreement between theory and experiment for the alignment angle γ, where theory passes through almost every data point. Experimental determination of γ has a high accuracy due to the number of measurements taken at each scattering angle, culminating in an accurate fit to data. Again, the new results at 65eV closely follow the trend at 55eV, although we do not have any calculations with which to compare these results. The results for P_{lin} also compare well to theory [6], apart from at the higher scattering angles taken with the MAC, where experiment indicates a lower value of this parameter than theory. The dip around 25° is not as pronounced as predicted by the calculation, but once again theory is clearly including the important collision physics in the interaction. Again, the new results at 65eV follow a similar trend to those at 55eV, with a deep minimum in P_{lin} being predicted at lower scattering angles, as well as a shallower dip around 90°.

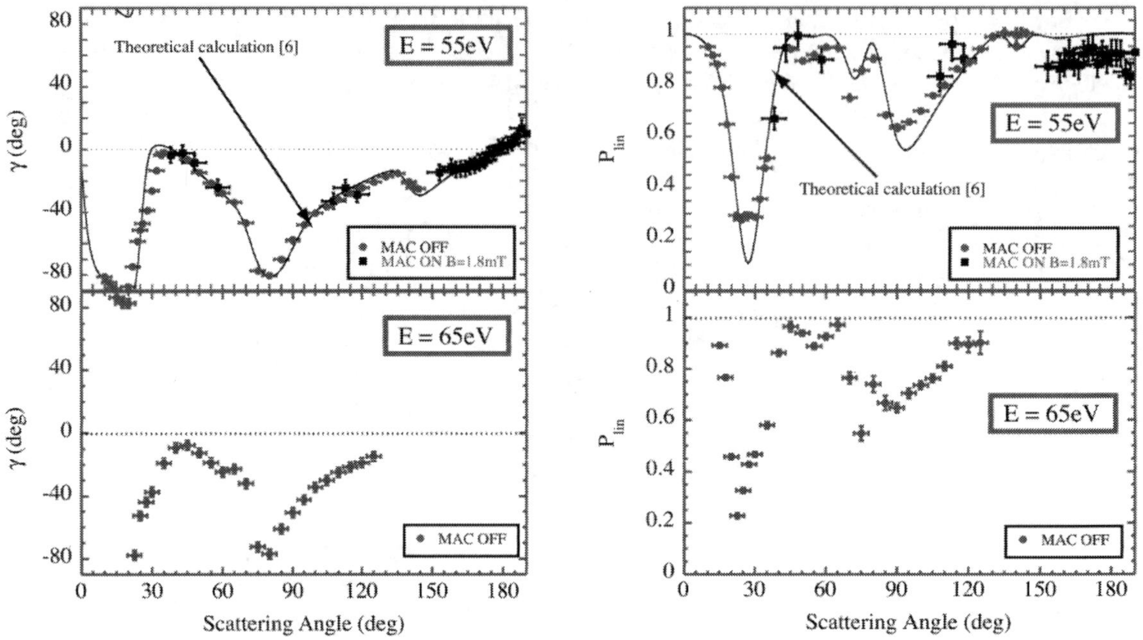

FIG 3. Results for γ & P_{lin} at 55eV & 65eV incident energies, compared to theory at 55eV[6].

4. Conclusions.

We have demonstrated the advantage that a MAC device brings to the super-elastic scattering process, and have shown that measurements can now be obtained for the atomic collision parameters over a wide range of scattering angles. The impressive agreement noted between theory and experiment clearly indicates that the important collision processes have been included at these intermediate energies, and we are confident the new data at 65eV will also show agreement once calculations are carried out at this energy.

The QED model derived for the interaction of laser radiation with atoms immersed in a magnetic field has been used to predict the effects on the super-elastic scattering process, and close agreement has been found between these predictions and experiment as seen above. This model is now being used in other work involving magnetic fields, including atom trapping and manipulation experiments where these effects are important. As such, this model is expected to have wide applicability in different fields.

The apparatus is currently being modified to allow data to be collected at energies approaching the threshold for excitation. These are challenging experiments due to the low scattering yield and due to the difficulty working with calcium and low energy electron beams. It is in this region that differences between various excitation models have been predicted [7], and it will be particularly interesting to see if the high accuracy data produced by the super-elastic technique can distinguish between different models. Experimental results from these studies are expected to be produced in the near future.

References

[1] M Hussey *et al*, J Phys B At Mol Opt Phys **41** 055202 (2008)
[2] F H Read and J M Channing, Rev Sci Inst **67** 2372 (1996)
[3] A J Murray *et al*, Phys Rev A **77** 013409 (2008)
[4] A J Murray *et al*, Meas Sci Tech **16** N19 (2005)
[5] A J Murray, Meas Sci Tech **13** N12 (2002)
[6] R K Chauhan et al, J Phys B At Mol Opt Phys **38** 2385 (2005)
[7] O Zatsarinny et al, J Phys B At Mol Opt Phys **40** 4023 (2007)

Journal of Physics: Conference Series **185** (2009) 012026 doi:10.1088/1742-6596/185/1/012026

Quantum random walk with Rb atoms

Kia Manouchehri and Jingbo Wang

School of Physics, The University of Western Australia, Australia

E-mail: wang@physics.uwa.edu.au

Abstract. Quantum random walks are known to have non-intuitive dynamics, which makes them an attractive area of study for devising quantum algorithms for long-standing classical problems as well as those arising in the field of quantum computing. In this paper we describe a new implementation scheme for quantum random walks, using ultra cold rubidium atoms trapped in optical lattices.

1. Introduction

Quantum random walks represent a generalised version of the well-known classical random walk, which can be elegantly described using quantum information processing terminology [1]. Despite their apparent connection however, dynamics of quantum random walks are often non-intuitive and far deviate from those of their classical counterparts [2,3]. Among the differences, the superior propagation properties of quantum random walks are particularly noteworthy, making them an attractive area of study for devising efficient quantum algorithms. Areas of study using quantum random walks have included connectivity and graph theory [3,4], as well as quantum search algorithms [5,6].

If quantum walks are to be taken seriously as a viable platform for quantum information processing, it is essential to find an implementation that is experimentally feasible, readily scalable and not limited to problems with specific connectivity criteria. In our previous work [7], we have provided a comprehensive list of theoretical as well as experimental efforts to construct quantum walks using a broad range of physical systems including nuclear magnetic resonance, cavity QED, ion traps, classical and quantum optics, as well as quantum dots. In this paper, we present a new implementation scheme for performing quantum random walks using ultra cold Rubidium atoms trapped inside optical lattices, which satisfies the above-mentioned criteria and is based on existing laboratory techniques reported in recent literature [8, 9, 10].

2. Theoretical model

Quantum random walks follow a similar idea to classical random walks, but incorporate a more complex definition of nodes, coins and steps. In a one-dimensional discrete-time quantum random walk for instance, each node n is made up of two sublevels, conventionally denoted by ↑ and ↓, and the probability of being present at a node $P(n) = |\psi(\uparrow,n)|^2 + |\psi(\downarrow,n)|^2$, where $\psi(\uparrow,n)$ and $\psi(\downarrow,n)$ are complex amplitudes associated with each sublevel at site n. A quantum coin $\hat{C}$ is defined as a 2 × 2 unitary matrix which in some way mixes the sublevel amplitudes within each node. A commonly used unbiased quantum coin, for instance, is defined as:

$$\hat{C} = \frac{1}{\sqrt{2}}\begin{pmatrix} 1 & -i \\ -i & 1 \end{pmatrix},$$

which mixes the two sublevels in the form of

$$\begin{pmatrix} \psi(\uparrow,n) \\ \psi(\downarrow,n) \end{pmatrix}_{new} = \frac{1}{\sqrt{2}}\begin{pmatrix} 1 & -i \\ -i & 1 \end{pmatrix}\begin{pmatrix} \psi(\uparrow,n) \\ \psi(\downarrow,n) \end{pmatrix}_{old}.$$

As shown in figure 1, each step in such a quantum random walk consists of

1. Mixing the sublevels using a quantum coin $\hat{C}$;

2. Shifting the amplitudes of $\psi(\uparrow,n)$ and $\psi(\downarrow,n)$ from node n to nodes $n-1$ and $n+1$, using the translation operator $\hat{T}_\uparrow(1)$ and $\hat{T}_\downarrow(-1)$ respectively.

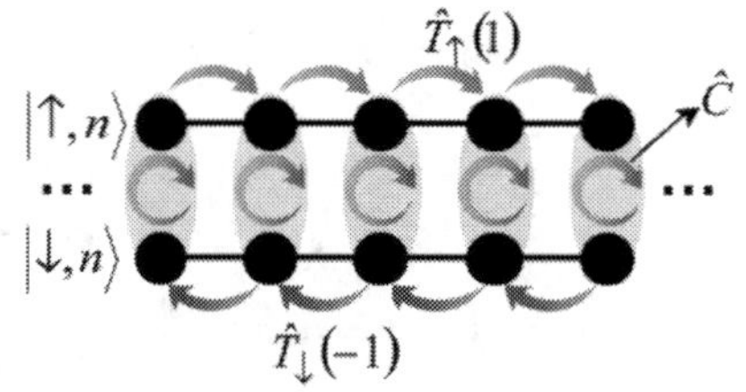

Figure 1: Illustration of a single step in a discrete-time quantum random walk. Green arrows represent sublevel mixing, which is followed by sublevel shifting as indicated by the red and blue arrows.

Repeating the above procedure over a number of steps leads to a characteristic probability distribution for quantum random walks, which is starkly different from that of a classical random walk, as illustrated in figure 2.

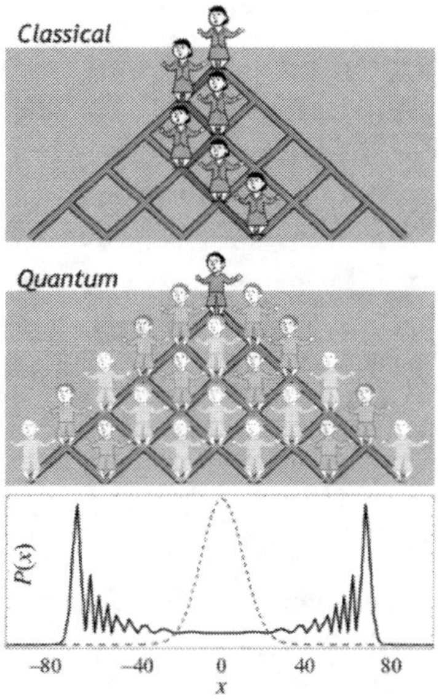

Figure 2. Upper panel: A classical walker making her way through a decision tree; middle panel: a quantum walker making his way through a decision tree; lower panel: probability distribution for a classical (dotted) vs. quantum (solid) random walk in one dimension.

3. Physical implementation using cold atoms

Trapping of neutral atoms in optical lattices have opened fascinating new experimental possibilities in quantum information processing. An optical lattice is formed by interacting a pair of counter-propagating lasers to produce a standing laser wave. When neutral atoms such as ^{87}Rb are placed in the resulting beam line, they will effectively see a spatially varying dipole potential. What we propose here is to implement the quantum walks as described above using cold ^{87}Rb atoms trapped inside an optical lattice. The position states are encoded using the individual trapping sites and the atom acts as the quantum walker with some initial distribution throughout the lattice sites. A series of specially tailored control laser operations are then introduced to address, manipulate and interact the atomic wave packets in individual sites, in a way that corresponds exactly to the action of the coin and translation operators along a row of lattice sites.

The ^{87}Rb atoms trapped inside the optical lattice experience a natural splitting of the 5S level into two hyperfine sub-levels, which are used to define the $|0\rangle$ and $|1\rangle$ basis state. These states can be manipulated using Microwave techniques to drive corresponding Rabi-Oscillations [8]. With wavelengths that are in the order of a few centimeters however, microwaves can not resolve any locations within the lattice. Instead we propose performing arbitrary unitary transformations on states $|0\rangle$ and $|1\rangle$ with the aid of a pair of three-photon Stimulated Raman Adiabatic Passage (STIRAP) operations [9], as illustrated in figure 3 (left panel). Each STIRAP requires the use of three control lasers applied in a counter intuitive order to transfer the atomic population in states $|0\rangle$ and $|1\rangle$, to and from an auxiliary state $|v\rangle$, via an intermediate upper state $|u\rangle$ that does not get populated during the transfer. The two-photon Λ STIRAP $|1\rangle \leftrightarrow |u\rangle \leftrightarrow |v\rangle$ transition has already been experimentally demonstrated using circularly polarized laser [10]. Our proposal extends this implementation through the addition of a third linearly polarized laser to facilitate the additional $|0\rangle \leftrightarrow |u\rangle$ transition. This operation allows for the implementation of a generalized quantum coin operator, controlled by the phases, amplitudes and timing of these laser pulses.

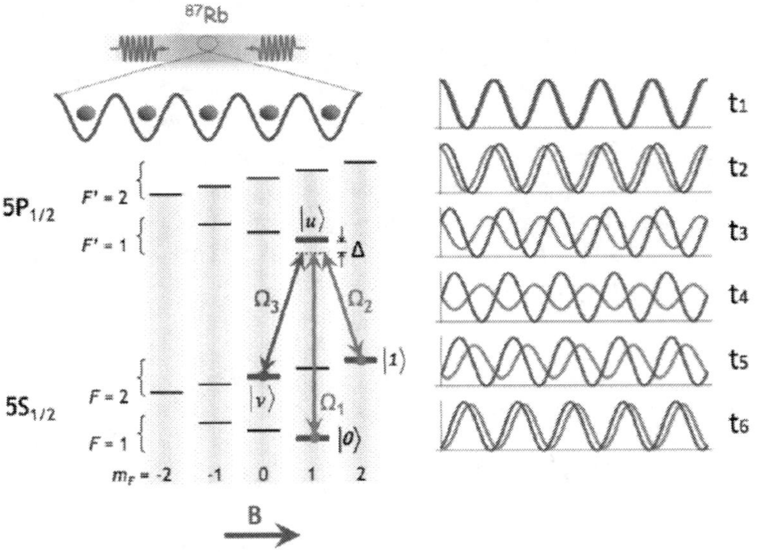

Figure 3. Left panel: schematic diagram of a three-photon STIRAP operation on the trapped ^{87}Rb atom; right panel: state dependent optical potentials moving in opposite directions.

To perform the translation operator $\hat{T}_{\uparrow}(1)$ and $\hat{T}_{\downarrow}(-1)$, we utilize a state-dependent transport scheme originally proposed in [11]. By setting the lattice wavelength at $\lambda_{lattice} = 785nm$, states $|0\rangle$ and $|1\rangle$ experience different optical potentials $V_{|0\rangle}(x,\theta) = V_+(x,\theta)/4 + 3V_-(x,\theta)/4$ and $V_{|1\rangle}(x,\theta) = V_+(x,\theta)$, respectively as illustrated in figure 3 (right panel). Here $V_\pm(x,\theta) = V_{\max}\cos^2(kx \pm \theta/2)$, $k = 2\pi/\lambda_{lattice}$, and θ is the relative polarization angle between the pair of counter-propagating lasers. Hence for an atom in the superposition state $\alpha|0\rangle + \beta|1\rangle$, increasing the polarization angle θ will lead to a split in the spatial wave packet of the atom, with the components α and β moving in opposite directions.

Figure 4 depicts the control operations involved in performing a single quantum walk step. In the proposed scheme adjacent lattice sites represent the two $\uparrow$ and $\downarrow$ sublevels within a node. The atom is initially in its internal ground state $|0\rangle$. All sites denoted as $\uparrow$ are then excited to the internal state $|1\rangle$ using a π control pulse. We utilize the state dependant transport mechanism to shift $\psi(\uparrow,n)$ to their adjacent lattice sites, still in the excited state $|1\rangle$, where they overlap with $\psi(\downarrow,n)$. A series of STIRAP operations (R_1, R_2, R_3,...) are then performed on the overlapping sites, effectively mixing the two states in accordance with the coin operator $\hat{C}$. Another state dependant transport is performed displacing $\psi(\uparrow,n)$ by an additional lattice wavelength such that $\psi(\uparrow,n)$ is adjacent to $\psi(\downarrow,n+1)$. Another π pulse brings the excited sites back to their ground state, completing one quantum walk step.

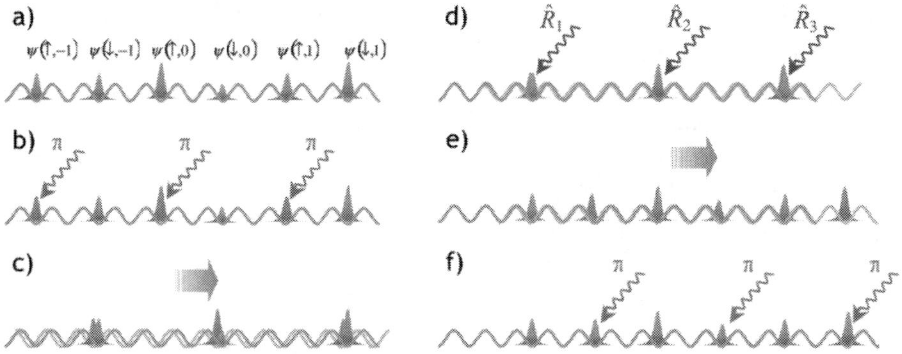

Figure 4. Basic procedure in a quantum walking step.

4. Conclusions

In this paper, we introduced a proposal for the physical implementation of discrete-time quantum random walks using ultra cold rubidium atoms trapped in an optical lattice. We demonstrate that, by tuning the polarization angles of a pair of counter-propagating lasers and utilizing the 3-photon STIRAP process, one can implement the action of the coin and translation operators and thus the specified quantum random walks.

References

[1] Y. Aharonov, L. Davidovich, and N. Zagury, Phys. Rev. A 48, 1687 (1993).
[2] P. L. Knight, E. Roldán, and J. E. Sipe, Phys. Rev. A 68, 020301 (2003).
[3] J. Kempe, Contemporary Physics 44, 307 (2003).
[4] B. L. Douglas and J. Wang, J. Phys. A: Math. Theor. 41, 075303 (2008).
[5] N. Shenvi, J. Kempe, K. B. Whaley, Phys. Rev. A 67, 052307 (2003).
[6] A. Childs and J. Goldstone, Phys. Rev. A 70, 022314 (2004).
[7] K. Manouchehri and J. Wang, J. Phys. A: Math. Theor. 41, 065304 (2008).
[8] O. Mandel, M. Greiner, A. Widera, T. Rom, T. W. Hansch, I. Bloch, Phys. Rev. Lett. 91, 010407 (2003).
[9] Z. Kis and F. Renzon, Phys. Rev. A 65, 032318 (2002).
[10] K. C. Wright, L. S. Leslie, N. P. Bigelow, Phys. Rev. A 77, 041601 (2008).
[11] D. Jaksch, H.-J. Briegel, J. I. Cirac, C. W. Gardiner, P. Zoller, Phys. Rev. Lett. 82, 1975 (1999).

Journal of Physics: Conference Series **185** (2009) 012027 doi:10.1088/1742-6596/185/1/012027

Measuring the Internal Energy Content of Molecules Transported Across the Liquid-Gas Interface

Olivia J. Maselli[1], Jason R. Gascooke[1,2], Warren D. Lawrance[2] and Mark A. Buntine[1,3*]

1. School of Chemistry and Physics, The University of Adelaide, SA 5005, Australia.
2. School of Chemistry, Physics and Earth Sciences, Flinders University, GPO Box 2100, Adelaide SA 5000, Australia
3. Department of Applied Chemistry, Curtin University of Technology, GPO Box U1987, Perth WA 6845, Australia

Email: m.buntine@curtin.edu.au

Abstract. Many details concerning the mechanism associated with the liberation of molecules from a liquid surface remain to be elucidated. We use the liquid microjet technique coupled with laser spectroscopy to measure the rotational and vibrational energy content of benzene spontaneously evaporating from a water-ethanol solution. These measurements provide molecular level insight into the mass and energy transfer processes associated with evaporation.

1. Introduction

While the thermodynamics of evaporation have been understood for over a century [1,2], the molecular-level dynamics of the evaporation process remain ill-defined. Recently, Molecular Dynamics (MD) simulations of evaporation have begun to provide insight into the physics of mass and energy transfer across the liquid-vapor interface in unprecedented detail. Most commonly, simulations of monatomic liquids have made predictions about the translational energy content of the evaporate after it has passed through the interfacial region [3-5]. However, MD simulations have rapidly evolved to the stage that they have begun to describe polyatomic liquid interfaces and have, within a rigid rotator approximation, included predictions of the rotational energy content of the evaporate [6,7]. Further extensions that explore the vibrational energy content of evaporating molecules have yet to be reported.

To date, the only experimental evidence against which to test the validity of predictions made by MD simulations have been measurements of the translational energy distributions of molecules from a liquid microjet (LμJ) evaporating into a vacuum [8,9]. There remains a critical need to experimentally characterize the nascent internal energy content of the evaporate to provide a much more rigorous test of the predictions made by MD simulations. In tackling this challenge, we have developed an experimental technique to probe the rotational and vibrational energy content of molecules evaporating from a liquid surface. Specifically, we report results for benzene evaporating from aqueous solution into vacuum using the LμJ approach coupled with laser spectroscopy.

The mechanism by which molecules leave a liquid surface can be classified as 'spontaneous' or 'stimulated' [4]. Spontaneous evaporation occurs only when molecules are liberated from the condensed phase in the absence of external forces; all the energy to leave the surface is supplied either from the bulk phase beneath the evaporating molecule or by the molecule's own internal energy.

Stimulated evaporation involves vapor-phase molecules above the liquid surface contributing to mass and energy transfer. Experimental studies describing spontaneous evaporation have traditionally been challenging due to the presence of this vapor. Collisions between the emerging molecules and the vapor-phase rapidly equilibrate the nascent energy distributions of the emerging molecules and thereby obscure interrogation of the spontaneous process. The LμJ approach overcomes this complexity by presenting the evaporating liquid surface directly into a vacuum, thus enabling direct interrogation of evaporation in the absence of competing processes.

2. Experiment

We have used a 7.5 μm radius liquid microjet. All experiments were conducted using 10^{-3} M benzene with ethanol in de-ionized water (25% v/v EtOH in H_2O) as the solvent. The aqueous solution was injected into the vacuum at a flow rate of 0.25 mL/min. Solute benzene molecules that evaporated from the liquid surface were ionized by a collimated and telescoped ultraviolet (UV) laser beam (10 μJ/pulse, 150 μm radius FWHM) located a fixed distance downstream of the LμJ nozzle aperture and propagating orthogonally to the liquid filament. The UV laser radiation ionized the solute benzene molecules by 1 + 1 resonance-enhanced multiphoton ionization (REMPI) via the $^1B_{2u}\leftarrow{}^1A_{1g}$ transition for the first absorption step. The UV laser did not directly irradiate the LμJ, but ionized molecules that had moved a fixed distance away from the liquid filament. Typically, the UV laser was located 300 μm from the LμJ. Resultant benzene ions were injected into a reflectron time-of-flight (TOF) mass spectrometer. Mass spectra were collected over 50 laser shots at each UV laser wavelength. Typically, 2-4 consecutive wavelength scans were averaged to generate the spectra used for spectral analysis.

The rotational and vibrational energy content of the benzene molecules are reported in terms of rotational and vibrational mode-specific 'temperatures'. Rotational temperatures in the vibrational ground state of benzene were determined by fitting the experimental 6^1_0 vibronic spectral band contours with ones simulated from known spectroscopic constants and transition line strengths [10-12] using a Levenberg-Marquardt non-linear least squares algorithm [13] and assuming a Boltzmann distribution of internal energies. Measurements of the temperatures of select vibrational modes were made by comparing the integrated intensity ratios of the relevant vibronic hot bands to the 6^1_0 cold band [14].

To test the spectral simulation algorithm, and to provide a benchmark against which the LμJ evaporation spectra can be interpreted, we recorded and analyzed the room-temperature REMPI spectrum of benzene in a static cell. The cell is equipped with a pair of 1 cm^2 electrodes separated by a distance of 1 cm and biased with a potential difference of 90 V. This spectrum was recorded at a pressure of 0.45 Torr by passing the resulting photocurrent from each electrode through a differential amplifier.

3. Results and Discussion

Measured rotational contours of the 6^1_0 vibronic band of benzene at room temperature and after evaporating from the LμJ are shown in Figure 1. The LμJ spectrum in Figure 1(b) was recorded with the UV laser set at a distance of 300 μm from the centre of the liquid filament and 1 mm downstream of the nozzle orifice. Comparison with the room temperature spectrum (Figure 1(a)) shows that the evaporation contour is significantly truncated at lower transition energies, revealing a considerably colder rotational energy distribution than found at room temperature. Overlaid on each spectrum is the best-fit spectral simulation. The simulated spectral contours, displayed as dashed lines in Figure 1, are convoluted with a Gaussian of 0.3 cm^{-1} (FWHM) to account for the laser bandwidth. Simulation of the gas cell spectrum in Figure 1(a) yields a best-fit rotational temperature of 295±6 K, where the uncertainty represents 1 standard error. This illustrates the accuracy of the rotational temperatures that can be extracted. The average of fitting nine spectra equivalent to that presented in Figure 1(b) yields a rotational temperature for the benzene evaporate of 206±4 K.

Extended spectra of benzene recorded at room temperature in the low pressure gas cell and under LμJ evaporation conditions are presented in Figure 2. Here, several additional vibronic bands are labeled according to the assignments of Stephenson *et al* [15]. The multi-peaked band between 38,515 cm^{-1} and 38,535 cm^{-1} is assigned to overlapping 6^2_1 and $6^0_1 1^2_0$ transitions. Although the peaks in this band involve mixed vibronic transitions, the initial vibrational level prior to laser excitation always includes 6_1, so the change in the band's integrated intensity under evaporation conditions relative to the equilibrated room-temperature gas cell allows us to determine the degree of cooling for the ν_6 vibrational mode. Such analysis yields a ν_6 vibrational temperature of 256±11 K following evaporation.

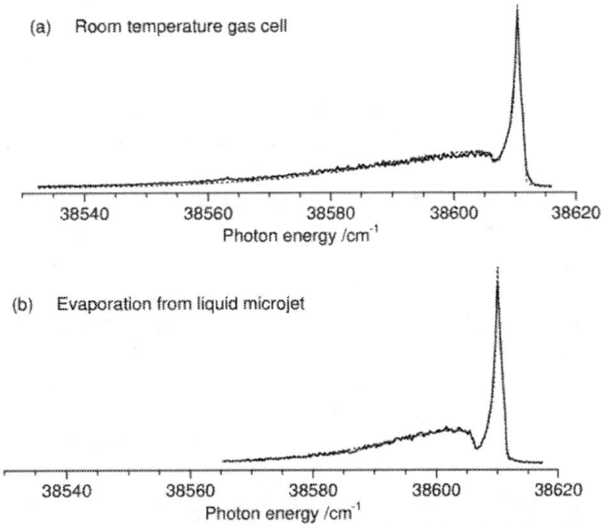

Figure 1 1+1 Resonance-Enhanced MultiPhoton Ionization (REMPI) spectra of the 6^1_0 vibronic transition of benzene. **(a)** under room-temperature gas cell conditions. **(b)** evaporated from the surface of a H_2O/EtOH liquid microjet. Overlaid on each spectrum is a dashed line representing the best-fit spectral simulation for the 6^1_0 transition, assuming a Boltzmann distribution of rotational states.

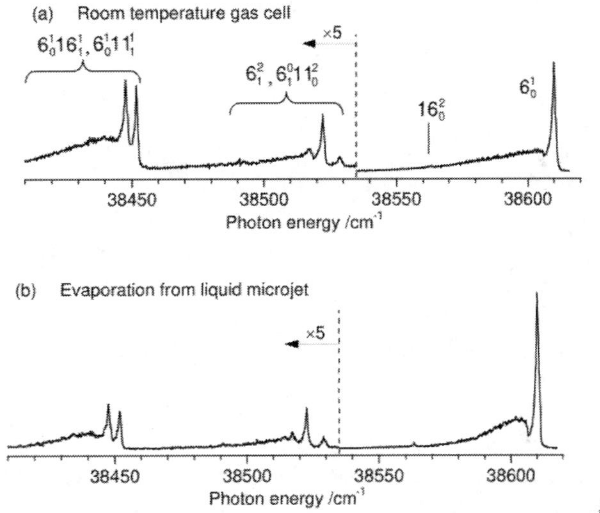

Figure 2 Extended 1+1 REMPI spectra of benzene to lower photon energy of the 6^1_0 vibronic transition. **(a)** under room-temperature gas cell conditions. **(b)** evaporated from the surface of a H_2O/EtOH liquid microjet. Intensities to lower photon energy of the dashed vertical line in each spectrum have been multiplied by a factor of five. Vibronic transitions discussed in the text are labeled. Vibrational temperatures are determined by the ratio of integrated band intensities for each hot band relative to that of the 6^1_0 transition.

To lower transition energy, the series of peaks around 38,450 cm^{-1} are assigned to the $6^1_0 16^1_1$ and $6^1_0 11^1_1$ vibronic transitions. Analysis of this band to determine vibrational temperatures is less straightforward because the overlapping transitions arise from molecules initially populating both ν_{11} and ν_{16} in the ground electronic state. However, the vibrational energies of ν_{11} and ν_{16} are 674 cm^{-1}

and 399 cm^{-1}, respectively [16,17], so assuming a Boltzmann distribution at the gas-cell temperature of 295 K as well as equal electronic transition probabilities, the population ratio between ν_{11} and ν_{16} is 0.13. As such, 87% of the intensity from this band, at least in the upper trace of Figure 2, arises from ground electronic state population in the lower-energy ν_{16} level. We therefore conclude that the contribution of ν_{11} to the spectral profile is minimal. Assuming an identical ratio of vibrational populations under evaporation conditions yields a ν_{16} vibrational temperature of 229±12 K. Because of its lower frequency, ν_{16} is expected to relax faster than ν_{11} so this temperature should be viewed as an upper limit [16,17].

It is important to ascertain that the measured temperatures represent the final values, *i.e.* collisional cooling is complete prior to laser ionization of the benzene evaporate. This was verified by measuring spectra at the larger distance of 1200 μm from the LμJ. Within experimental error there is no change in either the rotational or vibrational temperatures in moving from 300 to 1200 μm away from the liquid filament, indicating that the experiment probes the quantum state distributions of benzene after vapor-phase collisional energy transfer is complete.

The important results to emerge from this study are that (*i*) the rotational and vibrational temperatures are different, and (*ii*) the temperatures of ν_6 and ν_{16} are different. This is only possible if there have been insufficient collisions to establish equilibrium between the rotational and various vibrational degrees of freedom. The data allow for a comprehensive assessment of the number of collisions undergone by a molecule as it passes from the condensed phase, and thereby insight into the dynamics occurring at the liquid-vapor interface. This analysis is reported elsewhere [18].

We have directly measured the rotational and mode-specific vibrational energy content of benzene that has spontaneously evaporated into the vacuum from an aqueous liquid microjet. The rotational and ν_6 and ν_{16} vibrational temperatures of benzene are all different, pointing to non-equilibrated collisional energy transfer occurring within the liquid-vapor interphase region.

Acknowlegements
Financial support for this work was provided by the Australian Research Council and The University of Adelaide. OJM was the recipient of an Australian Postgraduate Award.

References

[1] Sadtler P B 1909 *J. Franklin Inst.* **166** 291-296
[2] Sadtler P B 1910 *J. Ind. Eng. Chem.* **1** 644-653
[3] Frezzotti A, Gibelli L and Lorenzani S 2005 *Phys. Fluids* **17** 012102
[4] Matsumoto M 1998 *Fluid Phase Equilib.* **144** 307-314
[5] Meland R, Frezzotti A, Ytrehus T and Hafskjold B 2004 *Phys. Fluids A - Fluid Dynamics* **16** 223-243
[6] Frezzotti A 2007 *Eur. J. Mech. B-Fluids* **26** 93-104
[7] Ishiyama T, Yano T and Fujikawa S 2004 *Phys. Fluids* **16** 4713-4726
[8] Faubel M and Kisters T 1989 *Nature* **339** 527-529
[9] Faubel M, Schlemmer S and Toennies J P 1988 *Z. Phys. D-Atoms Mol. Clusters* **10** 269-277
[10] Herzberg G 1966 *Molecular Spectra and Molecular Structure. III. Electronic spectra and electronic structure of polyatomic molecules*. (New York: Van Nostrand)
[11] Okruss M, Muller R and Hese A 1999 *J. Mol. Spectrosc.* **193** 293-305
[12] Wilson Jr. E B 1935 *J. Chem. Phys.* **3** 276-285
[13] Press W H, Flannery B P, Teukolsky S A and Vetterling W T 1986 *Numerical Recipes: The Art of Scientific Computing*. (New York: Cambridge University Press)
[14] Demtröder W "Visible and Ultraviolet Spectroscopy: Physical Aspects" In: G. Scoles, ed. *Atomic and Molecular Beam Methods*. New York: Oxford University Press 1992:231-233.
[15] Stephenson T A, Radloff P L and Rice S A 1984 *J. Chem. Phys.* **81** 1060-1072
[16] Hollenstein H, Piccirillo S, Quack M and Snels M 1990 *Mol. Phys.* **71** 759-768
[17] Page R H, Shen Y R and Lee Y T 1988 *J. Chem. Phys.* **88** 4621-4636
[18] Maselli O J, Gascooke J R, Lawrance W D and Buntine M A 2009 *J. Phys. Chem. C* **113** 637-643

Journal of Physics: Conference Series **185** (2009) 012028 doi:10.1088/1742-6596/185/1/012028

Frequency measurement of pure rotational transitions of D_2O using tunable terahertz spectrometer

F. Matsushima, K. Kobayashi, and Y. Moriwaki

Department of Physics, University of Toyama, Gofuku 3190, Toyama 930-8555, Japan

E-mail: matsushi@sci.u-toyama.ac.jp

Abstract. Precise frequency measurement of the rotational lines of D_2O in the terahertz (far-infrared) region is demonstrated as a case study of a high-precision far-infrared spectroscopy using a tunable radiation source. Frequencies of about 150 rotational lines of D_2O were measured in the 0.5 - 5 THz region. Measured frequencies provide an excellent frequency standard for the terahertz region together with our measurements on $H_2{}^{16}O$ (v=0 and v_2=1 vibrational states), $H_2{}^{17}O$, and $H_2{}^{18}O$. Molecular parameters of Watson's A-reduced Hamiltonian have been obtained to reproduce the observed frequencies.

1. Introduction

Precise spectroscopy of the H_2O molecule is important for researchers both in experiment and theory. Rotational spectra of water vapor, which have large density of lines and strong absorption intensity in the terahertz region, or in the other word, far-infrared (FIR) region, is a good calibration choice for the spectrometers in the terahertz region. Theoretical calculation of the energy levels of H_2O is still one of the challenging subject for theoreticians because the molecule is a typical example of light rotator with large centrifugal distortion effects.

In this paper, precise frequency measurement of the rotational lines of D_2O in the terahertz region is described as a case study of a high-precision far-infrared spectroscopy using a tunable terahertz spectrometer in Toyama. Other than D_2O, we measured frequencies of pure rotational transitions of $H_2{}^{16}O$ (v=0 and v_2=1 vibrational states), $H_2{}^{17}O$, and $H_2{}^{18}O$ in the frequency range from 0.5 to 5 THz [1-4].

Historically, low frequency rotational lines of D_2O have been measured by several groups using microwave spectrometers [5-13]. Up to 1985, the number of reported rotational lines was 66 in total and data were limited to the frequency range below 1100 GHz.

Frequencies of a number of rotational lines were measured by using Fourier transform (FT) spectrometers. Frequency table of D_2O was extended to 220 cm^{-1} by the FT study of Johns in 1985 [14]. He also reported the measurements of other isotopic species of water molecule and gave the sets of molecular parameters for each species. The table was extended to 420 cm^{-1} by Paso and Horneman in 1995 [15]. The spectral resolution of the FT spectrometer is typically about 0.0025 cm^{-1}.

In our work, rotational transitions of D_2O have been measured in the frequency range from 0.5 to 5.2 THz (16 to 173 cm^{-1}) with a frequency tunable terahertz spectrometer. Though

Journal of Physics: Conference Series **185** (2009) 012028 doi:10.1088/1742-6596/185/1/012028

the range of the measurement overlaps with that of FT measurements, the accuracy of the frequency measurement has been improved to better than 100 kHz which is comparable with that of microwave measurements.

2. Experimental

The frequency tunable terahertz spectrometer used in the present work (referred to as TuFIR as a contracted form of "tunable far-infrared") is based on a frequency synthesizing technique developed by Evenson and co-workers [16]. The principle of the spectrometer can be found in the previous papers [16-17]. A schematics of the spectrometer is shown in Fig.1. Two CO_2 laser radiations, frequency of each laser denoted as f_1 and f_2, are mixed in a metal-insulator-metal (MIM) diode to generate a far-infrared radiation of high spectral purity with the difference frequency of $|f_1 - f_2|$. Then a microwave radiation tunable up to 18 GHz is fed to the MIM diode through a whisker to generate tunable sidebands. By denoting the frequency of the microwave radiation as f_{MW}, frequency of the obtained FIR sidebands, f_{FIR}, are $f_{FIR} = |f_1 - f_2| \pm f_{MW}$. The present spectrometer can synthesize terahertz radiation with the frequency up to 5.3 THz by using two lasers with normal CO_2 lasing gas. Though the output of the FIR radiation is as small as the order of 100 nW, it is enough to observe strong rotational lines of neutral molecules such as water molecule. A liquid-helium-cooled silicon composite bolometer is used to detect the FIR radiation.

The sample cell is a Pyrex glass tube of 1.5 m in length and 12 mm in diameter with polypropylene windows of 1mm thickness at each end. The D_2O sample of 99.75 percent enrichment was used. Most of the lines were observed at a pressure of about 1 mTorr or less. The path of the FIR radiation outside the sample cell was purged with nitrogen gas to prevent the absorption of FIR radiation by atmospheric water vapor.

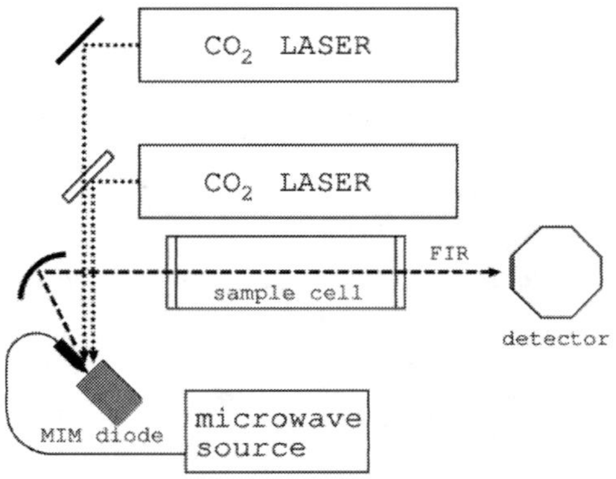

Figure 1. Schematic diagram of the tunable terahertz spectrometer.

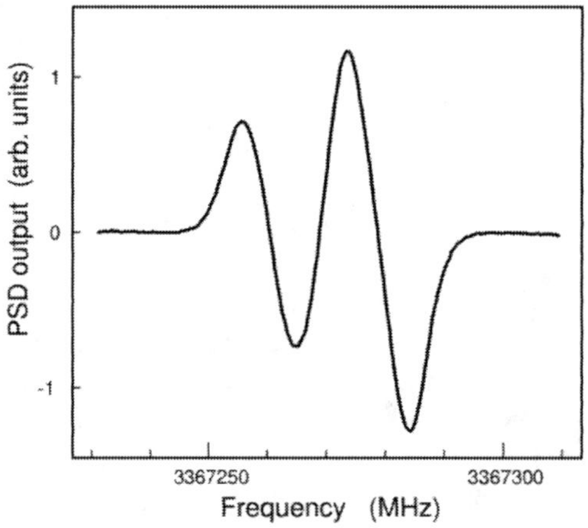

Figure 2. Accidentally overlapped lines are well resolved with Doppler line width.

3. Results and discussion

Though most of the rotational lines appear as an isolated single line, Figure 2. shows an unusual case of closely overlapped two lines near 3.367 THz. The lines in Fig. 2, $7_{70} \leftarrow 7_{61}$ and $7_{71} \leftarrow 7_{62}$, can not be resolved in the FT spectrometer but are now well resolved with Doppler line width.

Each spectral line shape was fitted by least squares program [18] to a theoretical Voigt profile to determine the line center. Two lines in Fig.2 were determined to be 3367260.541 (0.039)MHz and 3367279.136(0.037)MHz, respectively.

Journal of Physics: Conference Series **185** (2009) 012028 doi:10.1088/1742-6596/185/1/012028

We picked up the candidate lines of our measurement from the frequency tables obtained by FT spectrometers previously [14,15]. Frequency of about 150 rotational lines of D_2O were determined. Table of the measured frequency can be found in reference [4]. The estimated uncertainty is below 50 kHz for most of the lines. The uncertainties of measured line positions are contributed by (i) the uncertainty of the synthesized FIR frequency and (ii) that in determining the center frequency of the spectral line. The uncertainty (i) is estimated to be about 35 kHz assuming the uncertainty of the frequency of each laser line to be 25 kHz. The uncertainty (ii) is contributed by several factors such as the noise in the absorption signal which mainly comes from detector noise, a deformation of the spectral line, and that of the base line.

Calculation of the energy term of water molecule based on Taylor-series Hamiltonian in the Watson form [19] leads to large errors because of the large centrifugal distortion effect of the molecule. Researchers proposed theories to improve the calculation of water molecule [20-21]. Nevertheless, a calculation using Taylor series Hamiltonian is still a convenient method for giving precise term values for the levels with low rotational quantum numbers. So, we used a Hamiltonian based on Watson's A-reduction to reproduce our frequency data set. The Hamiltonian is shown as;

$$\begin{aligned} H \;=\; & \frac{1}{2}(B+C)J^2 + \Big(A - \frac{B+C}{2}\Big)J_z^{\,2} + \frac{1}{2}(B-C)J_{xy}^2 - \Delta_J (J^2)^2 - \Delta_{JK} J^2 J_z^{\,2} - \Delta_K J_z^{\,4} \\ & -2\delta_J J^2 J_{xy}^2 - \delta_K \{J_z^{\,2}, J_{xy}^2\} + H_J (J^2)^3 + H_{JK}(J^2)^2 J_z^{\,2} + H_{KJ} J^2 J_z^{\,4} + H_K J_z^{\,6} \\ & +2h_J (J^2)^2 J_{xy}^2 + h_{JK} J^2 \{J_z^{\,2}, J_{xy}^2\} + h_K \{J_z^{\,4}, J_{xy}^2\} + L_J (J^2)^4 + L_{JJK}(J^2)^3 J_z^{\,2} \\ & +L_{JK}(J^2)^2 J_z^{\,4} + L_{KKJ} J^2 J_z^{\,6} + L_K J_z^{\,8} + 2\ell_J (J^2)^3 J_{xy}^2 + \ell_{JK}(J^2)^2 \{J_z^{\,2}, J_{xy}^2\} \\ & +\ell_{KJ} J^2 \{J_z^{\,4}, J_{xy}^2\} + \ell_K \{J_z^{\,6}, J_{xy}^2\} + P_{KKJ} J^2 J_z^{\,8} + P_K J_z^{10} + p_J (J^2)^4 \\ & +p_{JJK}(J^2)^3 \{J_z^{\,2}, J_{xy}^2\} + p_K \{J_z^{\,8}, J_{xy}^2\} + T_K J_z^{12} \end{aligned}$$

where

$$J_{xy}^2 = J_x^{\,2} - J_y^{\,2}, \qquad \{A, B\} = AB - BA.$$

The molecular parameters obtained for D_2O are shown in Table 1.

Historically, the parameters have been based on the microwave spectroscopy of pure rotational transitions [5,13,14,22,23] or infrared spectroscopy of vibration-rotation transitions [24]. Calculated frequencies using previous parameters had typical discrepancies of several megahertz from frequencies observed in the present work. Because the precise frequency data comparable to that of microwave spectrometer were obtained by the present TuFIR spectrometer up to 5 THz, the molecular parameters in the ground state were improved so that the discrepancies between the calculated and observed values are mostly within 50 kHz.

Recently, Furtenbacher and Császár examined the accuracy and consistency of our TuFIR data of water molecule using their protocol MARVEL (standing for Measured Active Rotational-Vibrational Energy Levels) [25]. The data of $H_2^{16}O$ (ground state and v_2=1 state), $H_2^{17}O$, $H_2^{18}O$ and $D_2^{16}O$ were proved to have high accuracy and it was shown that the data can be candidates for becoming frequency standards in the FIR region of 15-170 cm^{-1} (0.5-5THz).

The present TuFIR spectrometer has been used to investigate many kinds of molecules other than water molecule such as neutral molecule or radical (LiH, KH, ^{18}OH), molecule with internal rotation (CH_3OH), molecular cation (HeH^+, NeH^+, ArH^+, KrH^+, XeH^+ (each including its isotopomers), H_2D^+), and molecular anion (OH^-, OD^-). These data obtained with the TuFIR spectrometer can be referenced by accessing the database "Toyama Microwave Atlas for spectroscopists and astronomers" which is maintained by the authors [26].

Table 1. Molecular parameters (in MHz) for the ground state of D_2O

A	462278.8077 (92)	$H_{KJ}\times10^2$	-7.629 (18)	$\ell_J\times10^7$	-2.05 (fixed)
B	218038.2178 (78)	$H_K\times10$	5.5552 (15)	$\ell_{JK}\times10^6$	-3.8 (fixed)
C	145258.0022 (67)	$h_J\times10^4$	9.6654 (70)	$\ell_{KJ}\times10^5$	1.59 (37)
Δ_J	9.29071 (30)	$h_{JK}\times10^3$	-1.936 (45)	$\ell_K\times10^4$	-7.27 (11)
Δ_{JK}	-45.71991 (99)	$h_K\times10$	1.1797 (34)	$P_{KKJ}\times10^7$	-7.21 (27)
Δ_K	277.4296 (14)	$L_J\times10^7$	-5.15 (51)	$P_K\times10^6$	5.813 (68)
δ_J	3.689546 (69)	$L_{JJK}\times10^6$	4.44 (76)	$p_J\times10^{10}$	-3.7 (fixed)
δ_K	10.4168 (26)	$L_{JK}\times10^4$	-1.322 (41)	$p_{JJK}\times10^8$	2.98 (32)
$H_J\times10^3$	1.9494 (68)	$L_{KKJ}\times10^4$	5.984 (96)	$p_K\times10^6$	1.258 (37)
$H_{JK}\times10^3$	-7.218 (61)	$L_K\times10^3$	-1.9338 (70)	$T_K\times10^8$	-1.229 (33)

4. Acknowledgments

This study was supported by a Grant-in-Aid from the Ministry of Education, Science and Culture of Japan and by National Astronomical Observatory of Japan.

References

[1] F. Matsushima, H. Odashima, T. Iwasaki, S. Tsunekawa, and K. Takagi, *J. Mol. Struct.* **352/353**, 371–378 (1995).
[2] F. Matsushima, H. Nagase, T. Nakauchi, H. Odashima, and K. Takagi, *J. Mol. Spectrosc.* **193**, 217–223 (1999).
[3] F. Matsushima, N. Tomatsu, T. Nagai, Y.Moriwaki, and K. Takagi, *J. Mol. Spectrosc.* **235**, 190–195 (2006).
[4] F. Matsushima, M. Matsunaga, G-Y. Qian, Y. Ohtaki, R-L. Wang, and K. Takagi, *J. Mol. Spectrosc.* **206**, 41–46 (2001).
[5] G. Steenbeckeliers, and J. Bellet, *J. Mol. Spectrosc.* **45**, 10–34 (1973).
[6] W.S. Benedict, S.A. Clough, L. Frenkel, and T.E. Sullivan, *J. Chem. Phys.* **53**, 2565–2570 (1970).
[7] C.K. Jen, D.R. Bianco, and J.T. Massey, *J. Chem. Phys.* **21**, 520–525 (1953).
[8] G. Erlandsson and J. Cox, *J. Chem. Phys.* **25**, 778–779 (1956).
[9] J. Verhoeven, H. Bluyssen, and A. Dymanus, *Phys. Lett.* **A26**, 424–425 (1968).
[10] G. Steenbeckeliers and J. Bellet, *Comp. Rend. B* **270** 1039–1041 (1970).
[11] J. Bellet and G. Steenbeckeliers, *Comp. Rend. B* **271**, 1208–1211 (1970).
[12] D.A. Stephenson and R.G. Strauch, *J. Mol. Spectrosc.* **35**, 494–495 (1970).
[13] J.K. Messer, F. C. De Lucia, and P. Helminger *J. Mol. Spectrosc.* **105**, 139–155 (1984).
[14] J.W.C. Johns, *J. Opt. Soc. Am. B* **2**, 1340–1354 (1985).
[15] R. Paso and V.-M. Horneman, *J. Opt. Soc. Am. B* **12**, 1813–1838 (1995).
[16] K.M. Evenson, D.A. Jennings, and F.R. Petersen, *Appl. Phys. Lett.* **44**, 576–578 (1984).
[17] I.G. Nolt, J.V. Radostitz, G. DiLonardo, K.M. Evenson, D.A. Jennings, K.R. Leopold, M.D. Vanek, L.R. Zink, A. Hinz and L.V. Chance, *J. Mol. Spectrosc.* **125**, 274–287 (1987).
[18] K.V. Chance, D.A. Jennings, K.M. Evenson, M.D. Vanek, I.G. Nolt, J.V. Radostitz and K. Park, *J. Mol. Spectrosc.* **146**, 375–380 (1991).
[19] J.K.G. Watson, in "*Vibrational Spectra and Structure*"(J.R.Durig Ed.), Vol.6, Elsevier, Amsterdam (1997).
[20] Vl. G. Tyuterev, *J. Mol. Spectrosc.* **151**, 97–129 (1992).
[21] L.H. Coudert, *J. Mol. Spectrosc.* **154**, 427–442 (1992).
[22] C.L. Lin and J.H. Shaw, *J. Mol. Spectrosc.* **66**, 441–447 (1977).
[23] N. Papineau, J.-M. Flaud, C. Camy-Peyret, and G. Guelachvili, *J. Mol. Spectrosc.* **87**, 219–232 (1981).
[24] R.A. Toth, *J. Mol. Spectrosc.* **162**, 41–54 (1993).
[25] T. Furtenbacher and A.G. Császár, *J. Quant. Spectros. Radiat. Transfer* **109**, 1234-1251 (2008).
[26] The web page address: http://www.sci.u-toyama.ac.jp/phys/4ken/atlas/

Journal of Physics: Conference Series **185** (2009) 012029 doi:10.1088/1742-6596/185/1/012029

Coincidence measurements of secondary ions and scattered ions in collision between slow Ar^{6+} and a GaN(0001) surface

K Motohashi[1]
[1] Department of Applied Physics, Tokyo University of Agriculture and Technology, 2-24-16 Naka-cho Koganei-shi, Tokyo 184-8588, Japan

E-mail: motohasi@cc.tuat.ac.jp

Abstract. An experimental apparatus has been developed to investigate interactions between highly charged ions (HCIs) and solid surfaces under off-normal incidence. The apparatus consists of a secondary ion mass spectrometer (SIMS) and a low-energy ion scattering spectrometer (LEIS) equipped with a Wien filter ($\boldsymbol{E} \times \boldsymbol{B}$ velocity filter). It allows us to carry out simultaneous SIMS and LEIS analysis. It successfully measured the time-of-flight SIMS spectrum in coincidence with scattered Ar^0 and Ar^+ in collisions between Ar^{6+} (15.0 keV) and a GaN(0001) surface. Protons and hydrocarbons, N^+, Ga^+ and GaN^+ ions were observed in the SIMS spectra taken in coincidence with the LEIS spectrum of Ar^0 and Ar^+. The kinetic energy of Ar^+ measured in coincidence with protons emitted from the surface was distributed over an energy range from 13.6 to 15.0 keV. This fact directly shows that kinetic processes play a significant role in proton emissions.

1. Introduction

Interactions between slow HCIs and solid surfaces have been attracting much attention because of their possible application to surface nanotechnology. There have been many reports on nanostructure formation as well as efficient secondary particle emission, which make it possible to realize the surface-sensitive chemical analysis. For example, some papers described sputtering yields almost independent of collision energy [1]. Nano-dot volume as a function of the collision energy showed similar energy dependence [2].

However, the details of the reaction processes between slow HCIs and solid surfaces have not been clarified yet because it is very difficult to observe the kinetic and potential processes separately. Secondary-electron emission measured in coincidence with no kinetic energy loss of incident ions was reported by Stöckl *et al.* [3]. It was the first observation of pure potential emission of secondary electrons induced with HCIs. Though it has been generally recognized that the potential energy of HCIs is very important, especially in slow collisions with solid surfaces, it is very difficult to reduce the kinetic energy enough to suppress kinetic processes in actual experiments.

The off-normal or glancing incidence has two important characteristics that are essentially different from those of normal incidence, especially in HCI collisions with solid surfaces. One is that, a lot of surface atoms positioned along the ion trajectory interact with the HCI in glancing collisions because the electron capture starts from a critical distance, which is almost proportional to the initial charge state of the incident HCI [4], and it continues until just in front of the surface. Thus, sputtering or desorption may also occur at multiple different atomic sites. Another important characteristic is, of course, a small kinetic energy supplied to the surface normal direction. The kinetic energy $E_\perp$ is given as follows [4],

$$E_\perp = E_0 \sin^2\theta + E_{im}, \tag{1}$$

where E_0 is the initial projectile energy, θ is the angle between the initial velocity vector and the surface and E_{im} is the image acceleration energy or energy gain. Although E_{im} depends on the charge state of the HCI [5], the potential energy of Ar^{q+} ($q \geq 3$) is greater than the calculated value of $E_\perp$ in collisions with metal surfaces with $E_0 = 2500 \times q$ eV and $\theta \leq 5°$.

A novel experimental apparatus allowing us to achieve multi-coincidence measurements among scattered atoms/ions, secondary electrons and secondary ions has been developed to clarify the sputtering or desorption processes due to electronic excitation induced by slow HCIs [4–7]. This paper will report the first experimental SIMS results measured in coincidence with scattered ions whose charge state and kinetic energy were selected in collisions between HCIs and the solid surfaces.

2. Experimental apparatus and methods

The details of the experimental apparatus were described elsewhere [7], so a brief description is presented here. Figure 1 shows a schematic illustration of the experimental setup. Ar^{6+} ions extracted from a compact electron beam ion source (EBIS) were accelerated by 15.0 keV. After their focus and trajectory are adjusted with an electrostatic lens (L) and four parallel plates ($P_{1\perp}$, $P_{1\parallel}$, P_2 and P_3), the ions collide with the target surface at an angle, $\theta \approx 5$–$10°$, tangential to the surface. The velocity and charge state of the scattered ions or atoms of $Ar^{q'+}$ ($0 \leq q' \leq 6$), that capture (6-q') electrons from the surface, were analysed with the Wien filter (WF) and the charge separator (CS), respectively. When each scattered ion or atom hit the detector, three signal waveforms from the position-sensitive anode (D_1) were recorded using a digital storage oscilloscope (DSO). One can analyse the position (y, z) (see the coordinates in Figure 1) by a simple arithmetic operation on the pulse heights of the waveforms. The WF works as a velocity-dispersion filter; i.e. the spacing of exit slits extends along the z direction. The displacement along the z-axis of the sensed position of a scattered ion with respect to that of a neutral atom reflects its charge state and velocity. Secondary ions emitted from the surface were detected with a multichannel plate (MCP, D_2). The difference between the time-of-flight of the secondary ion T_1 and that of scattered atom/ion T_2, or $t_{12} = T_1$-T_2, was also recorded with the DSO event by event.

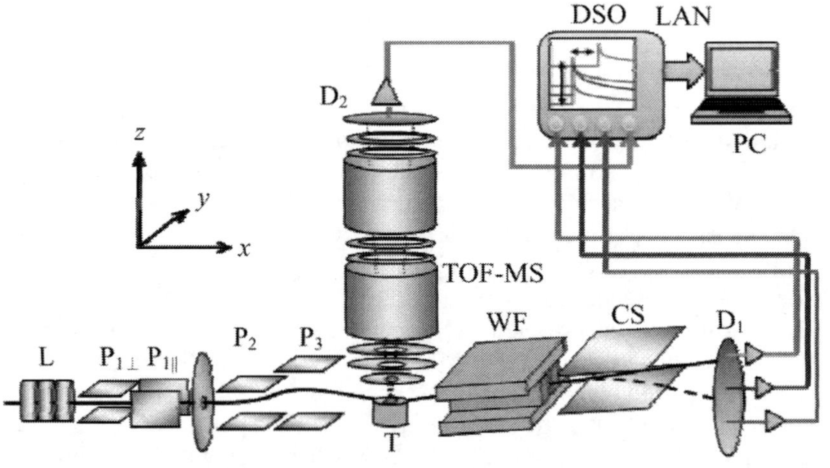

Figure 1. Schematic illustration of experimental setup
L, Einzel lens; $P_{1\perp}$, $P_{1\parallel}$, P_2 and P_3, Parallel plate deflectors; T, Target surface and its holder; WF, Wien filter; CS, Charge separator; D_1, Position-sensitive MCP detector of scattered ions and atoms; D_2, MCP detector; TOF-MS, Time-of-flight mass spectrometer; DSO, Digital storage oscilloscope; LAN, Local area network; PC, Personal computer

Figure 2 shows projectile-beam profile converted from the two-dimensional (2D) position images that were taken with the D_1 detector. The origin of the *z*-axis was set at the peak position of the 15.00 keV beam profile. The peak position of the 15.20 keV beam was shifted to the +*z* direction by 0.9 mm to maintain a width of ~2.5 mm (FWHM). The relationship between the energy and the displacement, i.e. the *E*-*z* calibration factor of the WF, was found to be 2.2×10^2 eV/mm. The energy resolution was evaluated to be 5.5×10^2 eV. The energy resolution was not so good because of the wide (1 mm) entrance slit of the WF. A few structures observed in both spectra may be due to the potential distribution at the ion trap region in the EBIS. Preliminary energy analysis of the primary ion beam will be required to achieve high-resolution measurements.

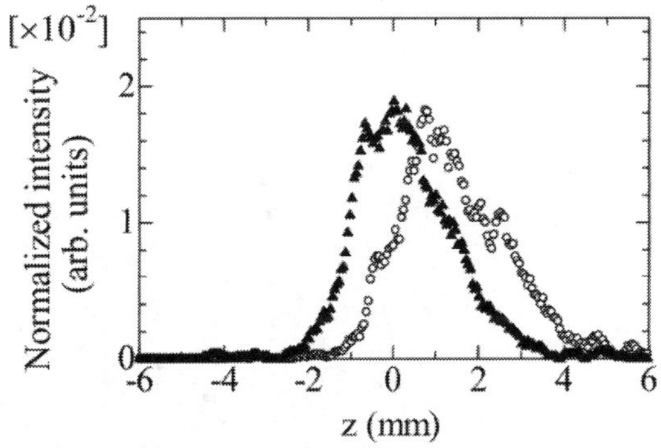

Figure 2. Beam profiles of Ar^{6+} ions measured with the position sensitive detector D_1
Triangles, 15.00 keV; Open circles, 15.20 keV

3. Results and discussion

Figure 3 shows the 2D-position image of Ar^+ scattered at the GaN(0001) surface upon incidence of Ar^{6+} (15.0 keV) at $\theta \approx 10°$. The vertical axis *z* can be converted to the kinetic energy of Ar^+ passing through the WF, as described above. The horizontal axis *y* may be related to the longitudinal scattering angle of Ar^+. However, the angle cannot be determined at present because its calibration has not been accomplished yet.

Figure 4 shows the kinetic energy distribution of Ar^+. The distribution, converted from Figure 3, corresponds to the LEIS spectrum. The width (FWHM) of the Ar^+ spectrum is ~2.6 times wider than that of the primary Ar^{6+} one. Thus, significant energy loss, up to ~2.0 keV, was found in the Ar^+ spectrum, and the loss peak was located around 0.6 keV.

Figure 5(a) shows a 2D-correlation plot between the displacement *z* of scattered $Ar^{q'+}$ (q' = 0, 1 and 2) and the time difference t_{12} (see section 2) of secondary ions. Figures 5(b) and (c) show the integrated intensity of Figure 5(a) along the vertical and horizontal axis, respectively. Figure 5(b) is almost the same as the usual SIMS spectrum, since T_1 is much longer than T_2. Very efficient proton emission is one of the most important characteristics of desorption and ionization due to electron capture by HCIs [1, 4–7]. Weak N^+, Ga^+ and GaN^+ peaks originating from substrate atoms were also observed, together with some hydrocarbon-ion peaks originating from contaminants. Figure 5(c) corresponds to the

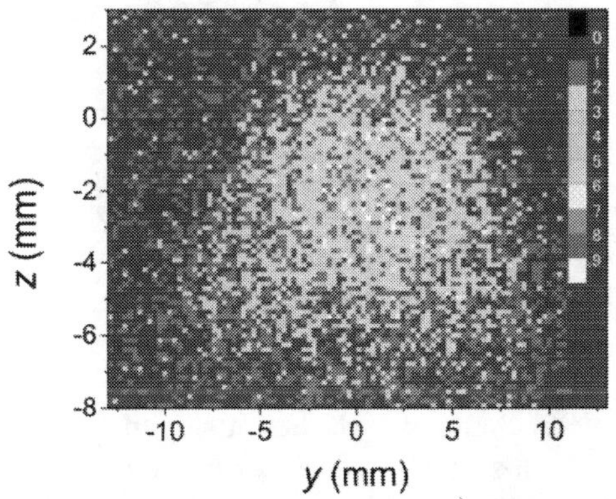

Figure 3. 2D-position image of Ar^+ scattered at a GaN(0001) surface under off-normal incidence of Ar^{6+} (15.0 keV)

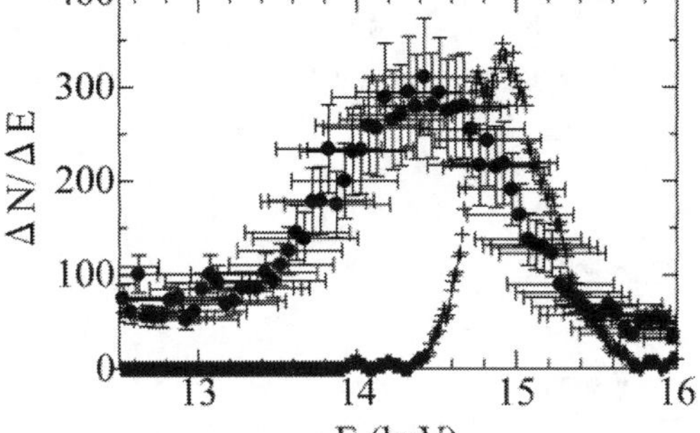

Figure 4. LEIS spectrum converted from the 2D-position image in Figure 3
Circles, Ar^+ scattered ions; Crosses, Ar^{6+} primary ions (The vertical axis of Ar^{6+} is arbitrary.)

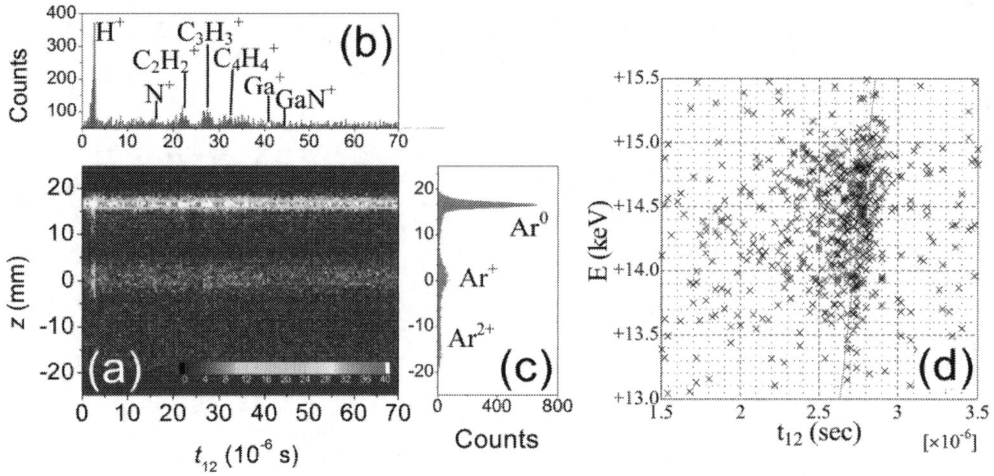

Figure 5. (a) 2D-correlation plot between the displacement z of scattered $Ar^{q'+}$ (q' = 0, 1 and 2) and the time difference t_{12} (see section 2) of secondary ions; (b) and (c) Distributions integrated along the vertical and horizontal axis of (a), respectively; (d) Correlation plot between the time difference t_{12} of protons and kinetic energy E of scattered Ar^+; Red line is provided to guide the eye.

charge state distribution of scattered $Ar^{q'+}$. The populations of the spectra with q' = 0, 1 and 2 were 66%, 23% and ≤ 11%, respectively. Figure 5(d) shows the 2D-correlation plot between the time difference t_{12} of protons and kinetic energy E of scattered Ar^+. The vertical axis of this figure was converted from the displacement z of Figure 5(a). The scale of the horizontal axis of figure 5(d) is limited to the time range corresponding to the proton peak of Figure 5(a) or (b). It was found that the E–t_{12} distribution slightly but clearly leaned, as shown by a red line. This fact confirms that t_{12} of protons changed with the kinetic energy of Ar^+, simply because T_2 depends on the kinetic energy. The energy loss of Ar^+ measured in coincidence with protons was distributed in the range of 0–2 keV, the same as in the LEIS spectrum of Figure 4. This finding directly shows that some kinetic processes of substrate atoms play a role in proton emission processes.

4. Conclusion

Simultaneous LEIS and SIMS analyses was achieved with a new experimental apparatus using collisions between Ar^{6+} and GaN(0001) at an off-normal incidence angle. It was found that some kinetic processes involving surface substrate atoms play a role in proton emission.

5. Acknowledgement

This work was partly supported by a Grant-in-Aid for Scientific Research (B) (19360016). The author would like to thank Prof. S. Tsurubuchi, Prof. A. Koukitu, Prof. Y. Yamazaki, Dr. Y. Kanai, and Dr. M. Flores for fruitful discussion.

References

[1] Tona M, Fujita Y, Yamada C and Ohtani S, 2008 Phys. Rev. B **77** 155427.
[2] El-Said A S *et al.*, 2008 Phys. Rev. Lett. **100** 237601.
[3] Stöckl J, Suta T, Ditroi F, Winter HP and Aumayr F, 2004 Phys. Rev. Lett. **93** 263201.
[4] Motohashi K and Tsurubuchi S, 2007 Nucl. Instrum. Meth. Phys. Res. B **264** 15.
[5] Motohashi K, Tsurubuchi S and Koukitu A, 2005 Nucl. Instrum. Meth. Phys. Res. B **232** 254.
[6] Motohashi K, Hosoya K, Imano M and Tsurubuchi S, 2007 Surf. Sci. **601** 5304.
[7] Motohashi K, 2009 e-J. Surf. Sci. Nanotech. **7** 21.

Journal of Physics: Conference Series **185** (2009) 012030 doi:10.1088/1742-6596/185/1/012030

Thomson scattering from the antiferroquadrupolar ordering phase in TmTe

Tatsuya Nagao

Faculty of Engineering, Gunma University, Kiryu, Gunma 376-8515, Japan

E-mail: tnagao@phys.sci.gunma-u.ac.jp

Abstract. On the basis of the recently established crystal electric field level scheme of $\Gamma_8 - \Gamma_6 - \Gamma_7$, we investigate the Thomson scattering intensity of the superstructure reflections $\mathbf{G} = (\frac{2\pi}{a})\left(\frac{h}{2}, \frac{k}{2}, \frac{\ell}{2}\right)$ with $h, k, \ell = 2n + 1$ in the Γ_3-type (O_2^2, O_2^0) antiferroquadrupolar (AFQ) ordering phase of TmTe. The intensities show strong dependence on **G** and the direction of the applied magnetic field **H**. These dependences on **G** and **H** similar to those found in the AFQ phase of CeB_6, another AFQ material with Γ_5-type (O_{yz}, O_{zx}, O_{xy}) order parameter having the Γ_8 ground states, may provide us a useful tool to distinguish the component of the AFQ order of Γ_3-type and Γ_5-type.

1. Introduction

Thulium telluride (TmTe) is a magnetic semiconductor crystallizing in the rocksalt structure $(Fm\overline{3}m)$ with the lattice constant being equal to 6.354 Å. This material is considered to show an antiferroquadrupolar (AFQ) order below $T_{\mathrm{Q}} = 1.8$ K[1] and, then, an antiferromagnetic (AFM) order below $T_{\mathrm{N}} \sim 0.43$ K. The thulium monochalcogenide family usually shows valence instability of Tm. The Tm ion in this material, however, is divalent in an ideal crystal.[2] Then, the ground multiplet becomes $^2\mathrm{F}_{\frac{7}{2}}$ in the configuration $(4f)^{13}$. Under a cubic point symmetry $(\mathrm{F}m\overline{3}m)$, the multiplet splits into two doublets Γ_6, Γ_7 and one quartet Γ_8. Since the overall level splitting does not exceed 15 K, we should include full multiplets into the analysis. Among the basic properties which have not achieve the consensus yet, the crystal electric field (CEF) level scheme is one of the most urgent issues to be addressed. Clementyev *et al.* performed inelastic neutron scattering measurements and proposed their data were well fitted by means of the CEF level schemes $\Gamma_8(0) - \Gamma_7(4.6\mathrm{K}) - \Gamma_6(10.7\mathrm{K})$ or $\Gamma_8(0) - \Gamma_6(4.9\mathrm{K}) - \Gamma_7(11.1\mathrm{K})$.[3] One the other hand, some claimed the ground multiplet might be Γ_7.[4]

Recently, through the investigation based on the magnetic phase diagram, the present author's group has demonstrated that the level scheme $\Gamma_8 - \Gamma_6 - \Gamma_7$ is the most promising candidate of the CEF level scheme.[5] Their analysis has shown that the component of the AFQ order parameter is O_2^2 in zero field while in the presence of the applied field, their components depend on the direction of the field. The O_2^0 and O_2^2 states are stabilized by the field along [001] and [110] directions, respectively. These findings are consistent with the results predicted by the mean field analysis on a classical model [6, 7] and those confirmed by the neutron diffraction experiments.[8, 9, 10] As a possible direct observation of the AFQ order, we have shown that the azimuthal angle dependence of the resonant x-ray scattering intensity at the Tm M_5 edge can be employed to identify the order parameter and CEF.[5]

Journal of Physics: Conference Series **185** (2009) 012030 doi:10.1088/1742-6596/185/1/012030

In this paper, we investigate the Thomson scattering intensity from the AFQ phase with the Γ_3-type (O_2^2, O_2^0) of TmTe assuming the $\Gamma_8 - \Gamma_6 - \Gamma_7$ level scheme and using the model Hamiltonian defined in Ref. [5]. We show that its dependence on the scattering vector and the applied magnetic field exhibits characteristic behavior and that the situation is similar to that found in another AFQ phase material with the Γ_5-type (O_{yz}, O_{zx}, O_{xy}) order of CeB_6.[11, 12, 13, 14, 15] Although both materials experience the AFQ phase with Γ_8 ground states under the cubic symmetry, we discuss that the difference of the component of the AFQ order may be distinguished by investigating the Thomson scattering data.

Sections 2 and 3 report the theoretical preparation in evaluating the Thomson scattering intensity and our findings for it from the AFQ phase in TmTe, respectively. A comparison between the results for TmTe and those for CeB_6 will be shown in section 3, too. A summary is given in section 4.

2. Theoretical framework

The Thomson scattering amplitude measured in units of the classical electron radius r_0 at the j-site $\mathbf{R}_j$ is written as

$$f_j = (\boldsymbol{\epsilon}' \cdot \boldsymbol{\epsilon}) e^{i\mathbf{G}\cdot\mathbf{R}_j} \langle \psi_j | e^{i\mathbf{G}\cdot\mathbf{r}} | \psi_j \rangle, \tag{1}$$

where $\mathbf{G}$ is the scattering vector. The polarizations of the incident and scattered photons are denoted as $\boldsymbol{\epsilon}$ and $\boldsymbol{\epsilon}'$, respectively. We restrict our interest in the $\sigma - \sigma'$ channel corresponding to $\boldsymbol{\epsilon}' \cdot \boldsymbol{\epsilon} = 1$. The eigenstate is represented by $|\psi_j\rangle$ which is expanded as $\sum_{\ell_z, s_z} A(\ell_z, s_z)|\ell_z s_z\rangle$ within the space spanned by the total angular moment $J = \frac{7}{2}$. With the help of Rayleigh expansion, we can express the matrix element as

$$\langle \ell_z s_z | e^{i\mathbf{G}\cdot\mathbf{r}} | \ell_z' s_z' \rangle = \delta_{s_z, s_z'} \sqrt{4\pi} \sum_{\ell=0}^{\infty} i^\ell \sqrt{2\ell+1} \langle j_\ell \rangle Y^{\star}_{\ell, \ell_z - \ell_z'}(\theta_G, \varphi_G) c^\ell(3\ell_z; 3\ell_z'), \tag{2}$$

$$\langle j_\ell \rangle \equiv \int_0^\infty r^2 R_{4f}^2(r) j_\ell(|\mathbf{G}||\mathbf{r}|) dr, \tag{3}$$

where j_ℓ is a spherical Bessel function of order ℓ. Radial part of the $4f$ wave function is denoted as $R_{4f}(r)$, which is evaluated by the Hartree-Fock approximation in the following calculation.[16] The Gaunt coefficient is represented as $c^\ell(3\ell_z; 3\ell_z')$. The arguments θ_G and φ_G of the spherical harmonics Y_{ℓ, ℓ_z} mean the spherical coordinate of $\mathbf{G}$.

The eigenstates are evaluated by solving the model Hamiltonian defined in Eq. (1) of Ref. [5] which includes the CEF term, the Zeeman term, and the terms describing the multipole interactions. The parameters needed to define the Hamiltonian are chosen as the same values as those used in Ref. [5]. As for the applied magnetic field, we consider they are along [001] and [110] directions with $|\mathbf{H}| = 4$ T.

3. Numerical results

In the fcc structure, four K-domains are given rise to with the magnetic modulation vectors $\mathbf{K}_1 = (\frac{1}{2}\frac{1}{2}\frac{1}{2})$, $\mathbf{K}_2 = (\frac{\bar{1}}{2}\frac{1}{2}\frac{1}{2})$, $\mathbf{K}_3 = (\frac{1}{2}\frac{\bar{1}}{2}\frac{1}{2})$, and $\mathbf{K}_4 = (\frac{\bar{1}}{2}\frac{\bar{1}}{2}\frac{1}{2})$ where $\mathbf{K}_i$ are measured in units of $\frac{2\pi}{a}$. Different K-domains are connected to the different magnetic Bragg reflections in the fcc structure. For instance, the intensities from $\mathbf{G} = \left(\frac{2n+1}{2}, \frac{2n+1}{2}, \frac{2n+1}{2}\right)$ and $\left(\frac{2n+1}{2}, \frac{2n+3}{2}, \frac{2n+3}{2}\right)$ are brought about by $\mathbf{K}_1$ and $\mathbf{K}_2$-domains, respectively. In the absence of the applied magnetic field, each K-domain has three S-domains where the choices of the quantized axis are different. The one is defined by the crystal z-axis as a quantization axis of J. We denote it as S_1-domain. The other two, named as S_2 and S_3, are obtained by rotating S_1 through angles $\pm 120°$ about the modulation vector $\mathbf{K}_i$ in each K-domain. In zero field, all the three S-domains contributes independently. On the other hand, in the presence of the applied field only K-domain consideration is needed. In the present treatment, we expect every domain is equally populated.

First, we summarize the results for a series of $\mathbf{G} = \left(\frac{2n+1}{2}, \frac{2n+1}{2}, \frac{2n+1}{2}\right)$. No intensities are detected in zero field and $\mathbf{H} \parallel [110]$ due to the symmetrical reason. When the magnetic field is applied along [001] direction, however, relatively weak none-zero intensity is observed with a single-peak structure as a function of n. Its maximum occurs around $n = 9$ and the dominant and secondary contributions are from the terms proportional to $\langle j_4 \rangle$ and $\langle j_6 \rangle$, respectively, in Eq. (2).

As for the detectability, it would be better to turn our attention to another series of $\mathbf{G}$ which gives larger intensities than that of $\left(\frac{2n+1}{2}, \frac{2n+1}{2}, \frac{2n+1}{2}\right)$. Figure 1 shows the results of Thomson scattering intensities for $\mathbf{G}$ being a series of $\left(\frac{2n+1}{2}, \frac{2n+3}{2}, \frac{2n+3}{2}\right)$. Intensities are measured in units of r_0^2. In this series of $\mathbf{G}$, the intensities in zero field and $\mathbf{H} \parallel [110]$ exhibit similar behavior as a function of n as shown in Fig. 1 (a) and (c). After achieving the maximum around $n = 2$ or 3, the intensities decay rapidly when $|\mathbf{G}|$ exceeds 1 Å^{-1} since the dominant contribution is due to the term proportional to $\langle j_2 \rangle$. It seems reasonable since the primary order parameter is O_2^2 in both cases. The qualitative difference may be attributed to the induced dipole and octupole moments emerged in the presence of the applied field. When the field is applied in the [001] direction, the intensity shows two-peak structure as a function of n as shown in Fig. 1 (b). Such a characteristic bahavior may be easy to detect in an actual experiment. The intensity includes a considerable contribution from the term proportional to $\langle j_2 \rangle$ which is absent for $\left(\frac{2n+1}{2}, \frac{2n+1}{2}, \frac{2n+1}{2}\right)$.

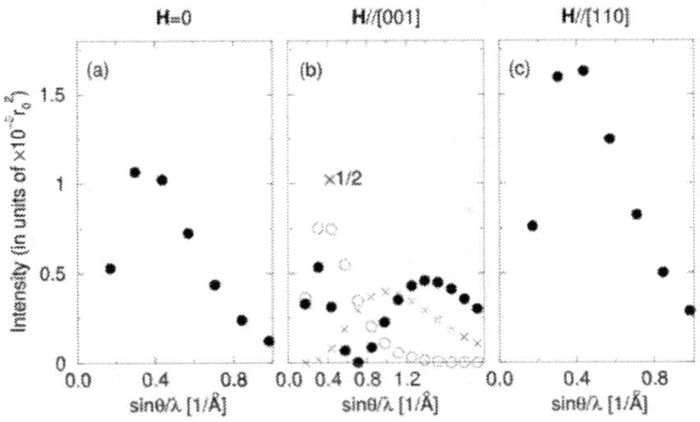

Figure 1. Thomson scattering intensities from TmTe with $\mathbf{G} = \left(\frac{2n+1}{2}, \frac{2n+3}{2}, \frac{2n+3}{2}\right)$ with $n = 1 \sim 7$. (a) zero field, (b) $\mathbf{H} \parallel [001]$, and (c) $\mathbf{H} \parallel [110]$. Intensities are measured in units of r_0^2. Symbols ● , ○ , and × denote the full intensity, the terms proportional to $\langle j_2 \rangle^2$ and $\langle j_4 \rangle^2$, respectively.

It would be worthwhile to compare the present results with those of CeB_6 in the AFQ phase (phase II).[11, 17] In CeB_6, the ground multiplet is also the Γ_8 quartet with $J = \frac{5}{2}$ but is widely isolated (540 K). A notable difference between TmTe and CeB_6 relevant to the behavior of the Thomson scattering intensity is that the AFQ order is believed to be of Γ_3-type (O_2^2, O_2^0) and of Γ_5-type (O_{yz}, O_{zx}, O_{xy}), respectively. Kono *et al.* obtained zero intensity for CeB_6 in zero field with $\mathbf{G} = \left(\frac{2n+1}{2}, \frac{2n+1}{2}, \frac{2n+1}{2}\right)$.[15] This is the same as our present case of TmTe. On the other hand, a prominent difference is found from the work by Yakhou *et al.*[11] The relative intensities of Thomson scattering in the AFQ phase of CeB_6 in the absence of the magnetic field. at the superlattice spots $\mathbf{G} = \left(\frac{5}{2}, \frac{3}{2}, \frac{3}{2}\right)$, $\left(\frac{5}{2}, \frac{1}{2}, \frac{1}{2}\right)$, and $\left(\frac{7}{2}, \frac{1}{2}, \frac{1}{2}\right)$ were listed as about 1, 0.02, and 0.02, respectively. The corresponding values in the present case of TmTe are 1, 0.08, and 20.3, respectively. Note that these values are obtained by multiplying a geometrical factor $\cos(2\theta_B)$ in the $\pi - \pi'$ channel where θ_B is the Bragg angle. There are several candidates for the origin of this huge difference of the relative intensity at $\left(\frac{7}{2}, \frac{1}{2}, \frac{1}{2}\right)$. For instance, it may be due to the difference of the values of J, the inclusion or exclusion of the excited states $\Gamma_{7,6}$, and the difference of the Γ_3-type and Γ_5-type components. Although we cannot specify a single reason at the present, it is safe to say that such a huge difference does not disappear unless the values of the parameters might be modified considerably.

4. Summary

On the basis of the CEF scheme Γ_8–Γ_6–Γ_7 spanned by $J = \frac{7}{2}$, we have investigated the Thomson scattering intensities of TmTe in the AFQ phase. We have demonstrated that the intensity exhibits characteristic dependence on the scattering vector and the direction of the applied magnetic field. We have considered three case; (i)zero field, (ii)$\mathbf{H} \parallel [001]$, and (iii)$\mathbf{H} \parallel [110]$. The primary order parameters of cases (i) and (iii) are O_2^2 while that of case (ii) is O_2^0. Reflecting this difference, intensity of case (ii) shows different dependence on $\mathbf{G}$ compared to those of cases (i) and (iii), which are similar to each other.

The strong dependence of the Thomson scattering intensity on the direction of $\mathbf{G}$ similar to found in the present work are also known for CeB_6 in the AFQ phase.[12, 17] By comparing the present results and those of CeB_6, we can conclude that the experimental observation of Thomson scattering in TmTe has realistic possibility. If we pay attention to the difference of the relative intensities of Thomson scattering at superlattice reflections between TmTe and CeB_6 for the appropriate $\mathbf{G}$ spots, we may get a useful clue to distinguish the component of the AFQ order of Γ_3-type (O_2^2, O_2^0) and Γ_5-type (O_{yz}, O_{zx}, O_{xy}) realized in TmTe and CeB_6, respectively. The Thomson scattering experiment on TmTe in the AFQ phase may clarify the situation.

Acknowledgments

The author would like to thank J. Igarashi and R. Shiina for valuable discussions. This work was partially supported by a Grant-in-Aid for Scientific Research from the Ministry of Education, Science, Sports and Culture, Japan.

References

[1] Matsumura T, Nakamura S, Goto T, Amitsuka H, Matsuhira K, Sakakibara T, and Suzuki T 1998 *J. Phys. Soc. Jpn.* **67** 612
[2] Brewer W D, Kalkowski G, Kaindl G, and Holtzberg F 1985 *Phys. Rev.* B **32** 3676
[3] Clementyev E, Köhler R, Braden M, Mignot J. -M., Vettier C, Matsumura T, and T. Suzuki 1997 *Physica* B **230-232** 735
[4] Nikolaev A V and Michel K H 2001 *Phys. Rev.* B **63** 104105
[5] Shiina R and Nagao T 2008 *J. Phys. Soc. Jpn.* **77** 124715
[6] Shiina R, Shiba H, and Sakai O 1999 *J. Phys. Soc. Jpn.* **68** 2105
[7] Shiina R, Shiba H, and Sakai O 1999 *J. Phys. Soc. Jpn.* **68** 2390
[8] Link P, Gukasov A, Mignot J -M, Matsumura T, and Suzuki T 1998 *Phys. Rev. Lett.* **80** 4779
[9] Link P, Gukasov A, Mignot J -M, Matsumura T, and Suzuki T 1999 *Physica* B **259-261** 319
[10] Mignot J -M, Link P, Gukasov A, Matsumura T, and Suzuki T 2000 *Physica* B **281 & 282** 470
[11] Yakhou F, Plakhty V, Suzuki H, Gavrilov S, Burlet P, Paolasini L, Vettier C, and Kunii S 2001 *Phys. Lett.* A **285** 191
[12] Tanaka Y, Staub U, Katsumata K, Lovesey S W, Lorenzo J E, Narumi Y, Scagnoli V, Shimomura S, Tabata Y, Onuki Y, Kuramoto Y, Kikkawa A, Ishikawa T, and Kitamura H 2004 *Europhys. Lett.* **88** 671
[13] Tanaka Y, Katsumata K, Shimomura S, and Onuki Y 2005 *J. Phys. Soc. Jpn.* **74** 2201
[14] Lovesey S W 2002 *J. Phys.: Condense. Matter* **14** 4415
[15] Kono H N, Kubo K, and Kuramoto Y 2004 *J. Phys. Soc. Jpn.* **73** 2948
[16] Cowan R 1981 *The Theory of Atomic Structure and Spectra* (Berkeley, University of California)
[17] Nagao T and Igarashi J 2003 *J. Phys. Soc. Jpn.* **72** 2381

Journal of Physics: Conference Series **185** (2009) 012031 doi:10.1088/1742-6596/185/1/012031

Atom interferometer using a free oscillation of Bose-Einstein Condensate on an atom chip

Ken'ichi Nakagawa[1,2], Munekazu Horikoshi[1]*

[1]Institute for Laser Science, University of Electro-Communications, 1-5-1 Chofugaoka, Chofu, Tokyo, 182-8585 Japan

[2]CREST, Japan Science and Technology Agency, 4-1-8 Honcho, Kawaguchi, Saitama 332-0012, Japan

nakagawa@ils.uec.ac.jp

Abstract. We have demonstrated a novel atom interferometer which uses a free oscillation of ^{87}Rb BEC atoms in a magnetic harmonic potential on an atom chip. This interferometer can effectively suppress the dephasing due to the trap potential by applying two Bragg laser pulses with an time interval T equal to the oscillation period of the axial mode of the magnetic trap. We observed a high contrast (~30 %) interference fringe signal for T = 58 ms. When we increased the time interval T to 97 ms, an interference fringe was washed out by the random phase fluctuation. We attributed this phase fluctuation to the external vibrational noise from the floor.

1. Introduction

Recently many attentions have been paid to atom interferometers using Bose-Einstein condensates (BEC) in a trap or waveguide [1]. Atom interferometers using trapped atoms instead of free falling atoms can offer the future prospect of realizing longer interrogation time and thus higher sensitivity than those of the conventional atom interferometers with free falling atoms. However, the dephasing effects of condensate atoms in a trap due to the atom-atom interaction and the trap potential has limited the maximum coherence time to about 10 ms [1-2].

In this paper, we present a novel atomic interferometer using a free oscillation of atoms in a magnetic harmonic potential on a chip [3]. With this new interferometer, we can suppress the dephasing due to the trap potential and we can realize a long coherence time of more than 50 ms.

2. Experimental setup

We use an atom chip for the production and manipulation of condensate atoms [4]. The atom chip used in the experiment is shown in Fig. 1. Using standard photolithographic and electroplating techniques, a gold wire pattern for the micro magnetic potential is fabricated on a Si substrate. A magnetic wave guide potential is produced by the combination of the magnetic field of the current (I_z) through the Z-shaped wire (Z-wire) and the uniform bias field (B_0) (Fig. 1 (b)). A current through the crossed wire is additionally used to realize a harmonic potential at the center of the Z-wire.

Using this atom chip, we can produce a Bose-Einstein condensate (BEC) of about 10^4 ^{87}Rb atoms in the F=2, m_F=2 state with an evaporation time of about 5s. Condensate atoms are trapped around

Journal of Physics: Conference Series **185** (2009) 012031 doi:10.1088/1742-6596/185/1/012031

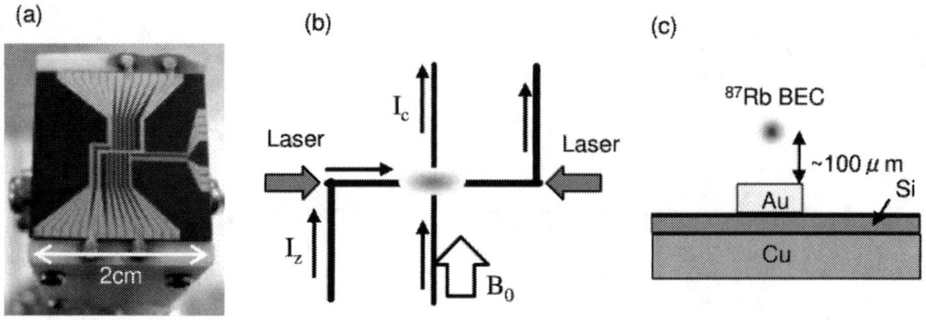

Fig. 1. Atom chip and magnetic potential. (a) Atom chip. (b) A wire configuration for the magnetic trap potential. (c) Trapped atoms in the magnetic potential above the chip.

100 μm above the wire surface (Fig. 1 (c)). In the following interferometer experiment, in order to split the condensate atoms into two components, counter-propagating two laser beams are incident on the condensate atoms (Fig. 2 (b)) along the axial direction of the trap.

3. Trapped BEC interferometer

A first BEC interferometer on an atom chip was demonstrated by Wang et al.[1]. They realized an atom Michelson interferometer using an optical standing wave light field as an atom beam splitter, and they observed the interference fringe signal for the interaction time up to 10 ms [1]. Following this first experiment, we realized a Mach-Zehnder atom interferometer using an optical Bragg diffraction as an atom beam splitter. We observed the high-contrast interference fringe signal for the interaction time up to 1 ms. However, the fringe contrast became degraded when we increased the interaction time and the trap frequency [2]. We attributed it to the dephasing effect due to the trap potential and the atom-atom interaction.

The dephasing effect due to the trap potential is caused by the velocity change of the condensate when the condensate move in the potential. In order to eliminate this dephasing effect, we have proposed a novel interferometer using a free oscillating BEC in a harmonic potential (Fig. 2). We apply two optical Bragg pulses to the BEC in a harmonic trap with a time interval T between two pulses. A first $\pi/2$ Bragg pulse splits a BEC in a trap into two components, one in $p = 0$, and the other in $p = 2\hbar k$ momentum state, respectively. Two BEC components freely evolve in a harmonic potential, and then the second $\pi/2$ pulse is applied to recombine two components after a time interval T from the first pulse. If T is exactly equal to the trap oscillating period $T_0 = 2\pi/\omega_0$, where ω_0 is the axial trap frequency, the velocity of the separated condensate is returned to the initial value at the first pulse, and the dephasing due to the trap potential is eliminated.

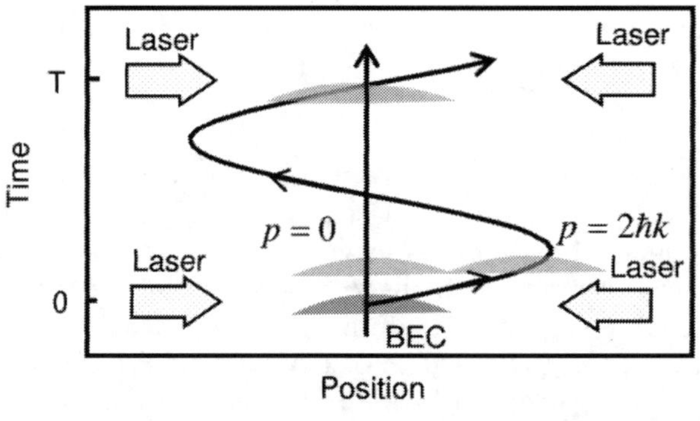

Fig. 2. Schematic drawing for the atom interferometer using a free oscillation of BEC.

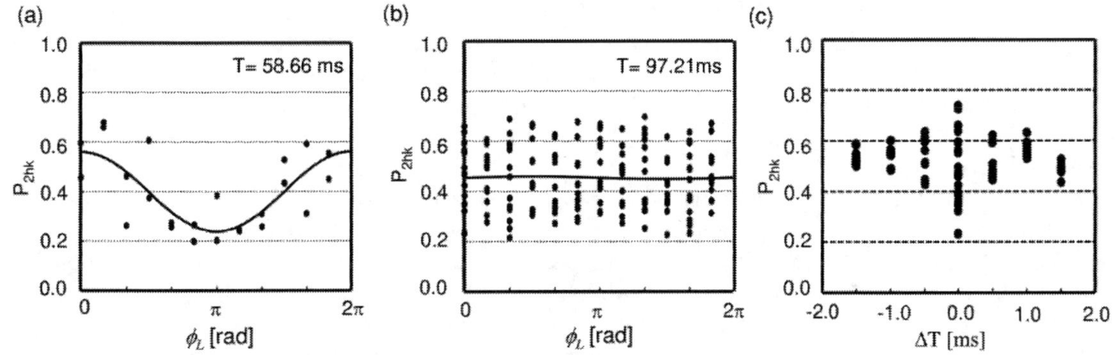

Fig.3. Observed interferometer signal. (a) T (= T_0) = 58.66 ms ($\omega_0/2\pi$ = 19 Hz). (b) T (= T_0) = 97.21 ms ($\omega_0/2\pi$ = 10 Hz). (c) T = T_0 + ΔT ($\omega_0/2\pi$ = 10 Hz).

Figure 3 shows the experimental results. We plot the fraction of atoms in the momentum state $p = 2\hbar k$ given by $P_{2hk} = N_{2hk}/(N_0+N_{2hk})$, where N_0 and N_{2hk} are the number of atoms in the momentum state $p = 0$, $p = 2\hbar k$, respectively. At first, we set the trap frequency $\omega_0/2\pi$ to 19 Hz. Thus the time interval T is set equal to the corresponding trap oscillation period $T_0 = 2\pi/\omega_0$ of about 59 ms. The observed signal shows a clear interference fringe with a fringe contrast of about 30% (Fig. 3 (a)). This result proves that the dephasing effect due to the trap potential is effectively reduced.

Next, we set the trap frequency $\omega_0/2\pi$ to 10 Hz, thus T = 97 ms. The observed signal is randomly distributed between 0.2 and 0.7, and the interference fringe is hardly seen in the signal (Fig. 3 (b)). We attribute this run-to-run fluctuation of the signal in Fig. 3 (b) to the internal and/or the external phase noise not to the dephasing effect. To confirm this assumption, we detune the time interval T from the trap oscillation period T_0. The amplitude of the signal fluctuation is maximum at T = T_0, while it decreases for $T \neq T_0$ (Fig.3 (c)). If $T \neq T_0$, the velocity of the condensate is not exactly returned to the initial one, and thus the dephasing is not completely eliminated.

One of the possible phase noise source is the atom-atom interaction [5,6]. When a BEC with N atoms is coherently split into two components, a relative atom number ΔN (= $N_1 - N_2$) is different from shot-to shot, where N_1 and N_2 are the number of atoms in each component. This shot-to-shot variation of ΔN induces the relative phase difference $\Delta\phi$ between two components due to the atom-atom interaction [5]. This phase fluctuation increases linearly with time, and it limits the maximum coherence time of the trapped condensates. The coherence time can be extended by using a number squeezed state of condensates [6]. For our experimental condition, the maximum coherence time is estimated to be 400 ms, which is still longer than 97 ms.

Another noise source is the external vibration from the floor. We analyzed the phase change of the interferometer signal due to the vibration of the experimental apparatus, and we found that this interferometer acts as a resonant-type accelerometer with a maximum phase sensitivity at the trap frequency of 10 Hz for T = 97ms [3]. Our experimental apparatus including the atom chip and the laser optics is put on the air-dumped optical table. The interferometer signal is largely affected by the large vibrational noise around the natural frequency of the optical table, which is about 2 Hz. We evaluated the phase fluctuation of the interferometer signal from the measured vibrational noise of the optical table, and it is nearly consistent with the experimental results. Thus we attributed the observed large shot-to-shot random phase fluctuation for T= 97 ms (Fig. 3 (b)) to the external vibration noise rather than the phase noise due to the atom-atom interaction.

The present experiment proved that the atom interferometer using a free oscillation of atoms in a harmonic trap is promising in terms of long interaction time of much longer than 100 ms which cannot be easily realized in the conventional atom interferometer using free falling atoms. Recently Burke et al. demonstrated an extension of the interaction time up to 0.91 s using the same method as ours, but they also observed the shot-to-shot random phase fluctuation due to the vibrational noise [7]. Thus if we employ an active stabilization of the optical table to effectively reduce the vibrational noise in the frequency region from 1 to 10 Hz, we will be able to realize a highly sensitive atom interferometer for the precision measurement of such as Newtonian gravitational constant [8].

4. Conclusion

We have demonstrated a novel interferometer based on a free oscillation of condensate atoms in a magnetic harmonic potential on an atom chip. It has been seen that the dephasing due to the trap potential could be effectively cancelled by setting the time interval between two Bragg pulses equal to the trap oscillating period, and an interference signal with a high fringe contrast of 30% was observed for the time interval of 58 ms. For a longer time interval of 97 ms, the dephasing effect was still eliminated, however, the interference fringe was washed out by the low frequency external vibrational noise.

Acknowledgments

This work was partly supported by a Grant in Aid for Science Research (No. 17340120) from the Ministry of Education, Science, Sports, and Culture, ''Ground-Based Research Program for Space Utilization'' promoted by Japan Space Forum.

* Present address: ERATO Macroscopic Quantum Control Project, JST, 2-11-16 Yayoi, Bunkyo-Ku, Tokyo 113-8656, Japan

References

[1] Wang Y, Anderson D Z, Bright V M, Cornell E A, Diot Q, Kishimoto T, Prentiss M, Saravanan R A, Segal S R and Wu S 2005 Atom Michelson Interferometer on a Chip Using a Bose-Einstein Condensate *Phys. Rev. Lett.* **94** 090405
[2] Horikoshi M and Nakagawa K 2006 Dephasing due to atom-atom interaction in a waveguide interferometer using a Bose-Einstein condensate *Phys. Rev. A* **74** 031602(R)
[3] Horikoshi M and Nakagawa K 2007 Suppression of dephasing due to a trapping potential and atom-atom interactions in a trapped-condensate interferometer *Phys. Rev. Lett.* **99** 180401
[4] Horikoshi M and Nakagawa K 2006 Atom Chip based fast production of Bose-Einstein condensate *Appl. Phys. B* **82** 363
[5] Castin Y and Dalibard J 1997 Relative phase of two Bose-Einstein condensates *Phys. Rev. A* **55** 4330
[6] Jo G -B, Shin Y, Will S, Pasquini T A, Saba M, Ketterle W, Pritchard D E, Vengalattore M and Prentiss M 2007 Long Phase Coherence Time and Number Squeezing of Two Bose-Einstein Condensates on an Atom Chip *Phys. Rev. Lett.* **98** 030407
[7] Burke J H T, Deissler B, Hughes K J, Sackett C A 2008 Confinement effects in a guided-wave atom interferometer with millimeter-scale arm separation *Phys. Rev. A* **78** 023619
[8] Lamporesi G, Bertoldi A, Cacciapuori L, Prevedelli M, Tino G M 2008 Determination of the Newtonian Gravitational Constant Using Atom Interferometry *Phys. Rev. Lett.* **100** 050801

Journal of Physics: Conference Series **185** (2009) 012032 doi:10.1088/1742-6596/185/1/012032

Spin dependence in resonance electron scattering from zinc atoms

S. A. Napier, D. Cvejanović, J. F. Williams, L. Pravica, S. N. Samarin, A. D. Sergeant, P. Guagliardo and P. Wilkie

ARC Centre for Antimatter-Matter Studies, Centre for Atomic, Molecular and Surface Physics, Physics Department, University of Western Australia, Perth 6009, Australia

E-mail: `jfw@cyllene.uwa.edu.au`

Abstract.
Spin effects in excitation of the $3d^{10}4s^24p\,^3P_{0,1,2}$ states of zinc via $3d$ core excited negative ion resonances were studied using a spin polarized electron beam. Energy dependence of the angle differential scattered electron intensity is reported for two scattering angles, $\theta = 30°$ and $\theta = 54°$. Observations indicate strong spin-up/down scattering asymmetries in the energy region where excitation proceeds via $3d^94s^24p^2$ negative ion resonances.

1. Introduction

Experiments with low energy electron scattering on zinc atoms have shown significant effects of electron correlation as soon as excitation of the inner 3d electron becomes energetically possible. Of special interest are negative ion resonances formed upon the excitation of the $3d$ core electrons. These were observed in excitation of valence states using detection of decay photons [1] or of energy selected scattered electrons [2, 3]. While photon studies include cascade effects and the intensity is summed over all scattering angles, the detection of an electron, scattered at a chosen scattering angle and with well-defined energy loss, is final state specific and the observed count rate is proportional to an angular differential cross section. When the cross section is expanded in angular-momentum dependent partial waves, the effects of resonance states may become more apparent and this is particularely so for elastic scattering [3]. The present study of excitation of the $3d^{10}4s^24s4p\,^3P_{0,1,2}$ state using spin polarized electrons extends that investigation to obtain spin-specific angular momentum information. Also the spin-dependent interactions in the formation and decay of the $3d$ core-excited negative ion resonances in zinc are probed.

The mechanisms responsible for spin dependence in electron-atom collisions and the unique information obtained with incident spin polarized electrons have been discussed [4, 5]. In zinc, with a closed $4s^2$ outer shell, spin effects in elastic scattering are caused by the spin-orbit interaction of a scattered electron in the field of an atom (Mott scattering). For the excitation process however, spin polarization phenomena can be introduced by both the exchange and the internal atomic spin-orbit interactions. If electron scattering proceeds via a short-lived, negative ion resonance, additional spin effects are possible and these characterize the spin-orbit interaction and electron exchange in a particular negative ion state. As the active electrons/holes have non-zero orbital momentum (4p and 3d), both the momentum coupling of the 3d core and the pair of outermost 4p electrons will influence resonance scattering.

The only previous measurement of the angular behavior of asymmetries (left/right) in low energy elastic scattering of spin polarized electrons on zinc [6], used a coarse energy mesh above the ionization threshold to report non-zero values and could not observe variations due to negative ion resonances. However, that investigation observed a strong effect on spin asymmetry for the 4.05 eV resonance in cadmium.

We observed previously the core-excited negative ion resonances in zinc [7] in the angle-integrated Stokes polarization parameters P_1, P_2 and P_3. That study used spin polarized electrons to probe exchange and spin-orbit interactions activated by promotion of one of the inner $3d$ electrons. Both the spin-orbit and exchange effects were observed and separated by measurement of the Stokes polarization parameters P_1, P_2 and P_3 of the $3d^{10}4s^24d\,^1D_2 \rightarrow 3d^{10}4s4p\,^1P_1$ decay photons. Here we report the spin dependence induced by the same group of core-excited resonances but now observed in angle differential scattering cross sections for excitation of the unresolved $3d^{10}4s^24p\,^3P_{0,1,2}$ states. The results are expressed in terms of spin up/down scattering asymmetries.

2. Experiment

The apparatus was based on crossed spin-polarized electron and unpolarized atom beams. The source of spin polarized electrons has been described in detail previously [8]. Briefly, photo-electrons are emitted from a GaAs crystal irradiated with circularly polarized laser light with spin vectors parallel or anti-parallel to the incident laser beam propagation direction depending on the helicity of the laser light. This initially longitudinal spin polarization was changed to transverse by deflecting electrons through 90°. Orientation of the spin vector with respect to the scattering plane, up ↑, or down ↓, can be changed by switching the helicity of the laser light from left to right using a liquid crystal variable retarder. The initial energy spread of electrons was reduced using an electrostatic 180° hemispherical monochromator. Spin polarization in the incident electron beam was 28%.

Electrons scattered on zinc atoms at a variable angle θ were energy selected by a 127° cylindrical analyzer and those exciting the $4s4p\,^3P_{0,1,2}$ states were detected using a channel electron multiplier. The incident electron energy and the scattered electron residual energy were scanned simultaneously for each spin direction to observe resonance structure and its energy and spin dependence. The scanning of electron energies, electron counters and the control of spin orientation were done using a PC-based data acquisition system. Repeated energy scans, with a change of spin orientation at each incident energy point, were taken until the required statistical accuracy was achieved. The scattered electron analyser can rotate to both sides of the incident electron beam direction which defines the zero of the angular scale. In this way measurements can be repeated and data compared for left/right and spin up/down scattering asymmetries. The incident electron energy scale was calibrated on known positions of negative ion resonances in zinc [1]. The reported studies were conducted with an energy resolution of 120 meV.

3. Results and discussion

Previous electron impact $3d^{10}4s4p\,^3P_{0,1,2}$ excitation studies [1, 2] showed a strong influence of negative ion resonances above the ionization threshold. The observed negative ions are created upon promotion of an inner 3d electron into the first unoccupied 4p orbital accompanied by the temporary attachment of the incident electron into the same 4p orbital. The assignment of negative ion states was discussed in terms of momentum couplings and related to the spin-dependent effects within the 3d ion core and the pair of outer 4p electrons [1, 3]. Previous work also outlines a comparison with the iso-electronic atom, i.e. autoionizing states of gallium atom with the same $3d^94s^24p^2$ electron configuration. This comparison gives an insight into the likely symmetry and also the energy ordering of different negative ion states. The three fine structure components of the $3d^{10}4s4p\,^3P_{0,1,2}$ decay channel are not resolved in the present experiment but

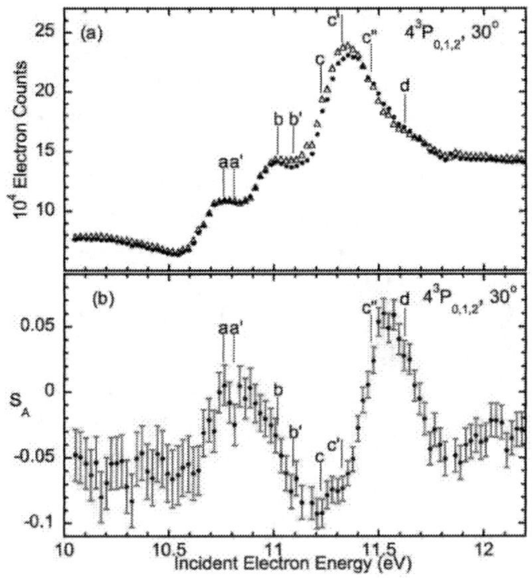

Figure 1. (a) Electron excitation functions measured $\theta = 30°$ with: •, spin up electrons; △, spin down. (b) Spin asymmetry.

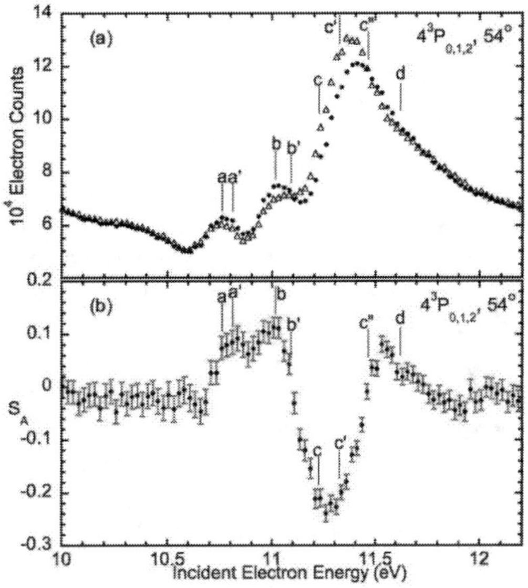

Figure 2. (a) Electron excitation functions measured at $\theta = 54°$ with: •, spin up electrons; △, spin down. (b) Spin asymmetry.

are well resolved from other scattering events at nearby energies, i.e. from elastic scattering and excitation of the $4s4p\,^1P_1$ state.

The results for spin-dependent angle differential electron scattering leading to the excitation of the $4s4p\,^3P_{0,1,2}$ states at scattering angles of $\theta = 30°$ and $\theta = 54°$ are shown in figures 1 and 2, respectively. For each scattering angle, the scattered electron intensities for the spin-up, $I^{\uparrow}(\theta)$, and spin-down, $I^{\downarrow}(\theta)$, are shown in parts 1(a) and 2(a). These intensities were used to evaluate the spin asymmetry function $S_A(\theta)$ using equation

$$S_A(\theta) = \frac{1}{P_e} \cdot \frac{I^{\uparrow}(\theta) - I^{\downarrow}(\theta)}{I^{\uparrow}(\theta) + I^{\downarrow}(\theta)}, \tag{1}$$

where P_e is the spin polarization of the incident electron beam. The spin asymmetry as a function of energy is shown in figures 1(b) and 2(b).

The energy positions of the a, a', b, b', c, c', c'' and d negative-ion resonance structures, determined from the photon [1] and electron excitation functions and electron transmission measurements [3, 9], are also indicated. As can be seen from figures 1 and 2 the energy behavior of the asymmetry function S_A shows a clear angle dependent, sharp variation in the vicinity of the indicated resonances. Also, a small, but statistically significant, non-zero asymmetry is measured outside the resonance region at some scattering angles. This is similar to the observation from angle integrated Stokes parameters P_1, P_2 and P_3 of the $3d^{10}4s4d\,^1D_2 \rightarrow 3d^{10}4s4d\,^1P_1$ decay photons that both spin-orbit and exchange were active in some of the $3d^94s^24p^2$ negative ion resonances [7].

Similar measurements to those presented here have been performed for mercury by Bartschat *et al* [10]. That study observed variations in the energy dependence of the asymmetry in excitation of the $6s6p\,^3P_{1,2}$ and $6s6p\,^1P_1$ at known resonance energies. However, important differences between mercury and zinc exist. The $6s6p\,^3P_{0,1,2}$ fine structure states were resolved in mercury, unlike the present study where the fine structure of the zinc $4s4p\,^3P_{0,1,2}$ states was not resolved. This is a consequence of the larger strength of the spin-orbit interaction in mercury which would have also a separate influence on momentum coupling and the multiplet structure relevant for both the negative ion spectrum and decay channels in the two atoms. Also some of the states with equivalent $5d^96s^26p$ electron configuration in mercury are bound ($^2D_{5/2}$) while some ($^2D_{3/2}$) are autoionizing. In conjunction with the strong spin-orbit interaction, this makes direct comparison of zinc and mercury not possible.

An explanation for the negative-ion induced asymmetries observed by Bartschat *et al* [10] was provided by Kessler [4]. When picking out the electrons of the dominant partial wave, i.e. orbital angular momentum l, we must take into account that transversely polarised electrons which are scattered to the left have total angular momenta j different from those of the electrons scattered to the right, since the orbital angular momenta are opposite to each other. In conclusion, for a given incident electron spin polarisation, either electrons scattered to the right or to the left, but not both, may have j suitable to form a compound state. Similar reasoning is applicable in the present measurements where only electrons scattered to the left were detected, but electrons in the incident electron beam have predominantly spin up or spin down. In compliance with conservation of angular momentum and parity, the incident electrons with spin-up or spin-down, but not both, may have j suitable for formation of the negative-ion resonance.

To demonstrate the effect of a negative-ion resonance on the spin asymmetry, let us consider a negative-ion resonance which enhances the cross section of the observed state, and is formed by incident electrons with orbital angular momentum l. If the observation is made at a positive scattering angle θ, and if this negative-ion resonance produces a positive spin-up/down asymmetry, it means that spin-up incident electrons are bound in the negative ion, which therefore has a total angular momentum $J = l + 1/2$. Likewise, if a negative spin-up/down asymmetry is observed, it means that spin-down incident electrons are bound and the negative ion has a total angular momentum $J = l - 1/2$. Obviously a negative-ion resonance generally will not enhance a cross section because it may interfere destructively, as well as constructively, with the non-resonance cross section. Whether a particular negative-ion resonance produces a positive or negative asymmetry will depend on its Fano profile, as well as its parity and J. In principle, spin asymmetry measurements can resolve overlapping negative-ion resonances with different J, even if because of their large natural width they cannot be resolved in unpolarised electron excitation function experiments. Therefore, the high-resolution spin asymmetry function measurements can also assist the assignments of negative-ion resonances in the autoionizing region of zinc.

Acknowledgments
This work was supported by the Australian Research Council and University of Western Australia. SAN is supported by an Australian Postgraduate Award.

4. References
[1] Napier S A, Cvejanovic D, Williams J F and Pravica L 2008 *Phys. Rev. A* **78** 032706
[2] Napier S A, Cvejanovic D, Williams J F and Pravica L 2007 *J. Phys. B: At. Mol. Opt. Phys.* **40** 1323
[3] Napier S A, Cvejanovic D, Burrow P D, Williams J F, Michejda J A and Pravica L 2009 *To be published*
[4] Kessler J (ed) 1985 *Polarized Electrons* (Springer-Verlag, Berlin)
[5] Hanne G F 1983 *Phys. Rep.* **95** 95–165
[6] Bartsch M, Geesmann H, Hanne G F and Kessler J 1992 *J. Phys. B: At. Mol. Opt. Phys.* **25** 1511
[7] Pravica L, Cvejanovic D, Williams J F and Napier S A 2007 *Phys. Rev. A* **75** 030701(R)
[8] Hayes P A, Yu D H and Williams J F 1997 *Rev. Sci. Instrum.* **68** 1708–1713
[9] Napier S A, Cvejanovic D, Williams J F and Pravica L 2009 *To be published*
[10] Bartschat K, Hanne G F, Wolcke A and Kessler J 1981 *Phy. Rev. Lett.* **47** 997–999

Measurement of K – shell x-ray intensity ratios in compounds of Barium

Phanisree KD Timmaraju and K PremaChand

Department of Nuclear Physics, Andhra University, Waltair, Visakhapatnam,

India – 530003

dr.phanisree@yahoo.com

Abstract. Intensity ratios of K X-rays generated by 59.6 keV gamma rays from Am-241 radioactive isotope are determined for Barium in two of its compounds using a high efficiency HpGe detector. The experimental results were compared with those of earlier investigators and theoretical estimates.

1. Introduction

X-ray intensity ratios have extensive applications in various areas of Physics and most predominantly in atomic, nuclear, molecular, radiation and health physics. An accurate knowledge of these ratios is very important in XRF techniques used for elemental analysis. Using these ratios, physical factors like the L/K electron capture ratio and photoionisation cross-section can be determined. They also find important applications in dosimetry, medicine and other irradiational processes. The measurement of these ratios is also important because, by comparison with theoretical predictions based on atomic models which stem from various inherent assumptions, there is a possibility of testing the validity of these models.

A large number of investigations dealing with evaluation of K X-ray intensity ratios for different elements have been reported earlier in the literature. The measured values were compared with theoretical evaluations based on Hartree-Fock theory and Hartree-Slater theory. The K_β/K_α yield ratio is essentially related to atomic potential. In the absence of multiple ionization of the target atom, the value of K_β/K_α is expected to be constant irrespective of the source of excitation used to create the vacancies in the K shell[1]. However in a number of recent studies, it has been demonstrated that the value of K_β/K_α depends on the method of X-ray generation [2-5] and chemical state of targets [6].

In the present investigations, X-rays are generated by exciting the samples using an intense source of gamma photons. Since efficiency plays an important role in the determination of fluorescent ratios, a detector whose efficiency characteristics are constant over a wide energy region is selected for the present investigations. A HpGe detector with constant efficiency characteristics is employed to study the variations in the value of K_β/K_α intensity ratios of Barium in two of its compounds namely Barium Sulphate and Barium Carbonate.

Journal of Physics: Conference Series **185** (2009) 012033 doi:10.1088/1742-6596/185/1/012033

2. Theory

In the present studies, X-ray intensity ratios were calculated using the following relation

$$\frac{I_{K_i}}{I_{K_j}} = \frac{N_{K_i}}{N_{K_j}} \frac{\beta_{K_j}}{\beta_{K_i}} \frac{\varepsilon_{K_j}}{\varepsilon_{K_i}}$$

where i and j corresponds to either α or β, N_{K_i} and N_{K_j} are the counts observed under the peaks corresponding to K_i and K_j peaks respectively, β_{K_i} and β_{K_j} are the target self – absorption correction factors and ε_{K_i} and ε_{K_j} are the efficiencies of the detector for the K_i and K_j series of X-rays. The value of β_{K_i} is calculated using the formula

$$\beta_{K_i} = \frac{1 - \exp\left[-\left(\frac{\mu_{inc}}{Cos\theta_1} + \frac{\mu_{K_i}}{Cos\theta_2}\right)t\right]}{\left(\frac{\mu_{inc}}{Cos\theta_1} + \frac{\mu_{K_i}}{Cos\theta_2}\right)t}$$

where μ_{inc} is the mass attenuation coefficient of the sample at incident energy, μ_{Ki} is the mass attenuation coefficient of the sample at the emitted K_i X-ray energy, θ_1 is the angle made by the incident photon with the normal to the target surface, θ_2 is the angle made by the emitted photon with the normal to the target surface and t is the mass thickness of the sample.

3. Experimental set up

The experimental arrangement employed for present studies is shown in Fig.1. It consists of an HpGe detector and an electronic system to process the pulses and provide digital data for storage and analysis. The equipments included under the head of electronic system are a pre-amplifier, a spectroscopy amplifier, a multi channel analyzer and a personal computer. Photons from an annular source of ^{241}Am are allowed to fall on the sample under investigation. These photons initiate the inner shell ionization process. The characteristic X-rays emitted as a result of filling up of vacancies in the inner shells by electrons from higher shells, are detected by the HpGe detector. The detected photons produce electronic pulses which are initially amplified by a pre-amplifier. The pulses thus amplified are processed by a spectroscopy amplifier and the output is passed on to a Multi channel analyzer (MCA). The MCA finally provides digital data for storage and further analysis. The digital data is analysed using a Quantitative X-ray Analysis System (QXAS) software package.

The effective incident photon flux ($I_0G\varepsilon$) over a wide energy range is determined and is plotted as shown in Fig - 2

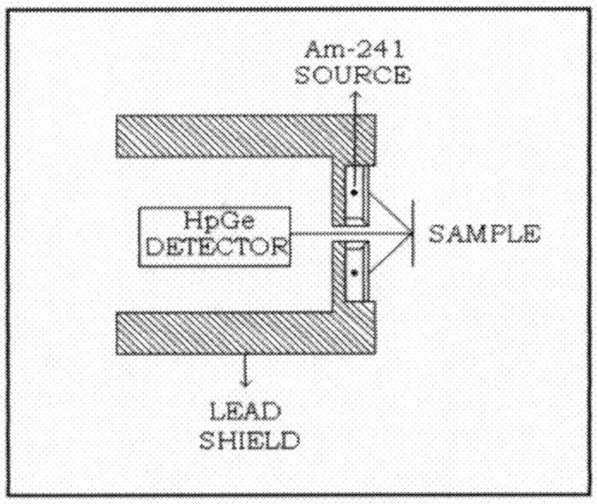

Fig : 1 : Experimental Set up

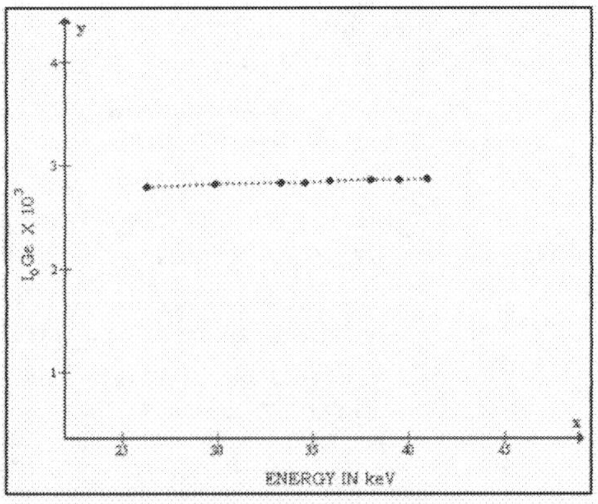

Fig : 2 : Efficiency Vs Energy graph

4. Results and Conclusions:

Since the term $I_0G\varepsilon$ is constant over the required energy range, the present experimental method eliminates the estimation of efficiencies of the detector for K_α and K_β X-rays. K X-ray intensity ratios of Barium were determined in two of its compounds namely Barium Sulphate and Barium Carbonate and the results are as shown in table – I.

TABLE – I: K_β / K_α intensity ratios of Barium

Compound	Present work	Sogut [7]	Theoretical values for Barium	
			Scofield [8]	Scofield [9]
$BaSO_4$	0.2307±0.0058	0.2365±0.019	0.237	0.243
$BaCO_3$	0.2331±0.0058	-	0.237	0.243

The results are presented with an experimental accuracy of 2.5%. For Barium Sulphate, the results are in agreement with the earlier experimental values of Sogut[7] within the experimental errors. However the values deviate with the theoretical estimates of Scofield[8,9] by 0.2% and 2.7% respectively. The intensity ratios of Barium determined in Barium Carbonate deviates with the theoretical values of Scofield[9] by 1.7%. These deviations are very minor to state that there is a necessity to determine X-ray intensity of elements in compound form.

References

[1] Middleman LM, Ford RL, Hofstadter R 1970 *Phys. Rev.* A **2** 1429
[2] Mukoyama T, Taniguchi K, 2000 *X-Ray Spectrometry* **29** 426
[3] Sawhney KJS, Lodha GS, Kataria SK, Kulsreshtha SK, 2000 *X-Ray Spectrometry* **29** 173
[4] Bruner G, Nagel M, Hartmann E, Arndt E, 1982 *J.Phys. B* **15** 4517
[5] Kucukonder A, Sahin Y, Buyukkasap E, Kopya E.1993 *J.Phys. B* **26**,101
[6] Kataria S.K Govil R, Saxena A, Bajpel HN, 1986 *X-Ray Spectrometry* **15** 49
[7] Sogut O, Seven S, Baydas E, Buyukkasap E, Kucukonder A 2001 *Spectrochemica Acta* **B 56** 1367
[8] Scofield J H 1969 *Phys. Rev.* **179** 9
[9] Scofield J H 1974 *Phys. Rev.* **A 9** 1041

Journal of Physics: Conference Series **185** (2009) 012034 doi:10.1088/1742-6596/185/1/012034

Investigating the fragmentation of C_{60} induced by nanosecond laser

D. B. Qian[1], X. Ma[1], D. C. Zhang[1], X. L. Zhu[1], B. Li[1,2], H. P. Liu[1], W. Q. Wen[1,2]
[1] Institute of Modern Physics, Chinese Academy of Sciences, Lanzhou 730000 China
[2] Graduate University of Chinese Academy of Sciences, Beijing 100049 China

E-mail: qiandb@impcas.ac.cn

Abstract. The C_{60} fragmentation induced by nanosecond (ns) laser was studied using a time-of-flight mass spectrometer. The average kinetic energies of fullerene ions C_{60-2n}^{+} ($1 \leqslant n \leqslant 14$) was extracted from the measured full width at half maximum of ion beam profiles. The measured data show approximately linear increase with the mass decreasing and little variation in the present range of laser fluences. To describe the primary formation of fullerene ions, a sequential decay chain including one ionization step and one or several C_2 emission is assumed and a second photo-absorption process in later part of the laser pulse is considered. The experimental results can be explained reasonably by the description of photoion formation.

1. Introduction

Photo-fragmentation of C_{60} has been studied after the discovery of fullerenes by ns laser excitation. Mass spectra of fragment ions C_n^{+} ($n \leqslant 58$) reveal a bimodal distribution with a minimum around n = 30 [1]. Above n = 30, only even-numbered fragment ions are observed. By contrast, both odd- and even-numbered fragment ions are found in the mass spectrum below n = 30. Kinetic energy release (KER) of fragment ions deserves a specific study as it brings additional information on the process of fragmentation. Most work on KER has focused on single dissociation of fullerene ion [2-4]. Until most recent experiment [5], the average kinetic energies of the global decay chain from C_{60}^{+} to C_{44}^{+} have been determined using velocity map ion-imaging technique, and the primary formation mechanism of fullerene ion is assumed to be a sequential decay chain including one ionization step and one or several C_2 emission. However, up to now, no work has been devoted to measuring and calculating the relation between the average kinetic energies of the global decay chain and laser fluences.

In the present paper, we measured the time-of-flight spectrum of fragment ions resulting from the photoexcitation of C_{60}. The relation between average kinetic energies for the fullerene ions C_{60-2n}^{+} ($1 \leqslant n \leqslant 14$) and laser fluences was determined (Section 2). A new theoretical description of fullerene ion formation was presented, and experimental result is in agreement with the theoretical description (Section 3). A general summary is given in section 4.

2. Experiment and results

The experimental set-up combines a vacuum chamber and a time of flight mass spectrometer (TOFMS). The C_{60} powder (purity > 99.5%) is sublimed in a resistively heated oven at 400℃. Optical excitation is achieved using the second harmonic of a Nd:YAG laser. The produced ions are immediately extracted by a static electric field and detected by a MCP detector. The ion signal is recorded with a digital oscilloscope. Fig. 1 displays the TOF spectrum of the fullerene ions produced in the different laser fluences.

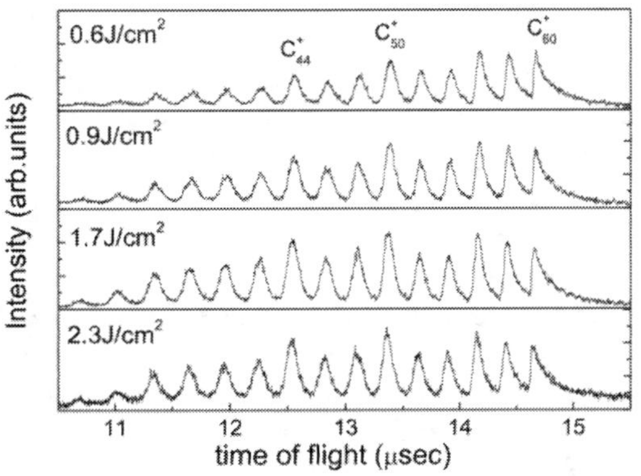

Figure 1. The TOF spectrum of the fullerene ions produced in the different laser fluences

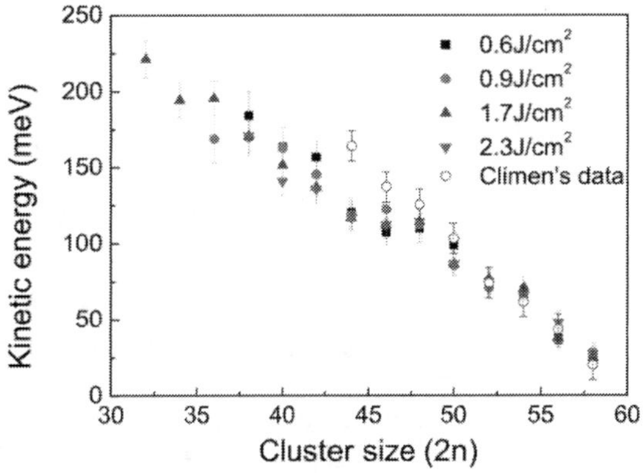

Figure 2. Average kinetic energies of the fullerene ions at various laser fluences

The average kinetic energy induced by fragmentation for each fullerene ion is expressed [6]:

$$\overline{\varepsilon} = \frac{(W_{FWHM}^{total} \times qE_s)^2 / m - 5.538(tg\alpha)^2 k_B T}{8} \tag{1}$$

where W_{FWHM}^{total} is the broadening induced by total initial kinetic energy of ions, T is the temperature of the C_{60} vapour, α is the half angle of divergence of the C_{60} beam, and E_s is the electric field of ionization region. The constants k_B, m and q are respectively the Boltzmann constant, the mass and the charge of ion.

The W_{FWHM}^{total} can be derived from experimental mass peak. The measured peaks shapes of fullerene ions $C_{60\text{-}2n}^+$ ($3 \leqslant n \leqslant 14$) are almost symmetrical as shown in Fig.1. It implies that almost all these fullerene ions are produced promptly. However the peaks shapes of fullerene ions $C_{60\text{-}2n}^+$ (n=1, 2) have obvious asymmetry due to delayed ionization. In order to obtain the broadening, the approach adopted by us is based on the assumption that the sharp mass peak part induced by promptly produced ions has a Gaussian shape [7]. The width of Gaussian fit is regarded approximately as the broadening induced by total initial kinetic energy of promptly produced ions.

The average kinetic energies for fullerene ions $C_{60\text{-}2n}^+$ ($1 \leqslant n \leqslant 14$) are shown in Fig. 2. To compare with other work, Climen's data [5] (open symbol) are also plotted. The two works give similar results. Formation of fullerene ions from C_{60} results from a sequential decay chain. Thus, it is not surprising that their average kinetic energies show approximately linear increase with the mass decrease. It can also be seen that the data show, within the experimental error, little variation with the laser fluences in the present range. This point will be discussed in Section 3.

3. Theoretical description and discussion

To explain the measured average kinetic energies, certain assumption regarding the formation mechanism of fragment ions is necessary. Our results for fullerene ions sustain the idea of formation from C_{60} result from a competition between sequential C_2 emission and delayed ionization. From fig.2 of ref. [5], one can see that C_{60} precursor with temperature about 5000K can decay to C_{54}^+. However the C_{60} lifetime until C_2 evaporation has been found on the time scale of picoseconds at this temperature [8]. That means C_{60} precursor already evaporates one or several C_2 to fullerene ions at certain time during laser pulse. Fullerene ions $C_{60\text{-}2n}^+$ will keep evaporating C_2 after a second photo-absorption from later part of the laser pulse.

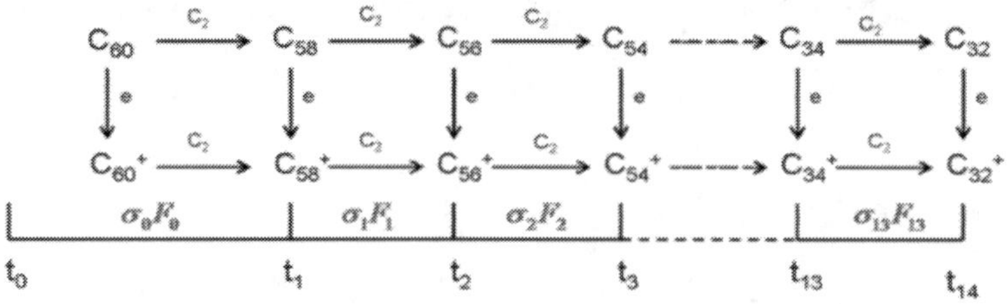

Figure 3. The formation chain of fullerene ion from photo-excited C_{60}

The formation chain of fullerene ion is shown in Fig. 3. There are many pathways to form fullerene ion C_{60-2n}^+. To simplify discussion, we consider here only one pathway corresponding to the ionization of C_{60}. The common characteristic of ns photo-excited C_{60} is that the electronic system is excited on a long time scale compared to the electron-phonon coupling time (ca. 250fs) [9, 10]. Therefore, a general statistical model based on detailed balance is used to describe relaxation processes [11, 12]. The average kinetic energy of C_{60-2n}^+ ($n \leqslant 14$) in the considered formation pathway is given by

$$\bar{\varepsilon}_{60-2n} = \sum_{m=1}^{n} \frac{3}{2} \frac{60-2n}{60-2m} \frac{2}{60-2(m-1)} k_B T_m \tag{2}$$

where T_m is the microcanonical temperature of intermediate ion C_{60-2m}^+ ($1 \leqslant m \leqslant n$). The internal energy of C_{60-2m}^+ is calculated from each step photo-absorption energy, the relevant dissociation and ionization energy. Assuming same photo-absorption cross section between C_{60} molecule and C_{60} ion, the internal energy of C_{60-2m}^+ can be written as

$$E_{C_{60-2m}^+} = \sum_{i=0}^{m-1} (\sigma_i F_i - D_i) - IP \tag{3}$$

and

$$F_i = \begin{cases} F & t_1 - t_0 > w \\ \frac{F}{w}(t_{i+1} - t_i) & t_{i+1} - t_0 \le w \end{cases} \tag{4}$$

where F and w are respectively the laser fluence and the laser pulse duration, σ_i is the photo-absorption cross section of C_{60-2i}^+, D_i and IP are respectively the dissociation energy of C_{60-2i}^+ and the ionization potential of C_{60}, t_0 is the time point of laser pulse arrival, t_i is the time point of formation of C_{60-2i}^+. In Eq. (4), t_{i+1}-t_i is regarded as the C_{60-2i}^+ lifetime until C_2 evaporation. Eq. (3) and (4) show that the internal energy of C_{60-2m}^+ is a function of F and t_{i+1}-t_i. To obtain the qualitative relation between the energy of C_{60-2m}^+ and laser fluence F, we discuss here only single dissociation $C_{60}^+ \rightarrow C_{58}^+ + C_2$. The internal energy of daughter ion C_{58}^+ corresponding to m=1 in Eq. (3) can be written by

$$E_{C_{58}^+} = \begin{cases} \sigma_0 F - D_0 - IP & \tau_0 > w \\ \sigma_0 \frac{F}{w} \tau_0 - D_0 - IP & \tau_0 \le w \end{cases} \tag{5}$$

where τ_0 is the lifetime of C_{60} precursor until C_2 evaporation. The photo-absorption cross section of C_{60} precursor for 532nm can be extrapolated from ref. [13, 14]. The caloric curve of C_{60} precursor in the range of T >1500K is derived from ref. [15], and the lifetime of C_{60} precursor as a function of temperature is derived from ref. [8]. Thus, the relation between internal energy of C_{58}^+ and laser fluence is obtained in Fig. 4.

One remarkable fact is that the energy value does not change significantly after the laser fluence exceeds $0.2 J/cm^2$. In other words, the kinetic energy release during C_2 evaporation from C_{60}^+ does not change significantly with increasing laser fluence after τ_0 is less than laser pulse duration. Internal energy of other intermediate ion C_{60-2m}^+ as a function of laser energy has the same trend because all fullerene ions have similar properties. Therefore, combining Eq. (2), the experimental result about the kinetic energies for fullerene ions C_{60-2n}^+ ($1 \leqslant n \leqslant 14$) regardless of laser fluence can be explained.

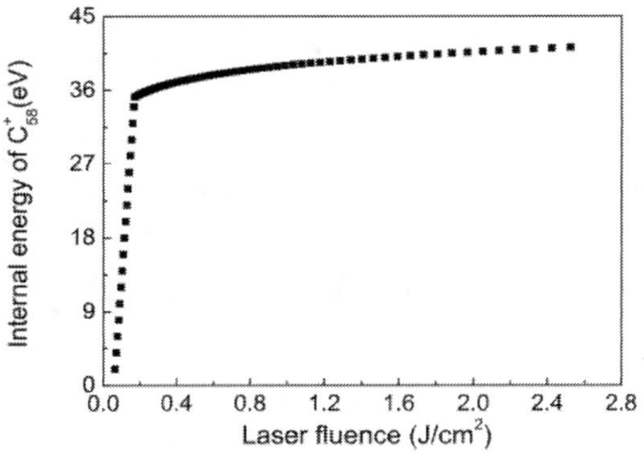

Figure 4. The internal energy of daughter ion C_{58}^+ by nanosecond laser excitation as a function of laser fluences

4. Summary

By carrying out detailed studies of the average kinetic energies for fullerene ions C_{60-2n}^+ ($1 \leqslant n \leqslant 14$) produced from C_{60} excited by nanosecond laser pulses, we have been able to provide convincing evidence that fullerene ion is formed by a sequential decay chain. Considering a second absorption process in later part of the laser pulse, the puzzle about the average kinetic energies of fullerene ions, which show little variation regardless of laser fluences, is solved reasonably.

Acknowledgments
This work is supported by West Light doctoral foundation of The Chinese Academy of Sciences and the National Natural Science Foundation (Grant No.10605034, 10434100, 10504037).

References

[1] P. Scheier, B. Duenser, T.D. Maerk, J. Phys. Chem. 99 (1995) 15428.
[2] K. Guch, S.M. Leubner, O. Echt, et al., J. Chem. Phys. 121 (2004) 2137
[3] K. Guch, S.M. Leubner, O. Echt, et al., Chem. Phys. Lett. 385 (2004) 449
[4] C. Lifshitz, Int. J. Mass Spectrom 198(2000) 1
[5] B. Climen, B. Concina, M.A. Lebeault, et al., Chem. Phys. Lett. 437 (2007) 17
[6] J.L. Franklin, P.M. Hierl, D.A. Whan, J. Chem. Phys. 47 (1967) 3148.
[7] E.E.B. Campbell, G. Ulmer, I.V. Hertel, Phys. Rev. Lett. 67 (1991) 1986
[8] L.A. Openov, A.I. Podlivaev, JETP Lett. 84 (2006) 68
[9] E.E.B. Campbell, K. Hansen, K. Hoffmann, et al., Phys. Rev. Lett. 84 (2000) 2128.
[10] K. Hansen, K. Hoffmann, E.E.B. Campbell, J. Chem. Phys. 119 (2003) 2513.
[11] J.U. Andersen, E. Bonderup, K. Hansen, J. Phys. B: At. Mol. Opt. Phys. 35 (2002) R1.
[12] F. Lépine, C. Bordas, Phys. Rev. A 69 (2004) 053201.
[13] A. Skumanich, Chem. Phys. Lett. 182 (1991) 486.
[14] A. Lassesson, A. Gromov, K. Mehlig, et al., J. Chem. Phys. 119 (2003) 5591.
[15] B. Tsipinyuk, A. Budrevich, M. Grinberg, et al., J. Chem. Phys. 106 (1997) 2449.

Journal of Physics: Conference Series **185** (2009) 012035 doi:10.1088/1742-6596/185/1/012035

Designing of Wave Packets by using Delayed Phase-locked Femtosecond Pulses of Different Durations

Krishna Rai Dastidar[1,a] **and Anindita Bhattacharjee**[2,b]

[1] Department of Spectroscopy, Indian Association for the Cultivation of Science, Kolkata 700032, India

[2] 1035 Aster Av., Apt. 2131, Sunnyvale, CA 94086, USA

E-mail: [a] spkrd@mahendra.iacs.res.in

E-mail: [b] anny_b_2000@yahoo.com

Abstract.

Ab-initio study on the quantum interference of two vibrational wavepackets on $A^1\Sigma^+$ state of NaH molecule excited by two delayed phase-locked 4 femtosecond pulses (pump and control) has revealed that coherent control over populating selective vibrational levels or selective product state can be achieved. We have investigated here how the quantum interference of two wavepackets can be controlled to design a wavepacket which resembles one or two vibrational levels on the excited bound state leading to large overlap with these vibrational levels and hence enhancing the cross section of excitation to selected vibrational levels wiping out the population flow to other nearby levels covered by the bandwidth of the pulses. It will be shown here that the designing of interfering wavepacket can be controlled by controlling the pulse duration and carrier frequency of two delayed phase-locked ultrashort pulses.

1. Introduction

Coherent control of quantum systems has potential applications in various fields like quantum computation and information, coherent cooling, chemical reaction, etc [1]. Wavepacket interferometry is widely used in studies of quantum control. In this technique wavepackets are generated on the excited state using ultrashort pulses and the interference between the wavepackets depend on the delay and relative phase difference between the pulses. By careful choice of the time delay and the phase difference between the pulses, population on the excited state can be enhanced or diminished. In recent experiments it has been shown that quantum interference can be used to extract both amplitude and phase information stored in the wavepackets [2]. Quantum interference between wavepackets can be probed by fluorescence measurement [2-4], ionization [5, 6, 7] or photodissociation [8].

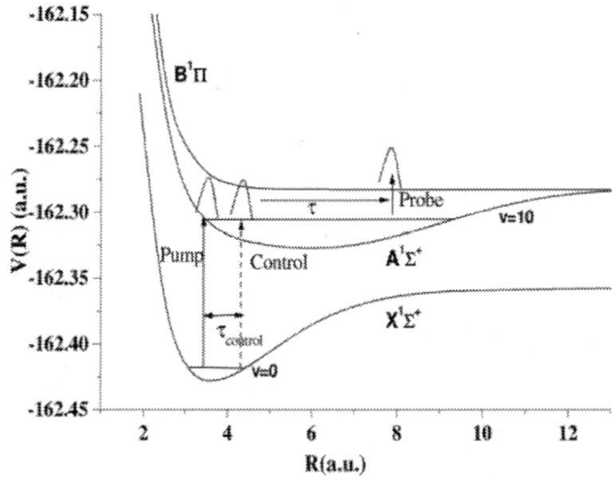

Figure 1. Potential energy curves for the $X^1\Sigma^+$, $A^1\Sigma^+$, $B^1\Pi$ electronic states of NaH molecule. Pump-control-probe scheme to study quantum interference of molecular vibrational wavepackets. See text for details.

In the present study, we have theoretically investigated the role of pulse duration and the carrier frequency on the quantum interference of wavepackets in NaH molecule. Scheme for the present study is shown in figure 1. Two phase-locked and delayed Gaussian femtosecond pulses (pump and control) sequentially excite two wavepackets on the $A^1\Sigma^+$ state of NaH molecule. Two wavepackets are allowed to propagate freely on the excited potential energy surface and to interfere. This interfering wavepacket oscillates on the potential energy surface. Previously we have shown that quantum interference probed by photodissociation is dependent on the delay time and phase difference between the pulses [1, 8]. It was also shown that for a selective delay only even or odd vibrational levels can be populated by changing the phase difference between pulses. In this work it will be shown that further control over the population in selective vibrational levels can be achieved by controlling the pulse duration and the carrier frequency of the ultrashort pulses.

2. Theory

Time dependent Fourier grid method [9, 10] is widely used to calculate total photodissociation cross section for one-photon excitation. We have extended this method to calculate two or three-photon dissociation cross section [8, 11] and vibrational excitation and de-excitation cross section for the bound-bound transition [1, 12]. The theory for vibrational excitation in pump-control method is given below.

By the first pulse, the molecule is excited from ground vibronic state to the intermediate $A^1\Sigma^+$ state. Wavepacket amplitude on the excited surface [1] at time t=T and at internuclear separation R is,

$$(i\hbar)\chi_{e1}(R,T) = \int_0^T dt_1 exp[\frac{-iH_e(T-t_1)}{\hbar}][D_{eg}(R).\epsilon(t_1)]exp[\frac{-iH_g t_1}{\hbar}]\chi_{gv}(R,0) \qquad (1)$$

where $\chi_{gv}(R,0)$ is the eigenfunction of the vibrational level 'v' of the ground state g at time t=0. H_g and H_e are the total Hamiltonian for the ground state 'g' and the excited state 'e' (bound $A^1\Sigma^+$ state) respectively. D_{ge} is the electronic dipole transition moment for transition from the ground state 'g' to the state 'e', T is the duration of the pump pulse. The ground vibrational

Journal of Physics: Conference Series **185** (2009) 012035 doi:10.1088/1742-6596/185/1/012035

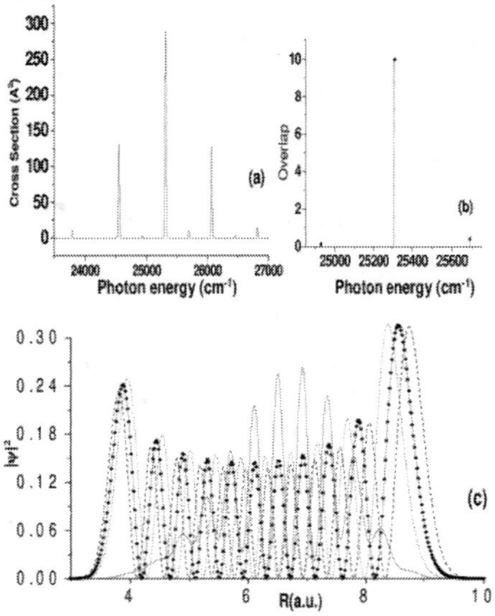

Figure 2. Vibrational excitation spectrum for excitation from $X^1\Sigma^+$ state to $A^1\Sigma^+$ state in NaH molecule, overlap of interfering wavepacket with v=10 level of $A^1\Sigma^+$ state and mod-square of the interfering wavepacket on $A^1\Sigma^+$ state for 10 *fs* pump and control pulses (delay=1.5 T_{vib}, phase difference $\phi = 3\pi/2$) are shown in 2a, 2b and 2c respectively. Scaled vibrational wavefunctions (mod-square) for v=9,10 and 11 vibrational levels (dashed, dotted and dash-dotted curves respectively) of $A^1\Sigma^+$ state have been plotted in figure 2c. Solid line curve shows Wavepacket (mod-square) in figure 2c. Carrier frequency is 25310 cm^{-1}, see text.

wavefunction $\chi_{gv}(R, 0)$ is obtained by using time-independent Fourier Grid Hamiltonian method. After a time interval t' where $t' = \tau_{control} + t_\phi$ where $\tau_{control}$ is the delay between the pump and control pulses and $t_\phi = \frac{\phi}{\omega}$ is the delay corresponding to the relative phase difference ϕ. The control pulse excites another wavepacket χ_{e2} on the $A^1\Sigma^+$ state surface of the same form as in equation (1), with T being replaced by T' (duration of the second pulse). The resultant wavepacket on the $A^1\Sigma^+$ state surface at time $T + t' + T'$ is the sum of these two wavepackets [1]. It is then allowed to propagate for a long time on the $A^1\Sigma^+$ potential energy surface and the excitation cross sections to different vibrationals levels are obtained by the fourier transform of the autocorrelation functions [1]. The propagated wavepackets on the excited state surface is calculated by using Chebyshev scheme [13]. Electric field at time t_i is given by $\epsilon(t_i) = g(t_i)exp[i\omega t_i]$ where $g(t_i)$ is the temporal profile of the pulse and ω is its carrier frequency. We have used Gaussian pulses. For details of calculation see [1].

3. Results and discussion

We have calculated the spectrum for excitation of NaH molecule from v=0 level of the ground $X^1\Sigma^+$ state to the $A^1\Sigma^+$ state by the pump and control pulses. Wavepackets on the $A^1\Sigma^+$ state formed by the pump and control pulses interfere to modify vibrational population in this state depending upon the relative phase (ϕ) as well as the delay ($\tau_{control}$) between these two pulses. It is found that only even vibrational levels (v=8,10,12) are excited by two delayed ($1.5T_{vib}$, T_{vib} = vibrational period of the molecule), phase-locked ($\phi = 3\pi/2$) 10 *fs* pulses (figure 2a). However by changing the temporal width of the pulses to 20 *fs* only v=10 level is excited with larger cross section (figure 3a). This enhancement of the cross section occurs due to the fact that different wavepackets (solid line curve in figures 2c and 3c) are constructed for different pulse durations leading to different overlaps of the composite wavepackets with v=10 level (figures 2b and 3b), although the wavepackets oscillate in phase with v=10 level (line with solid circle) and out of phase with v=9 and 11 levels (dash-dotted and dashed lines respectively)in both the cases. In

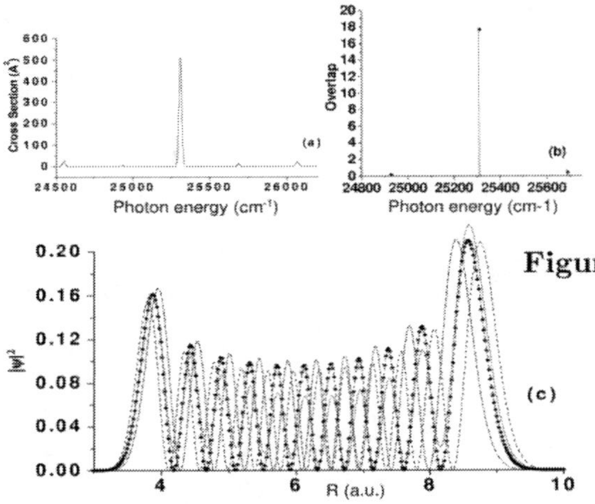

Figure 3. Same as in figure 2 for pulses of duration 20 *fs*

figures 2c and 3c the vibrational wave functions (for v=9,10 and 11 levels) are scaled to make these comparable in magnitude with that of wavepacket. It can be shown that the cross section depends also on the carrier frequency of the pulses.

4. Conclusions

It has been demonstrated theoretically (ab-initio) that a wavepacket can be constructed (by controlling the interference of two wavepackets excited by two delayed phase-locked ultrashort pulses) on an excited electronic state of a molecule (NaH) to resemble one or two vibrational levels of the excited state, thus leading to selectivity in vibrational excitations. This designing of wavepackets can be further controlled by changing the pulse duration and carrier frequency of pump and control pulses leading to enhancement or damping of excited populations in selective vibrational levels.

References

[1] Anindita Bhattacharjee and K. Rai Dastidar, J. Phys. Conf. Series, **80**, 012028 (2007) and references [1-21] therein.

[2] K. Ohmori, H. Katsuki, H. Chiba, M. Honda, Y. Hagihara, K. Fujiwara, Y. Sato and K. Ueda, *Phys. Rev. Lett.* **96**, 093002 (2006)
H. Katsuki, H. Chiba, B. Girard, C. Meier and K. Ohmori, *Science* **311**, 1589 (2006)

[3] N. F. Scherer, R. J. Carlson, A. Matro, M. Du, A. J. Ruggiero, V. Romero-Rochin, J. A. Cina, G. R. Fleming and S. A. Rice, *J. Chem. Phys.* **95**, 1487 (1991)

[4] K. Ohmori, Y. Sato, E. E. Nikitin and S. A. Rice, *Phys. Rev. Lett.* **91**, 243003 (2003)

[5] M. W. Noel and C. R. Stroud, Jr., *Phys. Rev. A* **77**, 1913 (1996)

[6] V. Blanchet, M.A. Bouchene and B. Girard, *J. Chem. Phys.* **108**, 4862 (1998)

[7] C. Petersen, E. Pronne, J. Thgersen, H. Stapelfeldt and M. Machholm, *Phys. Rev. A* **70**, 033404 (2004)

[8] A. Bhattacharjee and K. Rai Dastidar, *Phys. Rev. A* **72**, 023419 (2005)

[9] G.G. Balint-Kurti, R.N.Dixon and C.C. Marston, *Int. Rev. Phys. Chem.* **11**, 317 (1992) and references therein

[10] A. Bhattacharjee and K. Rai Dastidar, *Phys. Rev. A* **65**, 022701 (2002)
A.Bhattacharjee and K.Rai Dastidar, *Pramana* **58**, 569 (2002)

[11] A. Bhattacharjee and K. Rai Dastidar, *J. Phys. B* **36**, 4467 (2003)

[12] A. Bhattacharjee and K. Rai Dastidar, *Mol. Phys.* **104**, 2485 (2006)

[13] R. Kosloff, *J. Phys. Chem.* **92**, 2087 (1988)

Journal of Physics: Conference Series **185** (2009) 012036 doi:10.1088/1742-6596/185/1/012036

A New Way of Broadening the EIT Window: control over Subluminal Group Velocity

Krishna Rai Dastidar[1] **and Sulagna Dutta**[2]

Department of Spectroscopy and Raman Centre for Atomic, Molecular and Optical Sciences, Indian Association for the Cultivation of Science, Kolkata 700032, India

E-mail: [1] `krishna_raidastidar@yahoo.co.in`

E-mail: [2] `sulagna_dutta@yahoo.co.in`

Abstract.

The broadening of the Electromagnetically induced transparency (EIT) window can be achieved in a three level closed Λ system by controlling the strength of incoherent pumping. It has been shown that the presence of homogeneous/inhomogeneous decay facilitates the EIT window broadening. Moreover control over the subluminal group velocity associated with EIT can be achived by manipulating the number density of the molecules. This effect has been demonstrated in a heteronuclear molecule (e.g. LiH).

Keywords: Spontaneously Generated Coherence, EIT, Broadening of EIT Window

Electromagnetically induced transparency (EIT) has drawn tremendous attention both theoretically [1, 2] and experimentally [3, 4] due to its potential applications in many areas of quantum optics. In Λ-type three-level system the spontaneously generated coherence (SGC) due to the interaction with the vacuum bath of the radiation field has been investigated by Javanainen [5]. In the Λ systems observation of EIT can be explained with the help of a concept called coherent population trapping (CPT) [6, 7]. In our previous work, we have shown a new way of achieving EIT due to the destructive interference of spontaneously generated coherence and dynamically induced coherence, when the contribution to the probe response from two coherences are equal and opposite at resonance [8, 9, 10]. We have also shown that the broadening of the EIT window can be achieved by controlling the incoherent pumping rate [11].

In the present work we have shown that the broadening of the EIT window is further enhanced in presence of the homogeneous/inhomogeneous decay and the subluminal group velocity of the probe light for a particular value of the EIT window width can be controlled by manipulating the number density of the molecules.

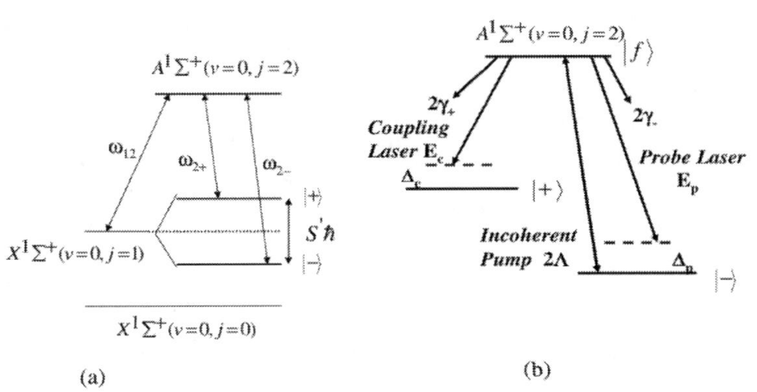

Figure 1. (a) Generation of dressed states ($|+\rangle$ and $|-\rangle$) by using a driving(dressing) laser field to couple the $X^1\Sigma^+(v = 0, j = 0)$ and $X^1\Sigma^+(v = 0, j = 1)$ states. (b) The system with upper state $A^1\Sigma^+(v = 0, j = 2)$ ($|f\rangle$) coupled to the lower dressed states $|+\rangle$ and $|-\rangle$ by the probe and the coherent lasers respectively behaves as a Λ system with antiparallel dipole transition moments.

1. Theory

Our aim is to create a Λ system with parallel or antiparallel dipole moments. Let us apply a laser field between $X^1\Sigma^+(v = 0, j = 0)$ and $X^1\Sigma^+(v = 0, j = 1)$ states of LiH molecule. It then produces dressed states $|+\rangle = \sin\psi|j = 0\rangle + \cos\psi|j = 1\rangle$ and $|-\rangle = \cos\psi|j = 0\rangle - \sin\psi|j = 1\rangle$ with $\cos\psi = \frac{1}{2} + \frac{\Delta_L}{2S'}$, where Δ_L is the detuning of the laser frequency from the molecular transition frequency. The two dressed states are separated by an interval $S'\hbar = \hbar\sqrt{S^2 + \Delta_L^2}$, where S is the Rabi frequency of the transition. For LiH molecule the separation between $X^1\Sigma^+(v = 0, j = 0)$ and $X^1\Sigma^+(v = 0, j = 1)$ states is of the order of 10^2 GHz [12]. Therefore to couple these two levels a maser emitting around 10^2 GHz energy (some water masers, HCN maser [13]) should be used. The energy separation between $|+\rangle$ and $|-\rangle$ levels should be greater than the bandwidth ($<< 1$ GHZ) of the maser and it should be much less than 10^2 GHz i.e. the separation energy between $X^1\Sigma^+(v = 0, j = 0)$ and $X^1\Sigma^+(v = 0, j = 1)$ states.

If $|\mu_{ij}|$ be the dipole moment matrix element for the transition, summed over the three perpendicular directions in space [14], we have $|\mu_{ij}|^2 = \mu^2\frac{(j+1)}{(2j+1)}$ for the transition $j + 1 \leftarrow j$ and $|\mu_{ij}|^2 = \mu^2\frac{(j+1)}{(2j+3)}$ for the transition $j + 1 \rightarrow j$. For $X^1\Sigma^+(v = 0, j = 0)$ to $X^1\Sigma^+(v = 0, j = 1)$ transition, $|\mu_{01}| = \mu$, where μ is the electric dipole moment of the LiH molecule ($\approx 5.866D$) at equilibrium nuclear distance [15].

Let us consider $A^1\Sigma^+(v = 0, j = 2)$ state of LiH as the upper excited level of a Λ system and the dressed states $|+\rangle$, $|-\rangle$ are coupled with the upper excited level by a coherent field and a probe field respectively (see fig.1). We find that the dipole matrix elements between the dressed states and the upper excited level are $d_{2+} = d_{12}\cos\psi$ and $d_{2-} = -d_{12}\sin\psi$, where d_{12} is the dipole transition moment between $A^1\Sigma^+(v = 0, j = 2)$ and $X^1\Sigma^+(v = 0, j = 1)$ states. Thus, the system with upper state $A^1\Sigma^+(v = 0, j = 2)$ and the lower dressed states $|+\rangle$, $|-\rangle$ behaves as a Λ system with antiparallel dipole moments. The spontaneous decay widths (γ_+ and γ_-) can be calculated from the dipole transition moments (d_{2+} and d_{2-}) and the frequencies of transition [10]. This system has the advantage that the magnitudes of the dipole transition moments (d_{2+} and d_{2-}) and the energy separation between the dressed states (S') can be controlled by the

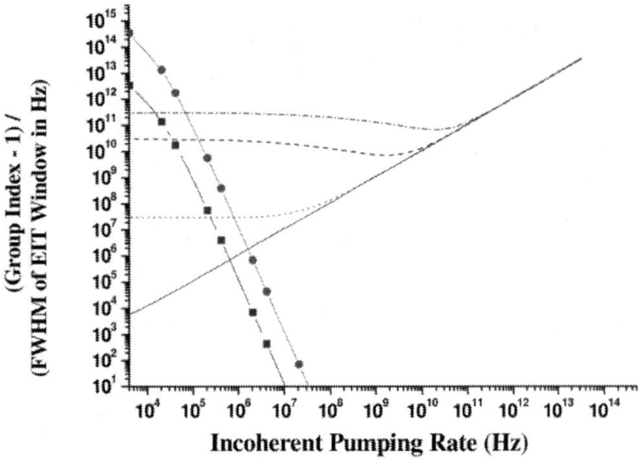

Figure 2. Dependence of FWHM (Full Width at Half Maximum) of the EIT window and $(n_g - 1)$ on incoherent pumping rate for $\Omega = 10^{-13}$ a.u. For FWHM curves, ——: $\gamma_0 = 0$, $\cdots\cdots$: $\gamma_0 = 10^{-8} a.u.$, - - - -: $\gamma_0 = 10^{-7} a.u.$ and — · —: $\gamma_0 = 10^{-6} a.u.$. For $(n_g - 1)$ curves, ■: N=10^{16}/ cm^3 and ● : N=10^{18}/ cm^3

Rabi frequency (S) and the detuning of the driving laser field (Δ_L).

Here the density matrix equations can be derived from the master equation without any restriction over the direction of field polarizations. These density matrix equations for closed three level Λ system with probe field, coherent field and the bidirectional incoherent pumping (which acts on $A^1\Sigma^+(v = 0, j = 2)(|f\rangle) \rightarrow |-\rangle$ transition) have been solved exactly for the coherences and populations in the steady state limit keeping all the higher order terms for the system parameters and without neglecting any term or imposing any restrictions over the values of system parameters [8, 9]. This will facilitate the realization of such scheme in realistic (atomic/molecular) systems. The exact analytical expression for the probe field response at $\Delta_p = \Delta_c = 0$ has been shown in our previous work to show how EIT can occur due to the interplay between two types of coherences [10, 11]. Hence the exact analytical expressions for the susceptibility ($\chi = \chi' + i\chi''$), group velocity (v_g) and group index ($n_g = c/v_g$) can be obtained by using the exact analytical derivatives of coherences (see equations 14-24 in ref. [10]).

2. Results and Discussions

Fig.2 shows the variation of FWHM (full width at half Maximum) of the EIT window and $(n_g - 1)$ as a function of incoherent pump rate in LiH molecule. The solid line indicates the variation of FWHM in the absence of homogeneous/inhomogeneous decay for $\Omega = 10^{-13}$ a.u and G= 1.7323×10^{-13}. FWHM increases linearly with the increase in the incoherent pumping rate in absence of homogeneous/inhomogeneous decay indicating the lossless and distortionless transmission of probe light. It is well known that the condition for the distortionless pulse propagation is that the spectral width of the probe pulse should be contained within the transparency window of the medium [16]. If the pulse is short i.e. its spectrum is broad relative to the transparency window of the medium, absorption and also the higher order dispersion need to be taken into account. But if the transparency window is broad, short pulses with large bandwidth can be propagated without dispersion and absorption. For further increase in γ_0 (see dotted, dashed and dashed-dot curves), G should be increased accordingly to get EIT, e.g. for $\gamma_0 = 10^{-8}$ a.u, $\Omega = 10^{-13}$ a.u and G= 6.6797×10^{-9}. For a fixed value of the incoherent pumping rate, FWHM increases with γ_0. It has been found that the

EIT window is wider in presence of the homogeneous/inhomogeneous decay than that in the absence of it. This indicates that the presence of the inhomogeneous/homogeneous broadening facilitates the widening of the EIT window. A closer inspection reveals that initially the FWHM curves are almost flat in presence of the homogeneous/inhomogeneous decay. However with further increase in incoherent pumping rate, all the curves merge together with that obtained in absence of the homogeneous/inhomogeneous broadening and increase with the incoherent pumping rate together. It happens due to the fact that the incoherent pumping dominates over the homogeneous/inhomogeneous decay which leads to the suppression of the effects of these decays. The solid lines with circle and square show the variation of $(n_g - 1)$ (where n_g is the group index) at the resonance as a function of the incoherent pump rate for two different value of the number density of molecules. These curves indicate that with the increase in the incoherent pump rate, the subluminal group velocity of probe light increases and approaches to the velocity of light, but does not exceed it. However by controlling the number density of the molecules, the superluminocity of the probe light can be increased for a fixed value of the incoherent pumping rate. As the homogeneous/inhomogeneous dacay width is much larger than the collisional decay width, the inclusion of collisional broadening has no effect on the results in these cases.

3. Conclusion

It has been shown that the SGC in heteronuclear molecule (LiH) can be invoked by an external field and this scheme can be applied to achieve broadening of the EIT window by controlling the incoherent pumping rate. This broadening is further enhanced in presence of the homogeneous/inhomogeneous decay for the moderate values of the incoherent pumping rate. Moreover it has been shown that by increasing the number density of molecules, the subluminocity of the probe light can be increased for a particular value of the width of the EIT window, thus allowing a lossless transmission of the probe light, slower than the velocity of light.

References

[1] Scully M O 1991 *Phys. Rev. Lett* **67** 1855
[2] Marangos J P et al 1998 *J. Mod. Opt.* **45** 471
[3] Kasapi A, Jain M, Yin G Y and Harris S E 1995 *Phys. Rev. Lett.* **74** 2447
[4] Boller K J, Imamolu A and Harris S E 1991 *Phys. Rev. Lett.* **66** 2593
[5] Javanainen J 1992 *Europhys. Lett.* **17** 407
[6] Alzetta G, Gozzini A, Moi L and Orriols G 1976 *Nuovo Cimento B* **36** 5
[7] Arimondo E 1996 *Progress in Optics*, edited by E.Wolf (North Holland, Amsterdam) **35** 257
[8] Dutta S and Rai Dastidar K 2006 *J.Phys.B: At.Mol.Opt.Phy.*, **39**,4525-38
[9] Dutta S and Rai Dastidar K 2007 *J. Phys.: Conf. Series* **80** 012030
[10] Dutta S and Rai Dastidar K 2007 *J.Phys.B: At.Mol.Opt.Phys.* **40**, 4287
[11] Rai Dastidar K and Dutta S 2008, *Euro. Phys. Lett* **82** 54003
[12] Bhattacharjee A, Sanyal S and Rai Dastidar K 2005 *J. Mol. Spect.* **232** 264
[13] www.en.wikipedia.org/wiki/Maser, www.britannica.com/eb/article-4678/maser
[14] Townes C H and Schawlow A L 1955 Microwave Spectroscopy, McGraw-Hill Book Company, INC.
[15] Haley Lawrence V. and Hameka Hendrik F 2004 Internation Journal of Quantum Chemistry, **11** 733
[16] Dey T N and Agarwal G S 2003 *Phys. Rev* A **67** 033813

Journal of Physics: Conference Series **185** (2009) 012037 doi:10.1088/1742-6596/185/1/012037

Two-dimensional fluorescence spectroscopy for the identification of discharge intermediates

NJ Reilly[1], M Nakajima[1], TP Troy[1], DL Kokkin[1], N Chalyavi[1], K Duncan[1], K Nauta[1], RG Sharp[2], TW Schmidt[1] and SH Kable[1,3]

[1] School of Chemistry, University of Sydney, Sydney, NSW, 2006, Australia

[2] Anglo-Australian Observatory, PO Box 296, Epping, NSW, 2710, Australia

E-mail: s.kable@chem.usyd.edu.au

Abstract. A protocol is developed for the identification of discharge intermediates from complex electronic spectra. The protocol involves resonant ionisation mass spectrometry to determine the mass of the unknown carrier followed by a two-dimensional fluorescence spectrum which is used as a "map" to guide higher resolution techniques. High resolution dispersed fluorescence provides accurate ground electronic state vibrational frequencies which are then compared with theoretical calculations of frequencies and structures of a range of isomers consistent with the observed mass. The protocol is demonstrated in the first identification of the *cis* and *trans*-1-vinylpropargyl radicals as a ubiquitous radical in hydrocarbon discharges.

1. Introduction

Energetic chemical environments, such as flames, discharges and plasmas, produce a wealth of reactive and metastable species. These transient molecules, radicals and ions participate freely in chemical reactions and are integral to the correct elucidation of complex reaction mechanisms. One motivation of our research program is to identify new free radical species that are important intermediates in combustion and discharge chemistry. We use two strategies to seek these new radicals: *i)* discharge simple precursors and let the chemistry dictate what species we observe, and *ii)* hypothesise which radicals might be important, then judiciously choose a related precursor that, upon mild discharge, is likely to provide the relevant radical. Frequently, as explained below, these two strategies are applied in tandem.

The complex menagerie of molecular species present in a discharge is largely preserved when quenched by supersonic expansion. The optical spectrum of the mixture is also complex, particularly throughout the visible and near-ultraviolet region of the spectrum, where many organic free radicals absorb. Small species such as C_2, C_3, and CH are very abundant. Their small partition function, large absorption cross-section and high fluorescence quantum yield means that they tend to swamp the spectral features of the more complex but less abundant species that we seek.

In 2006, we measured a new spectrum from a benzene discharge that piqued our interest due to its proximity to the diffuse interstellar band at 4762 Å. It took us nearly 18 months to positively identify and characterise the carrier of this spectrum as the 1-phenylpropargyl radical.[1] During this process we developed a combined experimental/theoretical protocol that captured the essential steps used in

[3] To whom any correspondence should be addressed.

this assignment. The protocol employs two 2D spectroscopic techniques, plus conventional fluorescence spectra and *ab initio* calculations to confirm the identity of the carrier.

In this paper we describe the protocol, which has now proven successful in the unambiguous identification of four other new radical species, all present in discharges. It has also allowed us to speed up the assignment process enormously.

2. Experimental

The experiments were carried out in two custom-made spectrometers, each of which has been described in detail previously.[1] In brief, each spectrometer had a pulsed nozzle with high voltage electrodes attached at the orifice. A sample comprising the room temperature vapour of an organic compound (benzene, toluene, 1-hexyne [Aldrich, as provided]), diluted in 2 – 5 bar of Ar, was expanded through the nozzle orifice and thence between the electrostatic plates. The plates were pulsed at 600 – 2000 V, for a short time within the gas pulse (typically 100 μs) to strike the discharge. The discharge chemistry was quickly quenched as the gas expand into a vacuum chamber (10^{-5} torr with nozzle pulsing). The discharge products could be controlled by varying the discharged conditions (pressure, voltage, length of discharge).

In the ionisation spectrometer, the free jet expansion was skimmed before entering another vacuum chamber (10^{-7} torr) equipped with a Wiley-McLaren time-of-flight mass spectrometer. Neutral discharge products were ionised by resonant 2-colour 2-photon ionisation (R2C2PI) using a tunable laser to resonantly excite a higher electronic state, followed by a fixed frequency (266/193/157 nm) laser to ionise from the intermediate state. The *m/z* ratio of the resultant ion was measured by accelerating it through the ToFMS. In the fluorescence spectrometer, the unskimmed, cooled discharge gases were probed by 2D fluorescence spectroscopy [2] or conventional laser induced fluorescence (LIF) excitation and dispersed fluorescence spectroscopy.

The tunable laser systems for both spectrometers were chosen from a YAG-pumped dye laser, YAG-pumped OPO or excimer-pumped dye laser. The fixed frequency ionisation laser was either an excimer laser (157/193 nm) or frequency quadrupled YAG (266 nm).

3. Results

3.1. Protocol for identification of unknown discharge intermediates

Over the past two years, we have developed a 5-step protocol that allows us to untangle the complex spectra and focus in on single species:

a) Measure a 2D fluorescence (2DF) spectrum of the discharge products. 2DF is a technique that we reported in 2006 for the specific purpose of intuitively separating emission features from different carriers [2]. It utilises a tunable excitation laser (usually a low resolution OPO) and a intensified CCD spectrograph to measure the dispersed fluorescence spectrum at every step of the laser. Electric discharge spectra are dominated by C_2 and C_3 bands, however this 2DF technique allows us to intuitively isolate features that are due to unidentified species. A similar technique, using laser Fourier transform spectroscopy instead of dispersed fluorescence was developed by Bondebey [3].

b) Find the matching spectrum in the R2C2PI-time-of-flight mass spectrometer. R2C2PI-ToF mass spectroscopy provides a separate resonant spectrum for every mass channel in the mass spectrometer. Measurement of the same resonant spectrum in the R2C2PI spectrometer as the fluorescence spectrometer provides the mass of the carrier. Steps *a)* and *b)* are independent and we have performed them in either order. It depends on where we first find a new and interesting spectrum.

c) Measure high resolution excitation and dispersed fluorescence spectra. Without the 2DF map in step *a)*, the spectral "space" occupied by separate high resolution techniques is huge and would take an enormous time to cover. However, using the 2DF spectrum as a map, we can concentrate on the unassigned fluorescence features using high resolution lasers, and high resolution spectrometers.

Dispersed fluorescence spectra provide ground state vibrational frequencies, which are crucial for distinguishing different isomers and for comparing with quantum chemical calculations.

d) Perform quantum chemical calculations of the candidate radicals. Knowledge of the mass usually allows us to determine the empirical chemical formula of the unknown radical, although not the structure. We perform either *ab initio* or density functional theory (DFT) calculations of the structure and vibrational frequencies of a series of isomers of the appropriate mass. *Ab initio* and DFT techniques are very reliable for ground state frequencies and structures. These frequencies, coupled with the experimental frequencies afforded by high resolution dispersed fluorescence spectra allows us to distinguish between the likely candidates.

e) Confirm the identity by mildly discharging a chemically similar precursor. Once a likely carrier is hypothesised, we select a chemically similar, closed-shell precursor – typically one that is simply hydrogenated at the radical site – and re-measure the spectrum. This mild discharge should abstract a H-atom to create the proposed radical in far greater abundance, thereby confirming the assignment.

3.2. Example – vinylpropargyl radical (C_5H_5)

While exploring a discharge of 1-hexyne, we observed features spanning 450-470 nm in the LIF spectrum. Following up by R2C2PI confirmed that the carriers of these bands had a mass corresponding to *m/z*=65. This mass-to-charge ratio corresponds to the empirical formula, C_5H_5, given the pre-cursor composition. Previous R2C2PI studies had assigned the carrier of some of these spectral features to the 3-vinylpropargyl radical [4] (see figure 1). However, other features at this mass were unassigned. To discover the identity of this unassigned isomer, and better determine which features belonged to which carrier we measured a 2DF spectrum which is shown in figure 2. As usual, strong C_2 and C_3 features are evident; these are circled in the figure. Vertical integration of

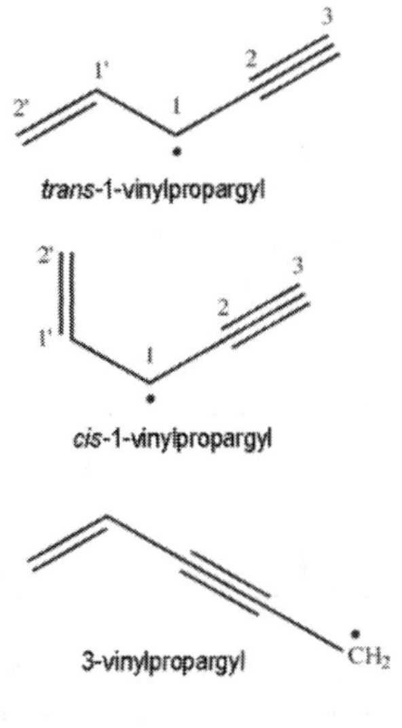

Figure 1. Chemical structure of several C_5H_5 isomers.

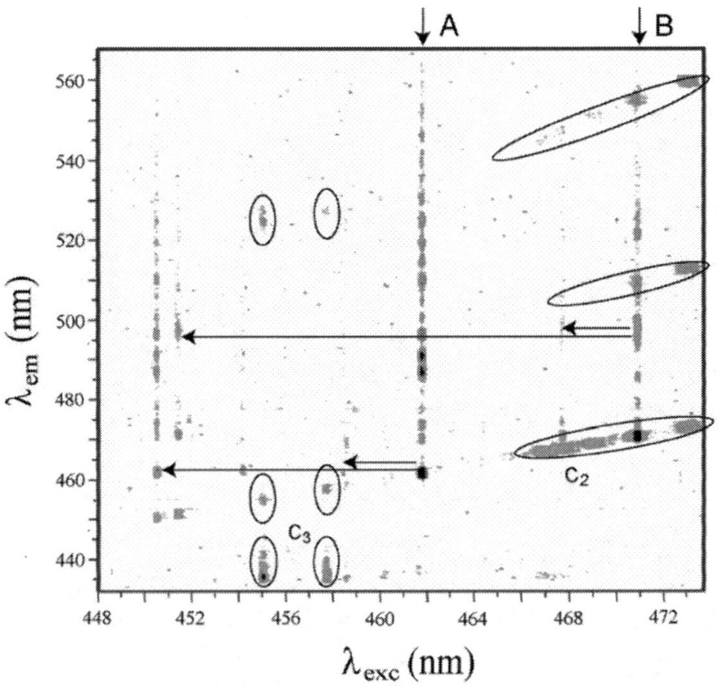

Figure 2. 2DF spectrum of a hexyne discharge. The circled structures are due to C_2 and C_3. The next brightest species are the *trans*- and *cis*-1-vinylpropargyl radicals (marked by A and B respectively).

this 2D spectrum yields the conventional fluorescence excitation spectrum, in which the very strong C_2 features completely obscure the finer features marked by "B" in the same region. Vertical slices of the 2DF spectrum correspond to dispersed fluorescence spectra at each excitation position. Using this 2D map as a guide, we rescanned the fluorescence excitation spectrum at high resolution (figure 3), choosing to record emission only between 480 and 500 nm. This removes all of the C_2 and most of the C_3 emission, leaving a relatively clean LIF spectrum of mostly the C_5H_5 carrier.

Even in the low resolution 2DF spectrum, it is apparent that the two strongest fluorescence spectra (A and B) are quite different. Scanned at higher resolution in figure 4, the dispersed fluorescence spectra after exciting these bands (A′ and B′ in figure 4) show quite different vibrational frequencies and intensities; clearly they belong to two different species. We performed DFT calculations of the vibrational frequencies of several C_5H_5 isomers, including the three in figure 1. To our surprise, the agreement between the previously proposed carrier, 3-vinylpropargyl, and the experimental frequencies from either spectrum A′ or B′ was poor. However, the agreement between the frequencies of A′ and *cis*-1-vinylpropargyl, and B′ with its *trans* isomer was excellent, particularly in the region of the characteristic low frequency modes. Analysis of the spectra is ongoing and we will report the full details of the calculations and spectral assignments in a forthcoming publication.

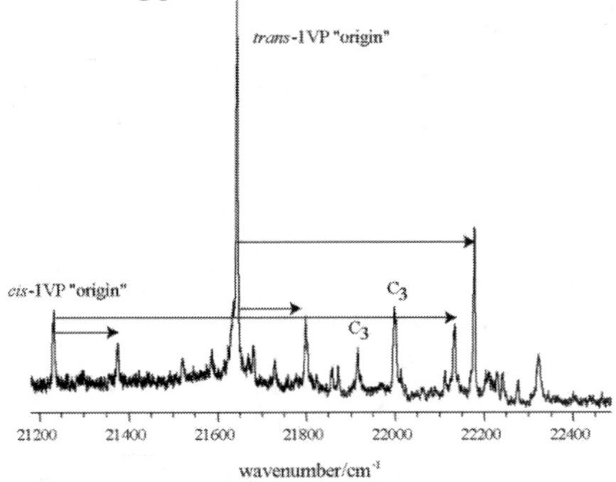

Figure 3. High resolution LIF spectrum. The arrows represent the same transitions as figure 2.

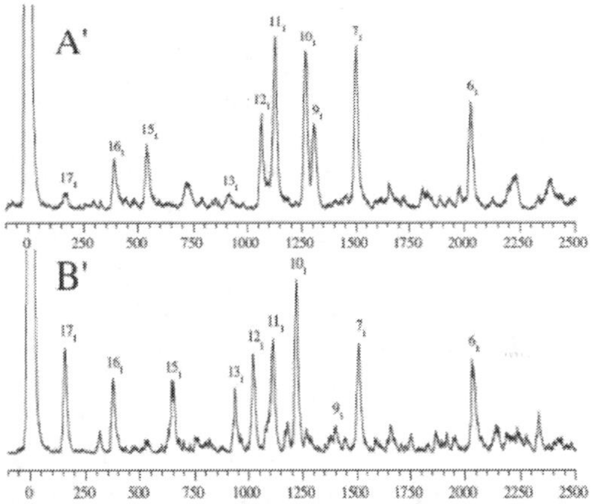

Figure 4. High resolution dispersed spectra after excitation of the *cis*- and *trans*-1VP origin bands.

Finally, we re-measured the spectrum using a tailored precursor. The same spectrum appears from the electric discharge of many pre-cursors showing that it is a ubiquitous component in hydrocarbon discharge chemistry. The signal is strongest in the pent-1-yn-4-ene discharge. Therefore, we assign these spectra to the *cis* and *trans* isomer of the 1-vinylpropargyl radical.

In summary, we have used the protocol described in this paper to assign, unambiguously, the electronic spectrum of the 1-vinylpropargyl radical for the first time. Use of the protocol meant that we were able to achieve this in much shorter time than we took for the phenylpropargyl radical, for which this protocol was developed. Since this work on vinylpropargyl, we have also identified the indanyl and naphthylmethyl radicals in hydrocarbon discharges, which are formed from proton addition to indene, and proton abstraction from methylnaphthalene.

[1] Reilly NJ, Kokkin DL, Nakajima M, Nauta K, Kable SH and Schmidt TW 2008 *JACS* **130** 3137
[2] Reilly NJ, Schmidt TW and Kable SH 2006 *J Phys Chem A* **110** 12355
[3] Bondybey VE, Smith AM and Agreiter J 1996 *Chem Rev* **96** 2113
[4] Ding H, Boguslavskiy A E and Maier J P 2004 *Phys Chem Chem Phys* **7** 888

Journal of Physics: Conference Series **185** (2009) 012038 doi:10.1088/1742-6596/185/1/012038

Fragmentation of positronium in collision with singly ionized lithium

S Roy and C Sinha[1]

Theoretical Physics Department, Indian Association for the Cultivation of Science, Jadavpur, Kolkata - 700032, India.

E mail: chand_sin@hotmail.com

Abstract. Fragmentation of ground state ortho Positronium (Ps) in collision with Li ion (Li^+) including the electron loss to the continuum (ELC) are studied in the framework of post collisional coulomb distorted eikonal approximation (CDEA) for the target elastic case . The present model takes account of the two center effect on the ejected e which is crucial for a proper description of the projectile ionization involving an ionic target, particularly for the explanation of the shifting of the ELC DDCS peak.

1. Introduction

Electron emission process in atom – atom or ion - atom collisions becomes particularly interesting at the same time complex when a structured projectile loses electron in collision with the target. Most of the earlier experiments [1, 2] and consequently theories [3] on the projectile ionization concentrated on bare , partially stripped or neutral heavy projectiles [1] for which a distinct signature of a cusp / peak was observed in the emitted electron energy spectrum at around $\vec{\upsilon}_e$ (velocity of the electron) $\approx \vec{\upsilon}_p$ (velocity of the positron). This peak was attributed to the electron loss from the projectile ion / atom into its low - lying continuum, usually referred to as the electron loss peak (ELP). With the advent of mono energetic energy tunable positronium (Ps) beams [4], attention is also being focused both experimentally [5] and theoretically [6,7] on the breakup process of the projectile Ps atom.

The present work addresses the extension of our earlier work [6] to a more complex system, e. g., the two electron (helium like) ionic target , Li^+ for the target elastic case ;

i.e., $$\mathrm{Ps}\,(1s) + \mathrm{Li}^+(\vec{r}_3, \vec{r}_4) \rightarrow e^+(\vec{r}_1) + e^-(\vec{r}_2) + \mathrm{Li}^+ \quad (1)$$

One major advantage of the Li^+ target (over He^+) is that the electrons of the former are much more tightly bound than the electron of the Ps atom and as such the probability of the electron loss from the projectile Ps is expected to be much higher than the ionization of the target. Further, in view of the large excitation energy of Li^+ as compared to the projectile Ps, we have neglected any virtual or real excitation of the Li^+ target during the fragmentation, i. e., only the target elastic case is considered.

2. Theory:

The prior form of the ionization amplitude for the process (1) is given as [6]:

$$T_{if}^{prior} = \left\langle \Psi_f^-(\vec{r}_1, \vec{r}_2, \vec{r}_3, \vec{r}_4) \right| H - E \left| \psi_i(\vec{r}_1, \vec{r}_2, \vec{r}_3, \vec{r}_4) \right\rangle \quad (2)$$

The prior interaction V_i is given by;

$$V_i = \frac{Z_t}{r_1} - \frac{Z_t}{r_2} - \frac{1}{r_{13}} + \frac{1}{r_{23}} - \frac{1}{r_{14}} + \frac{1}{r_{24}} \tag{3}$$

The initial asymptotic state ψ_i in Eq. (1) is given as:

$$\psi_i = \phi_{Ps}(\,|\,\vec{r}_1 + \vec{r}_2\,|\,)\, e^{i\,\vec{k}_i.\vec{R}}\, \phi_{Li^+}(\,\vec{r}_3,\ \vec{r}_4\,) \tag{4}$$

where $\vec{R} = \frac{\vec{r}_1 + \vec{r}_2}{2}$ and $\vec{k}_i$ is the initial momentum of Ps atom with respect to the target nucleus. The wave function of Li $^+$ (ϕ_{Li^+}) is chosen to be due to Morse et al [8] as well as due to Clementi & Roetti [9].

The final channel wavefunction Ψ_f^- in Eq. (1) satisfying the incoming wave boundary condition is constructed taking into consideration that the ejected electron is in the combined (attractive) fields of the two positive ions, e.g., its parent ion (e^+) and the target ion Li^+ and is approximated by the following ansaltz in the framework of Coulomb modified eikonal approximation [6]:

$$\Psi_f^-(\vec{r}_1,\ \vec{r}_2,\ \vec{r}_3,\ \vec{r}_4) = (2\pi)^{-3}\, \phi_{Li^+}(\vec{r}_3,\ \vec{r}_4)\, e^{i\,k_1.\vec{r}_1}\, e^{i\,k_2.\vec{r}_2}\ {}_1F_1(-i\alpha_2, 1,\ -i(k_2\, r_2 + \vec{k}_2.\ \vec{r}_2)) \times \exp$$

$$\left\{ i\eta_1 \int_z^{\infty} \left(\frac{1}{r_1} - \frac{1}{r_{12}}\right) dz' \right\} \tag{5}$$

with $\alpha_2 = -\frac{(Z_t - 2)}{k_2}$ and $\eta_1 = \frac{(Z_t - 2)}{k_1}$ $\vec{k}_1$, $\vec{k}_2$ being the final momenta of the scattered positron and the ejected electron respectively with respect to the target nucleus.

Following our earlier work [6], the break up amplitude T_{if} can be finally reduced to a two dimensional numerical integral [6, 10]. The triple (TDCS) and double (DDCS) differential cross sections for the process (1) are given as:

$$\frac{d^3\sigma}{dE_2\, d\Omega_1\, d\Omega_2} = \frac{k_1\, k_2}{k_i} \left| T_{if} \right|^2;$$
$$\frac{d^2\sigma}{dE_2 d\Omega_1 (d\Omega_2)} = \frac{k_1\, k_2}{k_i} \int \left| T_{if} \right|^2 d\Omega_2\ (d\Omega_1) \tag{6}$$

3. Results and Discussions

The fully (triple) differential (TDCS) and the double differential cross sections (DDCS) are computed for the process (1).

The TDCS in figure1 exhibits a broad electron loss peak (ELP) for forward emission (0^0) of both the e & the e^+ in contrast to the sharp ELP cusp around 0^0 for heavy ion impact [1] . This could be attributed to the probability of deflection of the light particle (e^+) to higher angles in contrast to the heavy projectile which is predominantly scattered in the forward direction (0^0). A prominent secondary peak appears particularly at $\theta_1 = 45^0$, that could be associated with the famous Thomas (p – n – e) mechanism [11] in charge transfer problems. The peak becomes more and more prominent with increasing incident energy, indicating the importance of the higher order effects at higher incident energies.

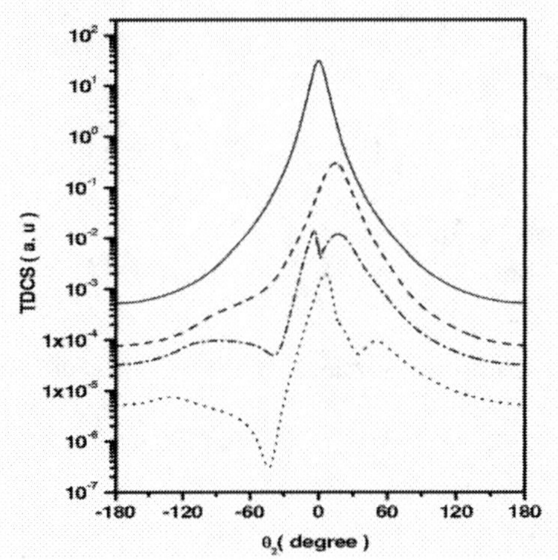

Figure 1. TDCS against the ejected e angle (θ_2) for different values of the scattered e^+ angle (θ_1). The incident energy (E_i) =50 eV, ejected e^- energy E_1 = scattered e^+ energy E_2=21.6 eV. Solid curve for $\theta_1 = 0^0$, dashed curve for $\theta_1 = 20^0$, dashed dot - dot curve for $\theta_1 = 30^0$, and the dotted curve for $\theta_1 = 45^0$.

In figure 2, the DDCS ($\sum \theta_1, \phi_1$) exhibits a broad peak, slightly shifted towards higher ejection energy with respect to half of the residual energy ($E_{res}/2$, $E_{res} = E_i - E_1 - 6.8$). However, this shift decreases and moves towards $E_{res}/2$ with increasing θ_2 and could be attributed to the post collisional two center effect [10,11].

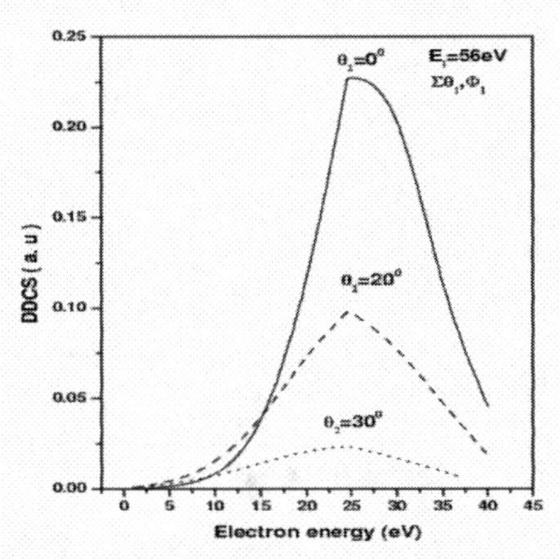

Figure. 2 The electron DDCS ($\Sigma\theta_1, \phi_1$) against the ejected e energy for different values of θ_2 at an incident energy E_i = 56 eV. The solid curve is for $\theta_2 = 0^0$, dashed curve is for $\theta_2 = 20^0$ and the dotted curve is for $\theta_2 = 30^0$.

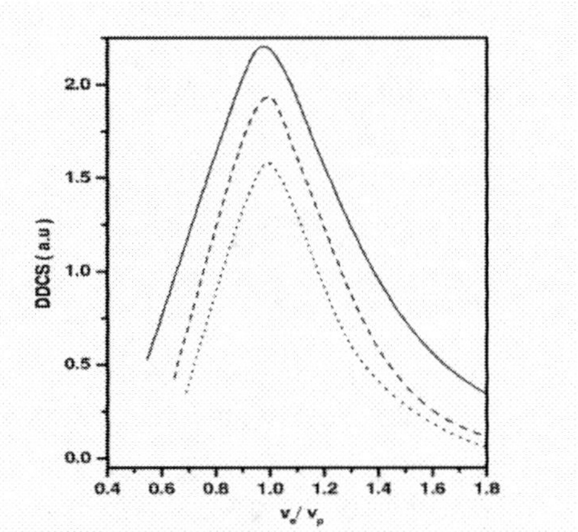

Figure 3. The positron DDCS ($\Sigma\theta_2, \phi_2$) against the ratio of $R = \vec{v}_e / \vec{v}_p$ for three different incident energies and for $\theta_1 = 0^0$. The solid curve is for E_i = 50 eV, dashed curve is for E_i = 75 eV and the dotted curve is for E_i = 100 eV.

Figure 3 demonstrating the e^+ DDCS ($\sum \theta_2, \phi_2$) against the ratio R = $\vec{\upsilon}_e / \vec{\upsilon}_p$ indicates that the shifting of the peak arising due to the post collisional two center effect moves gradually towards *R*=1 with increasing incident energy.

4. Conclusions

The angular distribution (TDCS) of the e exhibits a broad ELP at around $E_{res}/2$, unlike the sharp ELP cusp around 0^0 for heavy ion impact. The notable shift of the electron DDCS peak (towards higher momenta) from its standard position ($\vec{\upsilon}_e = \vec{\upsilon}_p$)) could be attributed to the post collisional two center effect. The e^+ DDCS shows exactly the reverse behaviour as is expected. The position of the $e^+(e)$ DDCS peak shifts gradually towards higher (lower) value of the ratio R = $\vec{\upsilon}_e / \vec{\upsilon}_p$ with increasing incident energy.

5. References

[1] Shah M B, McGrath C , Illescas Clara, Pons B , Riera A, Luna H., Crothers D S F , Rourke S F C O' and Gilbody H B 2003 *Phys. Rev. A.* **67** 010704 (R) .
[2] Misra D, Kelkar A H., Khadane U, Kumar A, Singh Y P, Tribedi L C, Fainsten P D 2007 *Phys. Rev A* **75** 052712.
[3] Barna I F, Gagyi-Palffy A C, Gulyas L, Tokesi K and Burgdorfer J, 2005 *Nucl. Instrum. Methods Phys. Res. B.* **233** 176 .
[4] Garner A J, Ö zen A and Laricchia G 2000 *J. Phys B.* **33** 1149 .
[5] Armitage S , Leslie D E, Garner A J and Laricchia G 2002 *Phys. Rev. Lett* **89,** 173402 .
[6] Roy S , Ghosh D and Sinha C 2005 *J. Phys. B*: **38** 2145.
[7] Starrett C, McAlinden Mary T and Walters H R J 2008 *Phys. Rev. A.* **77**, 042505.
[8] Morse P M, Young L A and Haurwitz E S 1935 *Phys. Rev* **48** 948.
[9] Clementi E and Roetti C, 1974 *Atomic Data and Nuclear Data Tables* (Academic, New York) **14** 268.
[10] Nath B and Sinha C, 2000 *J. Phys. B,* **33,** 5525.
[11] Thomas L H. 1927 *Proc.R. Soc. London, Ser. A* **114,** 561.
[12] Tribedi L C, Richard P, Gulyas L, Rudd M E, Moshammer R., 2001 *Phys. Rev A,* **63,** 062723.
[13] Misra D, Kumar Ajay, Kadhane U R, Fainstein P D, Tribedi L C 2006 *Radiation Physics and Chemistry* **75** 1723.

The 8th Asian International Seminar on Atomic and Molecular Physics IOP Publishing
Journal of Physics: Conference Series **185** (2009) 012039 doi:10.1088/1742-6596/185/1/012039

Inelastic ion-surface collisions: Understanding secondary emission of molecular ions

Biswajit Saha and Purushottam Chakraborty*

Surface Physics Division, Saha Institute of Nuclear Physics
1/AF Bidhannagar, Kolkata 700 064, India

E-mail: purushottam.chakraborty@saha.ac.in

Abstract. Electron transfer phenomena and inelastic ion-surface collisions lead to various processes of ionization: excitation of projectile/target atoms, secondary electron or atom/ion emission, etc. An important issue is the effect of electron exchange between the outgoing excited/ionized species and in particular, the effect of the chemical environment or surface composition on the outcome of the emission process. Secondary emission of molecular and/or cluster ions in the sputtering process has a large impact on the compositional analysis of materials. The problem arising out of 'matrix effect' in secondary ion mass spectrometry (SIMS) has largely been circumvented by the quantitative potential of the MCs^+-SIMS method. However, it generally suffers from a low useful yield of MCs^+ ions and in such case, MCs_2^+ molecular ions offers a better detection sensitivity. The present paper addresses the understanding of MCs_n^+ (n = 1,2...) molecular ions formed in inelastic ion-surface collision processes.

1. Introduction

The analysis of surfaces, interfaces and bulk materials through ion beam induced sputtering is a mature and well-established technique. The ejection of surface atoms or molecules under energetic ion impact (in the range of few hundred eV to several MeV) is estimated primarily by the 'sputtering-yield' (defined as the mean number of emitted atoms per incident ion) and in this process, the bombarding ion transfers energy in collisions to target atoms, which recoil with sufficient energy to generate other collisions. Some of the backward recoils in the generated 'collision cascade' impart enough energy to the surface atoms enabling them to overcome the surface binding energy and to escape from the solid. Therefore, the secondary sputtered species originate from the collision cascades that are created within the target lattice and from recoil sputtering.

In the linear cascade regime, the sputtering yield is proportional to the number of recoil-atoms that is proportional to the energy deposited per unit depth in nuclear processes (particle velocity much less than the Bohr velocity). Effective thickness of the sputtered layer arising out of a single collision cascade is of the order of an atomic dimension and its estimate can be obtained from a simple expression of the sputtering-yield involving information depth, typical average energy of the sputtered particles and nuclear stopping power [1]. As the ejected particles stem essentially from the outermost ($\sim$ 1-3) atomic layers, they carry information about the composition of the near-surface region. The varieties of the sputter-products constitute electrons, atoms, atomic clusters, molecules 'in tact' and distinctive molecular fragments. A very small fraction of the sputtered atoms or molecules may be ionized, either positively or negatively, and/or excited above their respective ground states. The atoms (or ions) sputtered in the excited states may subsequently go through two inelastic processes:

I. Radiative electronic transition via de-excitation of the excited neutral (or ionic) species leading to the emission of photons in the range ultra-violet to visible optics.

II. Non-radiative electronic transition leading to the emission of Auger electrons or to the ionization of the sputtered species via resonant-tunneling of electrons from these species to the valence band of the solid.

The above dynamic charge-transfer processes occurring between the escaping sputtered atom and the solid surface are further complicated by the strong electronic and structural perturbations from the ion bombardment.

2. Secondary ion emission and Matrix effect

Secondary emission of ions as well as photons in the sputtering process is an inelastic phenomenon in ion-surface interactions and has tremendous important applications in materials analysis. Analytically, one of the important sputtering techniques based on the above phenomenon is the "Secondary Ion Mass Spectrometry" (SIMS), in which the ionized fraction of the sputtered species (known as secondary ions) is detected by a mass spectrometer equipped with appropriate ion collection optics. Figure 1 shows the schematic representations of the technique. In this technique, the primary beam removes atomic (molecular) layers of the sample at a rate determined by the intensity, mass and energy of the bombarding ions and by the physical and chemical characteristics of the sample itself.

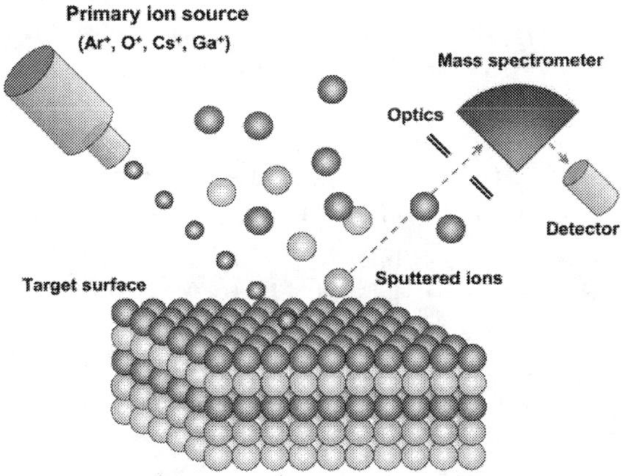

Figure 1. Schematic diagram of the SIMS technique.

Despite the highest detection sensitivity (~ part-per-billion level) and excellent depth-resolution (down to less than ~ 2 nm) of SIMS, the technique suffers from the "matrix effect" (variation of the ionization probability of a certain species due to the instantaneous change in local surface chemistry of the target containing the species) which hinders the quantification. A typical matrix effect was demonstrated through SIMS investigations of Pt/Cu and Au/Cu multilayers [2], where the large humps for Pt and Au signals across the interface positions in the respective multilayers are due to the preferential trapping of oxygen in these multilayer structures. Obviously, the matrix effect requires to be appropriately compensated while estimating the actual concentration of a certain species present in the host ('matrix'). The compensation is usually done by using "calibration standards", "relative sensitivity factor" (RSF) approach etc. The use of 'ion-implanted standards' has reached a high level of sophistication, allowing precision measurements in SIMS quantification. From an analytical point of view, motivation towards the reduction of the matrix effect has led to the development of another technique called "secondary neutral mass spectrometry" (SNMS). But basically the very fact that sputtered neutral signals are far less intense than the secondary ion signals has led most of the workers to concentrate on the secondary ion emission process for analytical work, despite the complexities in the ionization mechanism. Although quite sophisticated theoretical treatments of the ionization process have been developed for idealized models [3, 4], a quantitative theory that takes into account the dynamic evolution of the sputtering collision cascade and the chemical complexity of real materials is not yet available.

3. Molecular ion emission: possible mechanisms

3.1. MCs^+ ions

The most useful and promising technique to reduce the matrix effect in SIMS analysis seems to be the MCs^+ technique. Under Cs^+-bombardment, it has been shown that detection of the element M of interest by monitoring of MCs^+ molecular ions (instead of conventional M^- ions) is largely unaffected by changing matrix [5], thereby enabling this mode of analysis possible to quantify without the aid of calibration standards [6, 7]. Although the application of this technique is difficult for its poor sensitivity (MCs^+ molecular species are usually less abundant than the atomic ions), MCs^+ detection scheme is more favorable in cases where matrix effects are most severe, particularly for major components. This novel technique is referred to as Cs^+-assisted secondary ion mass spectrometry or simply as MCs^+-SIMS [8] and has been successfully been employed for various measurements. MCs^+-SIMS approach has been successfully utilized in a study of Pd-Rh interdiffusion which has a relevance in brachytherapy. Different Rh_xPd_y (from pure Rh to pure Pd) layers were deposited by PVD and were subsequently characterized by ToF-SIMS. A linear relationship between the relative $CsPd^+$ yields and Pd concentration into the Rh matrices enabled to calibrate the ToF-SIMS depth profiles of annealed Pd/Rh layers and hence to quantify the PdRh interdiffusion [9].

Formation mechanisms of MCs^+ secondary ions and the analytical applicability of MCs^+-SIMS technique have been examined by several workers [10, 11]. Based on some experimental observations, it was proposed that MCs^+ ions form via a combination of independently sputtered neutral M atoms and Cs^+ ions. Therefore, it is obvious that the binding energy between M and Cs (either charged or neutral) plays a very important role in the formation of MCs molecule and in that case, the polarizabilty of M plays the key role in the binding strength of the two species, since the interaction is basically between the electric field of Cs^+ and the ion-induced dipole moment of M. The interaction between Cs^+ and a polarizable neutral atom (or molecule) has been successfully modeled by applying a potential of the form

$$V(r) = Ar^{-12} - 0.5\alpha e^2 r^{-4}. \tag{1}$$

where r is the distance between the neutral atom and the Cs^+ ion, α is the polarizability of the former and A is a constant which characterizes the repulsive term. For a variety of semiconductors, MCs^+ yield was found to have a quadratic dependence on the polarizability of the atom M [12].

Secondary emission of atomic and molecular ions from elemental molybdenum surface has been studied under Cs^+ bombardment to explore the MCs^+ formation mechanism with changing Cs surface coverage [13]. Integrated count of $MoCs^+$ shows a monotonic increase with increasing primary ion energy. Change in $MoCs^+$ intensity is attributed to the variation of surface work function and cesium surface concentration c_{Cs} due to varying impact energies (inset of Fig. 2). Variation of c_{Cs} has been obtained from the expression, $c_{Cs} \propto 1/\ (1 + \mathrm{Y})$ where Y is the elemental sputtering yield. Systematic study of the energy distributions of all species emerging from Mo target has been done to measure the changes in surface work function. Changing slopes of the leading parts of Cs^+ energy distributions suggest a substantial depletion in surface work function with decreasing primary ion energies. As evidenced in (Fig. 2), surface work function showed a linear dependence on c_{Cs}. The maximum reduction in surface work function ($\sim$ 0.69 eV) corresponds to the highest value of $c_{Cs} = 0.5$. Emission of all atomic and molecular ionic species has been found to be consistent with the electron tunneling model. Furthermore $MoCs^+$ formation were found to consistent the recombination model.

3.2. MCs_2^+ ions

MCs^+ method suffers from a generally low useful yield, particularly for the electronegative elements. In such cases, detection of MCs_2^+ molecular ions offers a better sensitivity to that of MCs^+ ion, as the yield of MCs_2^+ ions has been observed much higher than that of MCs^+ ions [14]. Monitoring of MCs_2^+ could be important for the species where MCs^+ emission intensity is very small. The higher emission of MCs_2^+ is surprising as the yield of Cs_2^+ ions is significantly less than that of Cs^+ ions under identical Cs bombardment conditions. Holliger et al. [15] investigated the analytical usefulness of monitoring MCs_2^+ ions for semiconductor profiling of n-type dopants such as P, Ge or As with a primary (Cs^+ beam) impact energy of 1 keV. They indicated that the MCs_2^+ technique has an advantage over the MCs^+ technique of providing higher useful yields for electronegative elements by up to 10^2 and can therefore be applied to perform more precise junction measurements in terms of concentration, dose, decay length and detection limit.

Higher yield of MCs_2^+ compared to MCs^+ has been supported by an observation in case of Cu with varying impact ion energies [16] where the intensity ratio $I(MCs_2^+)/I(MCs^+)$ was found to be greater for lower impact energies. This is understandable from the fact that Cs surface-ion concentration becomes appreciable at low impact energies thereby leading to an enhanced formation of MCs_2^+ ions compared to that of MCs^+ ions. The fact that Cs concentration on the surface changes with impact Cs^+ ion energy was clearly evidenced from the kinetic energy distributions of MCs^+ ions [16].

Formation mechanism of MCs_2^+ is still not clear. Probable recombination mechanisms may be the following:

$MCs^0 + Cs^+ \rightarrow MCs_2^+$
$M^- + Cs^+ + Cs^+ \rightarrow MCs_2^+$
$M^0 + Cs_2^+ \rightarrow MCs_2^+$
$M^0 + Cs^+ + Cs^+ \rightarrow MCs_2^{++} \rightarrow MCs_2^+$

The first route indicates the recombination of a neutral MCs^0 molecule with one Cs^+ ion. The second mechanism suggests the vacuum recombination of sputtered M^- ion with two Cs^+ ions. These two mechanisms have been shown to be dominant for electronegative elements. The third mechanism indicates the recombination of sputtered M^0 atom (neutral) with a sputtered Cs_2^+ dimer [17]. The fourth route suggests the formation of MCs_2^{++} ion by the association of one M atom with two Cs^+ ions, followed by the conversion to MCs_2^+ ion via an electron capture. In this case higher Cs concentration at the surface should favor the production of MCs_2^{++} ions. In order to check the validity of this route, detection of MCs_2^{++} was attempted with Cu for 1 keV Cs^+ impact energy and the emission of $CuCs_2^{++}$ was evidenced although in a very small quantity [16]. The weak intensity of $CuCs_2^{++}$ ions could be due to the immediate conversion from $CuCs_2^{++}$ to $CuCs_2^+$.

Investigation on the formation of MCs_2^+ molecular ions under Cs^+ bombardment of silver surface in the SIMS process was reported under varying oxygen environments [18]. Energy distributions of MCs^+ and MCs_2^+ ions have shown a remarkable dependence on the changing oxygen environment. The changes in the intensities of the above molecular species are attributed to the changes in the resputtered Cs^+ intensity due to the decrease in the local surface work function. The MCs^+ molecular ion has been found to form via the recombination of a neutral M^0 atom with a Cs^+ ion and its formation probability remains unaffected by changing oxygen environment. A systematic study on the kinetic energy distributions of secondary Cs^+, Cs_2^+ , MCs^+ and MCs_2^+ ions and the estimation of the mean emission energies of various constituents participating in the respective formation processes confirm that the MCs_2^+ molecular ion formation via recombination of a neutral MCs^0 molecule with a Cs^+ ion is the most probable one [18].

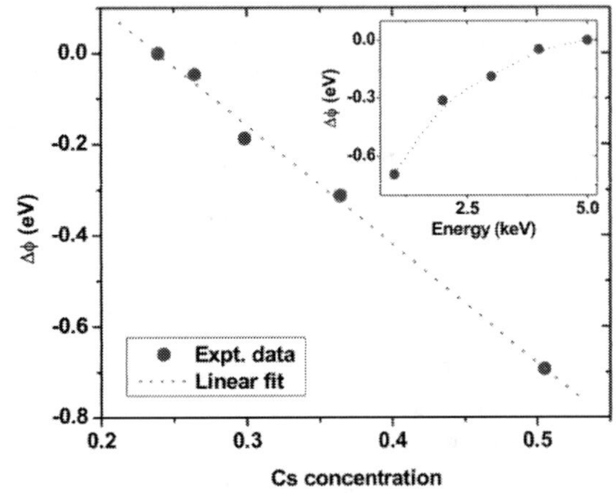

Figure 2. Change in work function for varying cesium surface concentration. Inset: $\Delta\phi$ as a function impact energy [13].

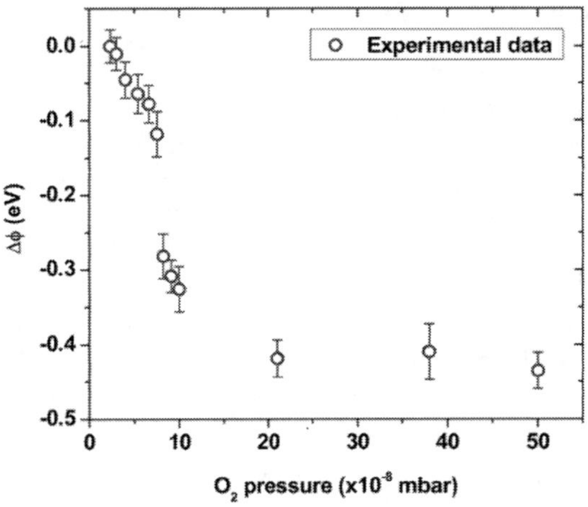

Figure 3. Change in surface work function $\Delta\phi$ as a function of O_2 pressure [22].

However, explanations towards the formation mechanism of MCs_2^+ ions according to the second route were given from the kinetic energy distributions of MCs_n^+ (n = 0,1,2,3, ., etc.) molecular ions for various metallic samples (Al, Cu, and Ag) under varying Cs^+ impact ion energies (1-5 keV) [19]. From the peak shifts of the energy distributions it was argued that the surface binding energy as well as the work function changes with the change in the surface Cs concentration arising out of the variation in the impact ion energy. The dependence of the measured integrated yields of the ions (M^+, MCs^+, and MCs_2^+ for individual elements) on the estimated work functions were found to be exponential. Results were interpreted in terms of the relative changes in the instantaneous local work function of the sputtered zone. The study was strongly supportive of the suggested formation mechanism of a MCs_2^+ molecular ion via a combination of M^- and two Cs^+ ions in the close proximity of the Cs-covered surface [19].

Recently, the ionization processes of elemental and cluster ions have been studied by co-sputtering cesium and xenon on silicon wafer with cesium beam concentration varying from the 0% (for pure xenon) to 100% (for pure cesium). The observations on MCs_n^+ signals were explained by using electron tunneling model [20]. Although the formation mechanisms of the MCs_2^+ clusters are unclear, the analysis of MCs_n^+ clusters on organic matter provides two potential applications. First, molecular information is preserved in the MCs_2^+ mode allowing organic depth profiling. Second, the charge state selection (neutral, singly or doubly negatively charged) occurring during the cluster formation may offer some new information carried by neutrals or multiple charged ions [21].

A recent study on the emission of MCs_n^+ (n = 1, 2, ...etc.) molecular ions from silver has been done under the joint influence of electropositive (cesium) and electronegative (oxygen) elements [22]. The kinetic energy distributions, measured for Cs^+, Cs_2^+ , $AgCs^+$ and $AgCs_2^+$ ions at different oxygen pressures exhibited changing slopes in their leading parts that hinted appreciable changes in the surface work function (Fig. 3). The maximum observed change in the surface work function was 0.44 eV. The integrated counts of Cs^+, Cs_2^+ , $AgCs^+$ and $AgCs_2^+$ ions exhibit an exponential dependence on the surface work function, supporting the electron tunneling model of secondary ion emission. Cs_2^+ ion intensity has been found to be consistent with recombination of neutral Cs^0 with Cs^+ ion. Study of $AgCs_n^+$ molecular ions also reveals the fact that the intensities of such molecular species are strongly affected by the oxygen environment. Variation in the Cs^+ intensity due to the change in oxygen environment has been found to be a key-factor behind $AgCs_n^+$ intensity variation. The linear dependence of $AgCs^+$ intensity on the ionization probability of Cs^+ ions strengthens the belief that an $AgCs^+$ is formed via combination of a neutral Ag atom with a Cs^+ ion in the near-surface region. Furthermore, a linear dependence of $AgCs_2^+$ intensity on the ionization probability of Cs^+ suggests that $AgCs_2^+$ molecular ion formation via the combination of neutral $AgCs^0$ molecule and Cs^+ ion is the most probable one in the present situation.

4. Summary

The problem arising out of 'matrix effect' in secondary ion mass spectrometry (SIMS) has largely been circumvented by the quantitative potential of the MCs^+-SIMS method. However, it generally suffers from a low useful yield of MCs^+ ions and in such cases detection of MCs_2^+ molecular ions offers a better sensitivity. Formation of MCs^+ is rather straight forward. The well accepted mechanism is basically via an interation between Cs^+ ion and the ion-induced dipole moment of a M^0 atom. Whereas, the formation mechanism of MCs_2^+ molecular ion is not yet clear. The two probable routes, one as a combination of two Cs^+ ions with one M^- ion and the other as a combination of MCs^0 molecule with a Cs^+ ion, both valid for electronegative elements, have been discussed as per literature survey.

5. References

[1] P. Chakraborty 2002 *Ion Beam Analysis of Surfaces and Interfaces of Condensed Matter Systems* ed. P. Chakraborty (New York, Nova Science)
[2] S. Sarkar and P. Chakraborty 2000 *J. Chinese Chem. Soc.* **48** 521
[3] A. Benninghoven, F.G. Rudenauer and H. W. Werner 1987 *Secondary Ion Mass Spectrometry: Basic Concepts, Instrumental Aspects, Applications and Trends* (New York, Wiley)
[4] Z. Sroubek 1987 *Proceedings of SIMS VI* eds A. Benninghoven et al (Wiley) p 17
[5] M. Haag, H. Gnaser and H. Oechsner 1995 *Fresenius' J. Anal. Chem.* **353** 565
[6] Y. Gao 1988 *J. Appl. Phys.* **64** 3760
[7] C. W. Magee, W. L. Harrington and E. M. Botnick 1990 *Int. J. Mass Spectrom. Ion Proc.* vol 103 p 45
[8] K. Wittmaack 1994 *Nucl. Instrum. Meth. in Phys. Res.* B **85** 374
[9] J. Brison, R. Hubert, S. Lucas and L. Houssiau 2006 *Appl. Surf. Sc.* **252** 7034
[10] J. Schroer, H. Gnaser and H. Oechsner 1993 *SIMS IX* eds. A.Benninghoven et al (New York, Wiley)
[11] U. Breuer, H. Holzbrecher, M. Gastel, J.S. Becker and H.H. Dietz 1995 *Fresenius' J. Anal. Chem.* **353** 372
[12] H. Gnaser and H. Oechsner 1994 *Surf. Sci. Lett.* **302** L289
[13] B. Saha and P. Chakraborty 2008 *Nucl. Instrum. and Meth. in Phys. Res.* B **266** 1386
[14] Y. Gao, Y. Marie, F. Saldi and H. -N. Megion 1995 *Int. J. Mass Spectrom. Ion Proc.* **143** 11
[15] P. Holliger, F. Laugier and J.C. Dupuy 2002 *Surf. Interface Anal.* **34** 472
[16] S. Sarkar and P. Chakraborty 2003 *Nucl. Instrum. Meth. in Phys. Res.* B **212** 364
[17] Y. Marie, Y. Gao, F. Saldi and H. -N. Migeon 1995 *Surf. Interface Anal.* **23** 38
[18] B. Saha and P. Chakraborty 2007 *Nucl. Instr. and Meth. in Phys. Res.* B **258** 246
[19] S. Sarkar, P. Chakraborty and H. Gnaser 2004 *Phys. Rev.* B **70** 195427
[20] J. Brison and L. Houssiau 2007 *Nucl. Instrum. Meth. in Phys. Res.* B **259** 984
[21] N. Mine, B. Douhard and L. Houssiau 2008 *Appl. Surf. Sci.* **255** 973
[22] B. Saha, S. Sarkar, P. Chakraborty and H. Gnaser 2008 *Surf. Sci.* **602** 1061

The 8th Asian International Seminar on Atomic and Molecular Physics IOP Publishing
Journal of Physics: Conference Series **185** (2009) 012040 doi:10.1088/1742-6596/185/1/012040

N-K Spectra of Adenine Amino Tautomers

Saumitra Saha[1] **and Feng Wang***

Centre for Molecular Simulation, Swinburne University of Technology, PO Box 218, Hawthorn, Melbourne, Victoria 3122, Australia

E-mail: *fwang@swin.edu.au

Abstract.
Adenine tautomeric process is not a small perturbation but a significant change of the inner shell configuration. For canonical adenine, ade n9, the inner-shell configuration in the ground electronic states (X^1A') is given by $1a'(N_{(9)})2a'(N_{(10)})3a'(N_{(7)})4a'(N_{(3)})5a'(N_{(1)})$ $6a'(C_{(6)})7a'(C_{(8)})8a'(C_{(4)})9a'(C_{(2)})10a'(C_{(5)})$. Although the five N1s orbitals in the core shell of all adenine amino tautomers are always given as $1a'2a'3a'4a'5a'$. The profound tautomer dependent inner-shell electronic reconfiguration has been largely masked by small net differences in their total electronic energies and subtle differences in their valence shell electronic structures. The present quantum mechanical spectral study demonstrates that inner-shell chemical shift of N1s orbitals of adenine amino tautomers is atomic site and tautomer specific: the orbital binding energy shifts of the amino N1s sites (-N-) are smaller than those of the imino N1s sites (-N=) in the same tautomer. The nitrogen $N^{@}$1s site (an amino site) which connects the mobile hydrogen exhibits the smallest energy shift within a tautomer and therefore, is an indicator of the specific tautomer. The preferential adenine (ade n9) protonation sites are the imino sites in the order of $5a'(N_{(1)})$, $4a'(N_{(3)})$ and $3a'(N_{(7)})$ which agrees with previous observations.

1. Introduction

Tautomers of DNA bases are vital for the understanding of chemical reactivity [1, 2] but experimental detection/identification of tautomers has been a challenge even in the gas phase [3, 4]. Molecules of biological relevance often are thermally labile and hardly be brought into the gas phase by sample heating, and for most of DNA bases the spectra of different tautomers superimpose and make the spectra congested [5, 6]. As a result, the study of DNA base tautomers has been largely based on computational methods [7, 8, 9, 10, 11, 12, 13]. Most of structural studies with respect to tautomerism of DNA bases, however, has concerned on the valence space [6, 7, 9, 11, 12, 14, 15], which hardly provides a thorough understanding of the role of possible proton transfer reactions without the recognition of inner-shell re-configuration process. For example, valence shell information has yet to interpret the reason why the preferential protonation sites [16, 14, 6] of ade n9 is in the order of $N_{(1)}$, $N_{(3)}$ and $N_{(7)}$ ($N_{(10)}$, $N_{(9)}$) andor why proton affinity of these sites is in such the order. Unfortunately, many DNA base tautomeric studies have treated proton transfer as small perturbations, and frozen core (FC) treatment [6] are popular in these studies, which miss the most important information on tautomerism.

[1] Present Address: Department of Materials Science and Engineering, University of Wisconsin, Madison, WI 53706, USA

2. Method and Computational Details

Adenine amino tautomers are slightly non-planar [17, 18, 19]. In the present study, adenine tautomers are treated as if they were planar molecules by imposing the C_s point group symmetry (σ-plane). This idealization to planarity not only simplifies the electronic calculations because the a' and a'' states are treated in separated and independent Hamiltonian matrices, but also provides insight into the chemical bonding by separating in-plane σ bonds from out-of-plane π bonds, as well as for other applications such as valence space studies [8, 19].

The planar geometries of the four planar amino tautomers of adenine are optimized respectively using the hybrid density functional theory (DFT) B3LYP/TZVP model. The TZVP basis set is a reasonably large triple zeta with valence polarized (TZVP) basis set [20], which has been proved reliable in our previous investigations of valence electronic properties of adenine amino tautomers [8], amino acids [21] and core shell of bicyclic-hydrocarbons [22] and DNA bases such as purine and adenine [23]. Based on the B3LYP/TZVP geometries of the tautomers, core ionization orbital energies of each tautomer are produced from single point calculations emplying the LB94/TZ2P model. All electronic calculations are performed using the GAUSSIAN03 [24] and Amsterdam Density Functional (ADF) [25] packages of computational chemistry programs.

3. Results and discussion

The inner-most five core orbitals, $1a'$, $2a'$, $3a'$, $4a'$ and $5a'$, of the tautomers are always the contributions from the N1s orbitals. Our simulated x-ray photoemission spectroscopy (XPS) of the canonical adenine (ade n9) using the LB94/T2ZP model has been proved accurate when compare to recent XPS experiments [23, 26]. As a result, this experimentally validated model is applied to study its amino tautomers theoretically in the present study.

The orbital energies change with respect to specific atom sites in the tautomers. The assignment of the inner-shell XPS of adenine tautomers will be completely different from each other. For example, the $N_{(9)}$ site, which locates in very different regions where a few electron volts are observed in the tautomers, is even larger than the total electronic energy differences among the tautomers. Figure 1 gives the simulated N-K spectra for the tautomers, based on the present LB94/TZ2P model. A theoretical resolution of 0.13 eV (FWHM) in the gaussian shape function is employed. The best experimentally available resolution used for adenine in gas phase [26] is 0.57 eV. The high theoretical resolution in the simulation is to resolve individual core states so that each spectral peak of adenine can be assigned to an individual N-K site of the molecule. The inner-shell spectra in Figure 1 indicates that the tautomers have significantly different inner-shell electronic structures, although they can be related in a certain manner. Although shifted by energy, the $N^{@}$-K site of a tautomer is always the highest energy peak in the spectra. When the $N^{@}$-K sites locate in the hexagon ring of the tautomer, such as ade n9 and ade n7, the chemical environment in energy term seems more similar as the three imino N-K peaks are closely positioned. When the $N^{@}$-K sites locate in the pentagon ring of the tautomer, such as ade n1 and ade n3, the N-K sites in the hexagon ring, $N_{(7)}$ and $N_{(9)}$, exhibit similar energy environment. In summary, the adenine tautomers are apparently different in their inner-shell N-K spectra.

4. Conclusions

The mobile proton causes significant reconfiguration in the inner-shell electronic structures in the adenine tautomeric processes. All N1s orbitals in the adenine tautomers exhibit energy shift in their binding energy spectra with respect to their corresponding free atoms. The electronic structural variations among adenine amino tautomers are **not** perturbations, but exhibit profoundly different tautomer dependent changes in inner-shell configurations. Therefore, studies based on valence shell alone and/or using the frozen core (FC) approximation unlikely produce reliable tautomer structural information of DNA bases and cannot interpret the order of

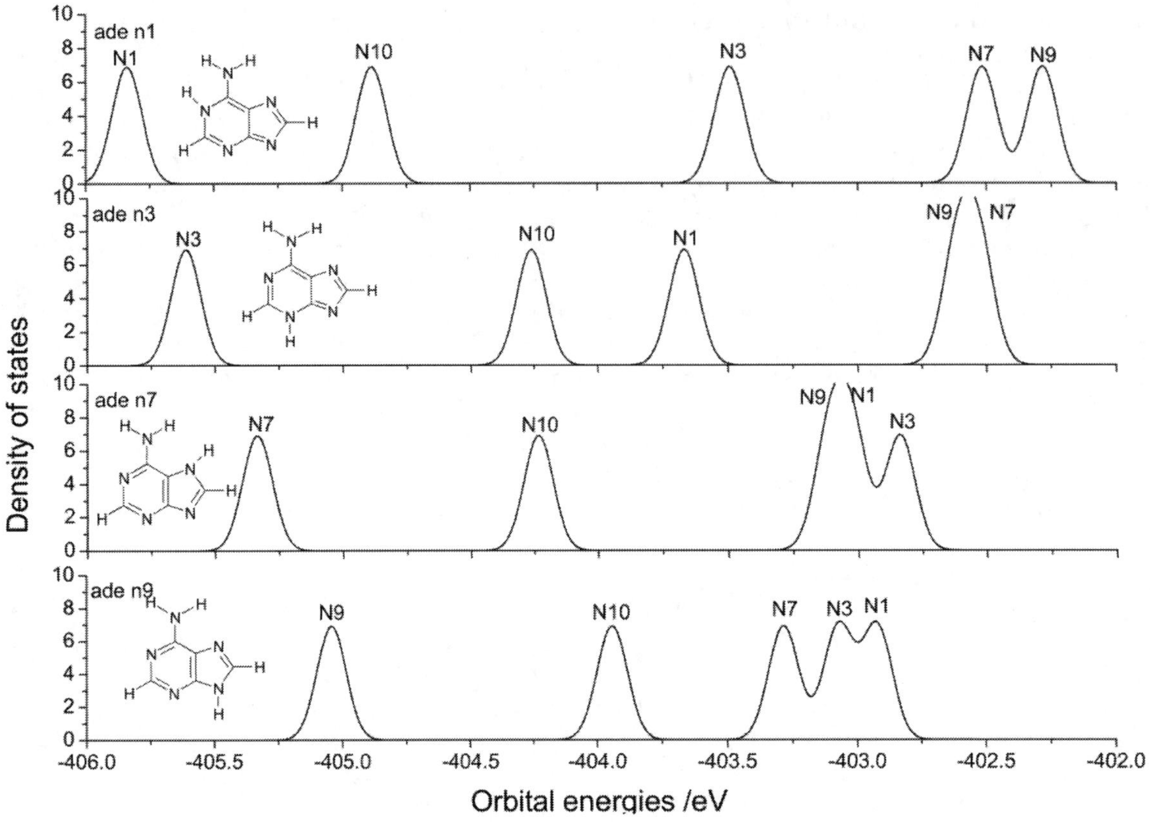

Figure 1: Density of states spectra of the N-K sites of adenine amino tautomers.

protonation sites in DNA bases. Core orbitals contain a dominant amount of energy in a molecule and cannot be ignored. Inner shell spectroscopy, when coupled with powerful synchrotron radiation sources and appropriate computational tools, is able to achieve the ultimated goal of identifying detailed electronic structures of DNA tautomers, their stabilities and properties, which will impose significant influences on biochemistry.

Acknowledgments
This work has been supported by an award under the Merit Allocation Scheme on the National Computational Infrastructure at the ANU. One of the authors, SS, wishes to acknowledge Swinburne University Postgraduate Research Award (SUPRA).

5. References

[1] Elguero J, Katritzky A R and Denisko O 2000 *Heterocyclic Chemistry* vol 76 ed Katritzky A R (San Diego: Academic Press)
[2] Harris V, Smith C, Cummins W, Hamilton A, Adams H, Dickman M, Hornby D and Williams D 2003 *J. Mol. Biol.* **326** 1389
[3] Brown R D, Godfrey P D, McNaughton D and Pierlot A P 1989 *Chem. Phys. Lett.* **156** 61
[4] Nir E, Kleinermanns K, Grace L and Vries M S D 2001 *J. Phys. Chem. A* **105** 5106
[5] Abo-Riziq A, Grace L, Nir E, Kabelac M, Hobza P and Vries M S D 2005 *Proc. Natl. Acad. Sci. USA* **102** 20
[6] Marian C, Nolting D and Weinkauf R 2005 *Phys. Chem. Chem. Phys.* **7** 3306
[7] Mochizuki Y, Koide H, Imamura T and Takemiya H 2001 *J. Synchrotron Rad.* **8** 1003

[8] Wang F, Downton M T and Kidwani N 2005 *Journal of Theoretical and Computational Chemistry* **4** 247–264
[9] Gu J and Leszczynski J 1999 *J. Phys. Chem. A.* **103** 2744
[10] Lührs D C, Viallon J and Fischer I 2001 *Phys. Chem. Chem. Phys.* **3** 1827
[11] Sukhanov O S, Shishkin O V, Gorb L, Podolyan Y and Leszczynski J 2003 *J. Phys. Chem. B* **107** 2846
[12] Dolgounitcheva O, Zakrzewski V and Ortiz J 2003 *Fundamental World of Quantum Chemistry* vol II ed Brandas E and Kryachko E (Kluwer Academic Publishers) p 525
[13] Piacenza M and Grimme S 2003 *J. Comput. Chem.* **25** 83
[14] Chen X, Syrstad E A, Nguyen M T, Gerbaux P and Turecek F 2004 *J. Phys. Chem. A* **108** 9283
[15] Chen X, Syrstad E A, Nguyen M T, Gerbaux P and Turecek F 2005 *J. Phys. Chem. A* **109** 8121
[16] Podolyan Y, Gprb L and Leszczynski J 2000 *J. Phys. Chem. A* **104** 7346
[17] Dong F and Miller R E 2002 *Science* **298** 1227
[18] Sponer J and Hobza P 2003 *Collection of Czechoslovak Chemical Communications* **68** 2231–2282
[19] Downton M and Wang F 2006 *Mol. Sim.* **(In Press)**
[20] Godbout N, Salahub D, Andzelm J and Wimmer E 1992 *Can. J. Chem.* **70** 560
[21] Falzon C and Wang F 2005 *Journal of Chemical Physics* **123** 000
[22] Wang F 2005 *Journal of Molecular Structure-Theochem* **728** 31–42
[23] Saha S, Wang F, MacNaughton J B, Moewes A and Chong D P 2008 *Journal of Synchrotron Radiation* **15** 151–157
[24] Frisch M J, Trucks G W, Schlegel H B, Scuseria G E, Robb M A, Cheeseman J R, J A Montgomery J, Vreven T, Kudin K N, Burant J C, Millam J M, Iyengar S S, Tomasi J, Barone V, Mennucci B, Cossi M, Scalmani G, Rega N, Petersson G A, Nakatsuji H, Hada M, Ehara M, Toyota K, Fukuda R, Hasegawa J, Ishida M, Nakajima T, Honda Y, Kitao O, Nakai H, Klene M, Li X, Knox J E, Hratchian H P, Cross J B, Bakken V, Adamo C, Jaramillo J, Gomperts R, Stratmann R E, Yazyev O, Austin A J, Cammi R, Pomelli C, Ochterski J W, Ayala P Y, Morokuma K, Voth G A, Salvador P, Dannenberg J J, Zakrzewski V G, Dapprich S, Daniels A D, Strain M C, Farkas O, Malick D K, Rabuck A D, Raghavachari K, Foresman J B, Ortiz J V, Cui Q, Baboul A G, Clifford S, Cioslowski J, Stefanov B B, Liu G, Liashenko A, Piskorz P, Komaromi I, Martin R L, Fox D J, Keith T, Al-Laham M A, Peng C Y, Nanayakkara A, Challacombe M, Gill P M W, Johnson B, Chen W, Wong M W, Gonzalez C and Pople J A 2004 Gaussian 03, Revision C.02
[25] Baerends E J, Autschbach J, Brces A, Bo C, Boerrigter P M, Cavallo L and Chong D P 2003
[26] Plekan O, Feyer V, Richter R, Coreno M, de Simone M, Prince K C, Trofimov A B, Gromov E V, Zaytseva I L and Schirmer J 2008 *Chemical Physics* **347** 360–375

Journal of Physics: Conference Series **185** (2009) 012041 doi:10.1088/1742-6596/185/1/012041

On the mechanism of spin-dependent (e,2e) scattering from a ferromagnetic surface.

SN Samarin[1], OM Artamonov[2], AD Sergeant[1], L Pravica[1], D Cvejanovic[1], P Wilkie[1], P Guagliardo[1], AA Suvorova[3] and JF Williams[1].

[1]ARC Centre of Excellence for Antimatter-Matter Studies, University of Western Australia, Perth, WA 6009, Australia,
[2]Research Institute of Physics, St. Petersburg University, St.Petersburg, Russia.
[3]CMCA, University of Western Australia, Perth, WA 6009, Australia

samar@physics.uwa.edu.au

Abstract. A simple model is suggested for a qualitative analysis of spin-dependent (e,2e) reaction on a ferromagnetic surface. The model is based on the scattering of the primary electron with the average spin projection $<s_1>$ by the valence electron with the average spin projection $< s_2>$. To test the model the energy distributions of correlated electron pairs are measured for parallel and anti-parallel orientations of the magnetic moment of the cobalt film and polarization vector of the incident beam. The proposed model explains qualitatively the spin-asymmetry of the measured binding energy spectrum.

1. Introduction

The exchange interaction in electron-electron scattering is a result of the fermionic (spin = ½) nature of the electrons. As a consequence, the wave function of an electronic system must be anti-symmetric which implies that an exchange of two electrons of the system results in the change of the wave function sign. To observe the exchange interaction one needs to monitor at least two interacting electrons with well-defined spin states. A spin-dependent scattering experiment includes preparation of the initial spin state of the two electron system and the measurement of the cross section of the scattering for two cases: i) the spin projections of both scattering electrons on the quantization axis are the same (differential cross section is denoted as $d\sigma\uparrow\uparrow$); ii) the projections have opposite signs (spin projection of one of the electrons is flipped while the second is unchanged, and the corresponding cross section is $d\sigma\uparrow\downarrow$). When such an experiment is performed on a ferromagnetic surface the spin state of the target electrons is defined by the magnetization of the sample, which is anti-parallel to the majority spin states of the sample. The spin state of the second (incident) electron is chosen to be parallel ("spin up") or anti-parallel ("spin down") to the majority spins of the target.

The two-electron spectroscopy in reflection with spin-polarized incident beam [1] offers the possibility to probe exchange interaction. The cross section of each of the two above mentioned channels ($d\sigma\uparrow\uparrow$ and $d\sigma\uparrow\downarrow$) is proportional to the number of electron pairs resulting from the interaction of the incident electron beam with the target at a fixed incident current. To test the model we measured (e,2e) spectra from 3 monolayer (ML) cobalt film on W(110) surface.

2. Model.

We present here a qualitative analysis of the spin-dependent electron scattering from a ferromagnetic surface based on a simple model of the ferromagnetic sample. The aim of this analysis is to evaluate the asymmetry parameter that can be measured experimentally and which is defined as:

$$A = (d\sigma_{\uparrow\uparrow} - d\sigma_{\uparrow\downarrow})/(d\sigma_{\uparrow\uparrow} + d\sigma_{\uparrow\downarrow}) \qquad (1)$$

Using a simplified free electron model for the ferromagnetic target one can assume that electrons within binding energy range ε to $\varepsilon + d\varepsilon$ have an average spin projection $\boldsymbol{\zeta}_2(\varepsilon) = \langle s_2 \rangle$. The incident electron beam with spin polarization **P** is characterized by an average spin $\boldsymbol{\zeta}_1 = \langle s_1 \rangle$. In a very simplified picture one can model our experiment by scattering of an electron with the average spin $\boldsymbol{\zeta}_1$ by another electron with average spin $\boldsymbol{\zeta}_2$. We will follow Landau [2] for solving this problem. The spin dependence in the cross section $d\sigma$ must be expressed by the term proportional to the dot-product $(\boldsymbol{\zeta}_1 \cdot \boldsymbol{\zeta}_2)$. We are looking for the general expression of the cross section in the form:

$$d\sigma/d\omega = a + b(\boldsymbol{\zeta}_1 \cdot \boldsymbol{\zeta}_2) \qquad (2)$$

To find parameters a and b in (2) we consider two particular cases of scattering: i) both electrons in average are unpolarized ($\boldsymbol{\zeta}_1 = \boldsymbol{\zeta}_2 = 0$) and ii) completely polarized beam ($\boldsymbol{\zeta}_1 = ½$) is scattered by identically completely polarized target ($\boldsymbol{\zeta}_2 = ½$). In the first case the second term of (2) is zero, and $d\sigma_i = a$. On the other hand it is known that for unpolarized electrons the cross section of scattering is the average of the singlet $d\sigma_s$ and triplet $d\sigma_t$ cross sections with their statistical weights:

$$d\sigma_i / d\omega = a = ¼\,(d\sigma_s / d\omega + 3d\sigma_t / d\omega). \qquad (3)$$

In the second case the two scattering electrons are in a triplet state and therefore:

$$d\sigma_{ii}/d\omega = d\sigma_t/ d\omega = a + b/4 \qquad (4)$$

From equations (3) and (4) one can extract a and b and substitute them in (2):

$$d\sigma = ¼\,(d\sigma_s + 3d\sigma_t) + (d\sigma_t - d\sigma_s)\,(\boldsymbol{\zeta}_1 \cdot \boldsymbol{\zeta}_2) \qquad (5)$$

Having the expression for the cross section (5) we can return now to the problem of the polarized electron beam scattering by a ferromagnetic target. The average spin of the incident electron is $\boldsymbol{\zeta}_1 = ½\,\mathbf{P}$, where P is degree of polarization of the incident electron beam. For our model of the target the average spin $\boldsymbol{\zeta}_2$ can be expressed using the majority $\rho_{maj}(E)$ and minority $\rho_{min}(E)$ density of states and the unit vector $\boldsymbol{m}$, which is parallel to the spin of majority of electrons (i.e. opposite to the magnetization direction of the sample):

$$\boldsymbol{\zeta}_2 = ½\,[(\rho_{maj} - \rho_{min})/(\rho_{maj} + \rho_{min})]\boldsymbol{m} \qquad (6)$$

The experimentally measured quantity of interest is the asymmetry of the scattering cross section, defined by (1) as: $A = (d\sigma_{\uparrow\uparrow} - d\sigma_{\uparrow\downarrow})/(d\sigma_{\uparrow\uparrow} + d\sigma_{\uparrow\downarrow})$, where $d\sigma_{\uparrow\uparrow}$ is the scattering cross section when the polarization of the incident beam is parallel to the spin of the majority electrons of the sample, and $d\sigma_{\uparrow\downarrow}$ is the cross section when the polarization vector of the incident beam is opposite to the spin orientation of majority electrons of the sample. These two cross sections differ from each other only by the sign of the second term in (5). Substituting $\boldsymbol{\zeta}_1$ and $\boldsymbol{\zeta}_2$ from (6) in (5) one can express the asymmetry as:

$$A = \{(\rho_{maj} - \rho_{min})/(\rho_{maj} + \rho_{min})\}\cdot\{(d\sigma_t - d\sigma_s)/(d\sigma_s + 3d\sigma_t)\}\cdot P \qquad (7)$$

The asymmetry normalized to the polarization of the incident beam is then:

$$A/P = \{(\rho_{maj} - \rho_{min})/(\rho_{maj} + \rho_{min})\}\cdot\{(d\sigma_t - d\sigma_s)/(d\sigma_s + 3d\sigma_t)\} \qquad (8).$$

This expression is similar to the one obtained in [3] from first principles. The term $\{(d\sigma_t - d\sigma_s)/(d\sigma_s + 3d\sigma_t)\}$in the product (8) is the so called "exchange asymmetry" and the $\{(\rho_{maj} - \rho_{min})/(\rho_{maj} + \rho_{min})\}$ is the spin asymmetry of the valence electrons. In symmetric kinematics (normal incidence and symmetric detection), when the momenta of both detected electrons are equal, the triplet cross section vanishes $d\sigma_t = 0$ because of the symmetry consideration [this statement is strictly correct only if the valence electron state involved in the scattering is even with respect to the mirror plane perpendicular to the scattering plane and to the surface]. Consequently, the asymmetry of the scattering cross section A is determined only by the asymmetry of the density of states at $k_{b\parallel} = 0$, where k_b is valence electron wave vector:

Journal of Physics: Conference Series **185** (2009) 012041 doi:10.1088/1742-6596/185/1/012041

$$A/P = -(\rho_{maj} - \rho_{min})/(\rho_{maj} + \rho_{min}) \quad (9).$$

The minus sign before the parenthesis means that if the majority density of states is larger at the binding energy where the valence electron is excited from, then the cross section of the (e,2e) reaction is larger for the "down" polarization of the incident electron. It simply means that the cross section for two-electron scattering is larger when the two electrons have opposite spin projections on the quantization axis.

3. Experiment.

The experiment was performed in the UHV conditions with the base pressure in the 10^{-11} Torr range. The residual magnetic field within the vacuum chamber was reduced to less than 5 mG using a combination of static and dynamic Helmholz coils. The W(110) substrate was cleaned using standard procedure including oxygen treatment at 1400 C° followed by high temperature flashes. The cleanliness of the sample was monitored by Auger Electron Spectroscopy and Low Energy Electron Diffraction. A 3 ML cobalt film was deposited using Omicron evaporator with the deposition rate of 0.5 ML per minute. The quality of the film was monitored by Low energy electron diffraction (LEED) and Auger electron spectroscopy. The LEED picture shows that the grown film was epitaxial with *hcp* structure and (0001) axis perpendicular to the sample surface.

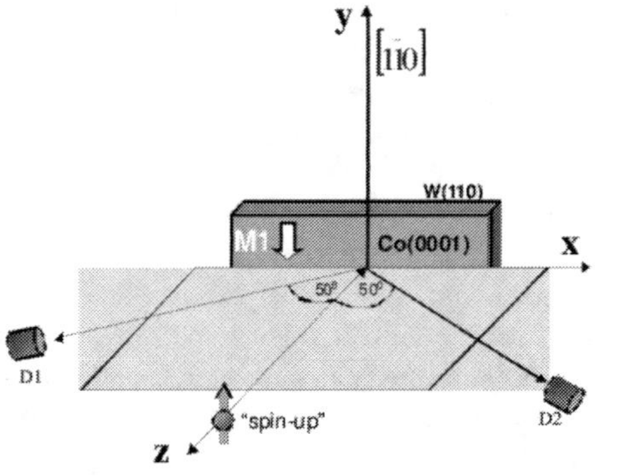

Figure 1. Geometry of the experiment. M1 – magnetization direction.

We applied spin-polarized two-electron spectroscopy in reflection geometry (referred to as (e,2e) spectroscopy) [4] for measuring energy distributions of correlated electron pair excited from the cobalt film by spin-polarized incident electrons. The geometrical arrangement of the experiment is shown in Figure 1. The 25 eV incident electron beam was spin-polarized with (60-70) % degree of polarization (measured in separate experiment).

4. Results and discussion

The energy conservation law in the (e,2e) reaction allows to determine the binding energy E_b of the valence electron involved in the scattering using measured energies of two correlated electrons: $E_b = (E_1 + E_2) - E_o$, where E_o is the incident electron energy, E_1 and E_2 are the two detected electrons energies. Two-dimensional energy distributions of correlated electron pairs from 3 ML cobalt film has been recorded with spin-up and spin-down polarization of the incident beam. As was mentioned in the model description for a particular selection of correlated electron pairs when both of the electrons are detected with equal energies and at symmetric angles the asymmetry of the binding energy spectrum represents the asymmetry (with "minus" sign) of majority and minority densities of states in the valence band. We plot this asymmetry and compare it with the asymmetry calculated using spin-up and spin-down densities of states from [5]. This comparison is presented in Figure 2.

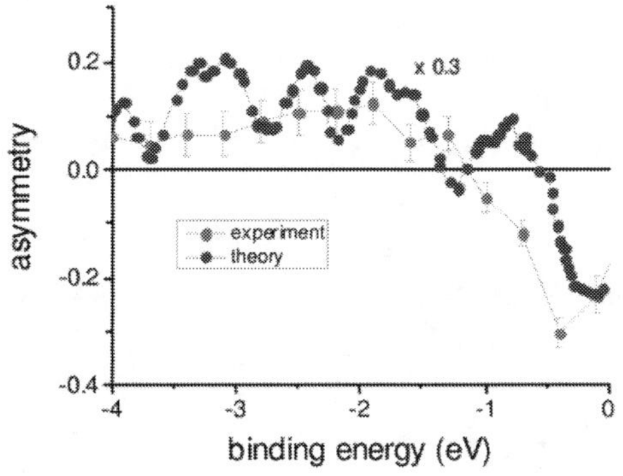

Figure 2. Comparison of measured and calculated asymmetries of binding energy spectrum.

The theoretical curve is scaled down by a factor of 3 to be compared with the experimental curve. Both curves show similar tendency: they have negative value just below the Fermi level (zero on the x axis) and then cross the x- axis and go positive. The theoretical curve oscillates and actually crosses the x- axis three times. Our experimental curve does not show such oscillations. This might be due to an insufficient energy resolution in the experiment that is estimated to be 0.5 eV and integration over the angle of 5^0 to have a reasonable statistics. The lower value of the experimental asymmetry than the theoretical one might be due to the misalignment of magnetic domains in the cobalt film.

4. Conclusions

A simple model is proposed to describe a spin asymmetry of the binding energy spectrum measured using spin-resolved (e,2e) spectroscopy on a ferromagnetic surface. We model our experiment by scattering of the primary electron with the spin projection $<s_1>$ by the valence electron with the spin projection $< s_2>$. One can assume that valence electrons within binding energy range ε to $\varepsilon + d\varepsilon$ have an average spin projection $\zeta_2(\varepsilon) = <s_2>$ defined by spin-asymmetry of the density of states within this energy range. The incident electron beam with spin polarization **P** is characterized by an average spin $\zeta_1 = <s_1> = \frac{1}{2}\mathbf{P}$. This model is applied to analyze the asymmetry of the measured binding energy spectrum of the cobalt film. The comparison of the measured and calculated spectra shows a qualitative agreement.

5. Acknowledgments

This research was supported by the Australian Research Council and the University of Western Australia. We thank S. Key, G. Light (UWA) for their technical support. One of the authors (O.M.A.) is grateful to the RFBR (◘ 06-02-17174-a) for financial support.

References

[1] Samarin SN, Berakdar J, Artamonov O and Kirschner J 2000 *Phys. Rev. Lett.* **85** 1746.
[2] Landau LD and Lifshitz EM 2003 *Quantum Mechanics (Non-relativistic Theory) Volume 3 Third Edition. Course of Theoretical Physics (Butterworth Heinemann)* p 574.
[3] Berakdar J 1999 *Phys. Rev. Lett.* **83** 5150.
[4] Morozov A, Berakdar J, Samarin SN, Hillebrecht FU, Kirschner J 2002 *Phys. Rev.* **B 65** 104425.
[5] Spišác D and Hafner J 2004 *Phys. Rev.* **B 70** 014430.

Journal of Physics: Conference Series **185** (2009) 012042 doi:10.1088/1742-6596/185/1/012042

Electron impact excitation of the $5p^5 7p$ levels of xenon from $5p^5 6s$ metastable states

Lalita Sharma[1], Rajesh Srivastava[1] and A. D. Stauffer[2]
[1]Department of Physics, Indian Institute of Technology, Roorkee 247667, India
[2]Department of Physics and Astronomy, York University, Toronto, ON, Canada M3J 1P3

Abstract
Relativistic distorted-wave (RDW) calculations have been carried out for the electron impact excitation from the lowest metastable states of xenon (the J = 0, 2 levels of the $5p^5 6s$ configuration) to the ten higher-lying fine-structure levels of the $5p^5 7p$ configuration. Integrated cross section results are reported for incident electron energies up to 300 eV and their analytic fits are also provided for obtaining results at higher energies.

1. Introduction

The importance of electron impact excitation cross sections from the lowest metastable levels of the inert gases, i.e. the $J = 0$, 2 levels of the np^5(n+1)s configuration to all ten fine-structure levels of the np^5(n+1)p and np^5(n+2)p configurations (with $J = 0$, 1, 2 and 3) has been repeatedly emphasized [1-4]. Experimental work to study excitations of the np^5(n+2)p configurations is in progress by the Wisconsin Group. They recently reported experimental results for argon [2] and we have carried out relativistic distorted-wave (RDW) calculations for these excitations [3]. In the present brief report we consider similar excitations in xenon in the light of further ongoing measurements. Results for such excitations in xenon have also been in demand as these contribute significantly to plasma modeling, in particular for plasma diagnostic of xenon-fed thrusters [4,5]. The role and effect of our calculated cross section in the plasma diagnostic of these thrusters is currently under way using a collisional radiative model (CRM).

We report here detailed results from our RDW calculations to obtain the integral cross sections (ICS) for the fine-structure transitions in xenon from the $5p^5 6$s metastable levels to the higher lying fine structure levels of $5p^5 7p$ configurations. The details of the theoretical calculations of the cross sections have been given in our earlier publications [1,3]. In the next section we describe the results for the atomic target wave functions of xenon. Section 3 presents our ICS results and their analytic fits.

2. Wave Functions

The ground state of xenon has a $5p^6$ outer shell in the nonrelativistic representation. Since we are using the relativistic j-j coupling scheme, the $5p^5$ core of the excited states is split into two separate configurations, $5\bar{p}5p^4$ and $5\bar{p}^2 5p^3$, with total angular momenta j_c of 1/2 and 3/2, respectively. The lowest lying metastable states of xenon with total angular momentum $J = 2$ and $J = 0$ (designated as $1s_5$ and $1s_3$, respectively, in the Paschen notation) have the configurations $5\bar{p}^2 5p^3 6s$ and $5\bar{p}5p^4 6s$ in the j-j coupling scheme. Thus the hole in the valence shell is different for these two metastable states.

We are considering the electron impact excitation to the ten individual fine-structure levels of the $5p^5 7p$ manifold. In the j-j coupling scheme, these states are linear combinations of the configurations $5\bar{p}^2 5p^3\, 7\bar{p}$ with $J = 1$, 2; $5\bar{p}^2 5p^3 7p$ with $J = 0$, 1, 2, 3; $5\bar{p}5p^4 7\bar{p}$ with $J = 0$, 1 and $5\bar{p}5p^4 7p$ with $J = 1$, 2. In the Paschen notation, the fine-structure levels are denoted as $3p_i$ where i varies from 1 to 10.

We have used the GRASP92 program [6] to obtain Dirac-Fock target-state wave functions for Xe which produced good agreement with the measured values of the oscillator strengths for these transitions where available. Our calculated dipole oscillator strengths for $1s_5$-$3p_6$ and $1s_5$-$3p_8$ transitions are 2.30×10^{-3} and 4.60×10^{-3} compared with the values available from NIST database (http://physics.nist.gov/PhysRef Data/ASD/ index. html) of 2.67×10^{-3} and 5.86×10^{-3} respectively. The estimated error in NIST values is 50% and oscillator strengths are available only for these two transitions. Table 2 shows our calculated excitation energies for different transitions as compared to the NIST values and can be said to be in reasonable agreement.

The contributions from the various configurations to the fine-structure levels of the final states are given in table 2. The $3p_8$ level with $J = 3$ has the single configuration $5\bar{p}^2 5p^3 7p$. Note that we have not calculated wave functions for the $3p_1$ level with $J = 0$ in Xe. This level with an excitation energy of 22.23 eV lies much above the first ionization potential of 12.13 eV for Xe and therefore accurate representation of this level cannot be achieved within a Dirac-Fock approximation without including mixing with the continuum. We observe that with the exception of the $3p_7$ and $3p_{10}$ levels these states are overwhelmingly dominated by a single configuration and that these two exceptions come from a single core with a $\bar{p}$ hole.

Table 1: Excitation energies of various transitions in Xe

Transition	GRASP	NIST
$1s_5$-$1s_3$	1.30104	1.13188
$1s_5$-$2p_{10}$	1.85745	2.58626
$1s_5$-$2p_9$	1.90877	2.63890
$1s_5$-$2p_8$	1.90529	2.65347
$1s_5$-$2p_7$	1.95456	2.68041
$1s_5$-$2p_6$	1.94552	2.68761
$1s_5$-$2p_5$	2.04625	2.69972
$1s_5$-$2p_4$	3.22909	3.94118
$1s_5$-$2p_3$	3.25242	3.96758
$1s_5$-$2p_2$	3.24499	3.96568

Table 2: Contributions of the various configurations to the $5p^5 7p$ states of Xe

Configurations	$5\bar{p}^2 5p^3 7\bar{p}$	$5\bar{p}^2 5p^3 7p$	$5\bar{p} 5p^4 7\bar{p}$	$5\bar{p} 5p^4 7p$
$J = 0$ levels				
$3p_5$		0.9976	0.0024	
$J = 1$ levels				
$3p_2$	0.0011	9.37×10^{-4}	0.0029	0.9950
$3p_4$	6.76×10^{-6}	4.25×10^{-4}	0.9966	0.0031
$3p_7$	0.4779	0.5219	2.76×10^{-4}	3.61×10^{-6}
$3p_{10}$	0.5210	0.4768	2.19×10^{-4}	0.0020
$J = 2$ levels				
$3p_3$	0.0017	4.36×10^{-5}		0.9998
$3p_6$	0.0421	0.9579		8.84×10^{-5}
$3p_9$	0.9577	0.0422		1.57×10^{-4}

3. Results and Discussions

We present cross sections for the electron impact excitation of xenon from the $1s_3$ and $1s_5$ metastable levels to the fine- structure levels of the $5p^5 7p$ manifold (designated as $3p_i$, i = 2 - 10). Figure 1 displays the cross sections for these transitions in the energy range up to 300 eV. In figures 1(a) and 1(b) we show the results for the dipole allowed transitions where the final states have respectively a $\bar{p}$ or p hole in the valence shell. A transition from the $1s_5$ level with configuration $5\bar{p}^2 5p^3 6s$ and $j_c = 3/2$ to a level whose wave functions are comprised almost entirely of configurations $5\bar{p} 5p^4 7p$ with $j_c = 1/2$ represents a two-electron transition. For such transitions cross section would be expected to be smaller compared to those from the $1s_3$ level with configuration $5\bar{p} 5p^4 6s$. This can be clearly seen in figure 1(a). Likewise, excitations of levels with wave functions dominated by the $5\bar{p}^2 5p^3 7p$ configuration with $j_c = 3/2$ from the $1s_3$ level would be expected to have smaller cross sections than those for excitations from the $1s_5$ state as shown in figure 1(b). Similar behaviour was shown to be the case for the excitation to the $np^5(n+1)p$ levels of Kr (n=4) and Xe (n=5) and to the $np^5(n+2)p$ levels of Ar (n=3) [1,3] where the wave functions are dominated by configurations with a single value of j_c.

Finally, figures 1(c) displays the results for the forbidden transitions. As compared to allowed transitions, these transitions have much smaller cross sections at higher energies though they have comparable magnitudes near threshold. Again, as discussed above, the cross sections for transitions between levels with the same core are larger than those where the core changes. These cross sections for all the forbidden transitions fall slightly faster than $1/E^3$ at higher energies. Note that the cross section for the transition from the $1s_3$ level to the $3p_8$ level with $J = 3$ is zero in our approximation from angular momentum coupling considerations.

We have fitted the ICS for the allowed transitions to the Bethe-Born form

$$ICS = \frac{1}{E}(c_0 + c_1 \ln(E))a_0^2 \tag{1}$$

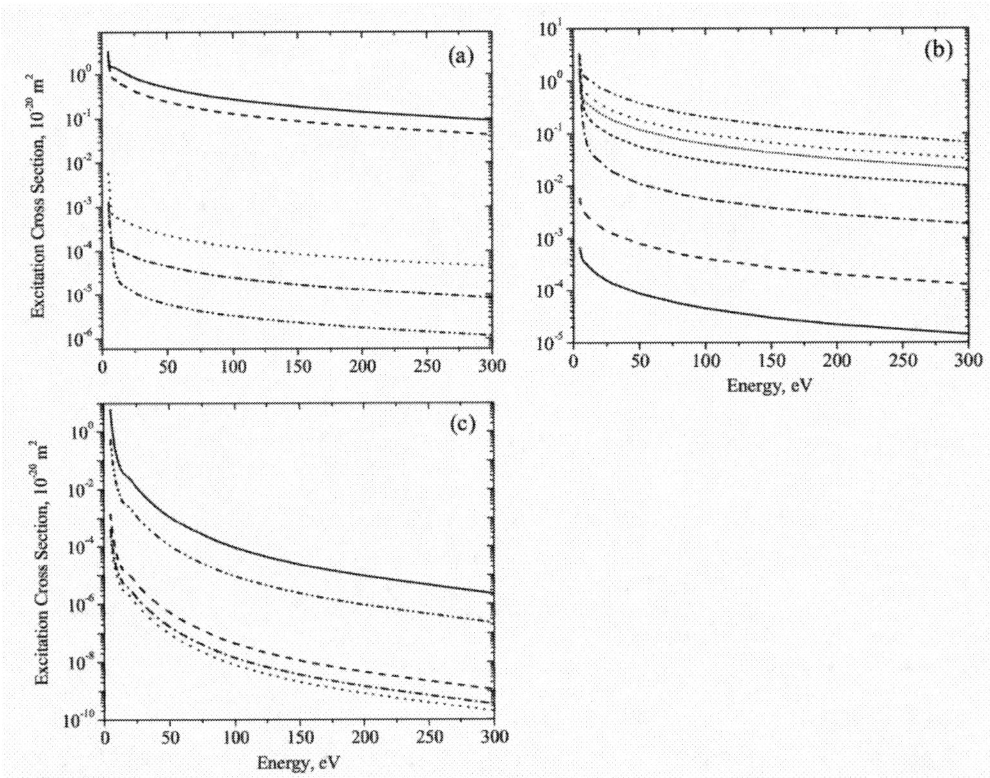

Figure 1:ICS for electron excitation of xenon: **(a)** the solid curve $1s_3$ - $3p_2$, the dashed curve $1s_3$-$3p_4$, the dotted curve $1s_5$-$3p_2$, the dashed-dot curve $1s_5$-$3p_3$, and the dashed-double-dot curve $1s_5$-$3p_4$ transitions; **(b)** the solid curve $1s_3$ - $3p_7$, the dashed curve $1s_3$-$3p_{10}$, the dotted curve $1s_5$-$3p_6$, the dashed-dot curve $1s_5$-$3p_7$, the dashed-double-dot curve $1s_5$-$3p_8$, the short-dashed curve $1s_5$-$3p_9$, the short-dotted curve $1s_5$-$3p_{10}$ transitions; **(c)** the solid curve $1s_3$ - $3p_3$, the dashed curve $1s_3$-$3p_5$, the dotted curve $1s_3$-$3p_6$ transitions, the dashed-dot curve $1s_3$-$3p_9$, and the dashed-double-dot curve $1s_5$-$3p_5$ transitions.

As well, we have found the cross sections for the forbidden transitions behave as

$$ICS = d_0 E^{d_1} a_0^2 \tag{2}$$

Journal of Physics: Conference Series **185** (2009) 012042 doi:10.1088/1742-6596/185/1/012042

at higher energies. The fitting constants c_0, c_1, d_0 and d_1 are given in table 3 and 4. The energy is in atomic unit. Since the excitation energies for transitions from the metastable states are less than 4 eV, these fittings can be utilized from 30 eV onwards. Note that the magnitude of c_0 and d_0 reflect whether or not there is a change in the core configuration as discussed above.

We have found that the cross sections for the forbidden transitions considered in Ar [3] behaved approximately as E^{-3} as E increases while in this case the power is slightly more negative at around -3.4.

Table 3: Values of the constant c_0 and c_1 in equation (1) for the allowed transitions

Initial State	$1s_5$		$1s_3$	
Final State	c_0	c_1	c_0	c_1
$3p_{10}$	0.74623	0.06233	4.89×10^{-3}	5.49×10^{-4}
$3p_9$	0.36277	0.02033	-	-
$3p_8$	2.33589	0.27723	-	-
$3p_7$	0.07017	2.01×10^{-3}	5.47×10^{-4}	3.71×10^{-5}
$3p_6$	1.11031	0.12797	-	-
$3p_4$	3.52×10^{-5}	5.74×10^{-6}	1.51459	0.11233
$3p_3$	2.78×10^{-4}	2.78×10^{-5}	-	-
$3p_2$	1.43×10^{-3}	1.41×10^{-4}	3.04581	0.37606

Table 4: Values of the constant d_0 and d_1 in equation (2) for the forbidden transitions

Atom	Xenon	
Transitions	d_0	d_1
$1s_3$ -$3p_9$	4.32×10^{-6}	-3.41409
$1s_3$ -$3p_8$	-	-
$1s_3$ -$3p_6$	2.43×10^{-6}	-3.41560
$1s_3$ -$3p_5$	1.32×10^{-5}	-3.40019
$1s_3$ -$3p_3$	3.02×10^{-2}	-3.45539
$1s_3$ -$3p_1$	-	-
$1s_5$ -$3p_5$	2.99×10^{-3}	-3.45280
$1s_5$ -$3p_1$	-	-

4. Conclusions

We have used RDW method to study electron excitation of the lowest lying metastable states of xenon to the higher lying fine structure states of the $5p^57p$ manifold. We have given a detailed analysis of the individual fine-structure transitions which are reflected in our results. Our present theoretical study will supplement ongoing experimental work in this field and provides reliable data for use in plasma modeling studies.

Acknowledgements

ADS is grateful to the NSERC of Canada for a grant in support of this research. RS would like to thank the Asian Office of Aerospace Research and Development ,Tokyo, Japan (Grant No. 064029) for support.

References

[1] Srivastava R, Stauffer A D and Sharma L, 2006 *Phys. Rev.* A **74** 012715
[2] Jung R O, Boffard J B, Anderson L W and Lin C C, 2007 *Phys. Rev.* A **75** 052707
[3] Lalita Sharma, Rajesh Srivastava and A D Stauffer, 2007 *Phys. Rev* A **76** 024701
[4] Chiu Y-H, Austin B L, Williams S, Dressler R A and Karabadzhak G F, J. 2006 *Appl. Phys.* **99** 113304
[5] Katsonis K and Berenguer Ch, Report LPGP-GA-**21** (2008)
[6] Parpia F A, Froese Fischer C and Grant I P 1996 *Comput. Phys. Commun.* **94** 24

Journal of Physics: Conference Series **185** (2009) 012043 doi:10.1088/1742-6596/185/1/012043

Collisions between anti-particles in presence of laser field

C Sinha[1] and S Ghosh Deb
Department of Theoretical Physics, Indian Association for the Cultivation of Science, Jadavpur, Kolkata-700 032, India

E-mail: chand_sin@hotmail.com

Abstract. Three body recombination (TBR) process for the formation of antihydrogen $(\overline{H})$ inside a trapped plasma of antiprotons $(\overline{p})$ and positrons (e^{+}), supposed to be the most efficient mechanism for the production of cold and trapped antihydrogen is studied quantum mechanically. Present study also involves a background laser field anticipating further enhancement of the $\overline{H}$ formation cross sections. Significantly large cross sections are obtained in the presence of the field at lower incident energies as compared to the field free ones. Cross sections are enhanced for higher excited states of $(\overline{H})$.

1. Introduction

The recent technological advances in the cooling and trapping mechanism of antiproton $(\overline{p})$ and positron (e^{+}) has now made it feasible to produce cold and trapped antihydrogen $(\overline{H})$ in the laboratories [1-6] which is essential to achieve the long term goal of performing high precision spectroscopic as well as gravitational studies in order to test the CPT.

The most efficient method for a significant production of cold $\overline{H}$ is supposed to be the three body recombination process of an antiproton $(\overline{p})$ and positron plasmas. Further, irradiation by laser light in the above plasmas is predicted to enhance the $\overline{H}$ production. The present work addresses to this process, e.g. the laser assisted $\overline{H}$ formation through the following three body recombination process:

$$\overline{p} + e^{+} + e^{+} \rightarrow \overline{H} + e^{+} \tag{1}$$

where the e^{+} (the third body) carries off the extra energy efficiently.

Recently the ATRAP [7] has also devised a laser assisted (LA) method of producing $\overline{H}$ that might give colder atoms which is now in the preliminary stage.

2. Theory

The prior form of the transition matrix element [8] for the LA process (1) is given by:

$$T_{if} = \int_{-\infty}^{\infty} dt < \psi_f^{-}(\vec{r}_1, \vec{r}_2)(1 + \vec{P}) | V_i | \psi_i(\vec{r}_1, \vec{r}_2) \rangle \tag{2}$$

$\vec{P}$ denotes the exchange operator, the initial channel perturbation $V_i = \dfrac{1}{r_{12}}$. The external laser field is chosen to be monochromatic, linearly polarised and spatially homogeneous represented by the vector potential (in the Coulomb gauge) by $\vec{A}(t) = A_0 \cos(wt + \delta)$ with $A_0 = c\varepsilon_0 / \omega$, δ being the initial phase and ε_0 is the field amplitude. The incident positrons are dressed by the laser field and are

[1] To whom any correspondence should be addressed.

represented by Coulomb Volkov wavefunction while the laser dressed final channel spectator positron e^+ are represented by Eikonal modified Volkov wavefunction [9].The dressed wavefunction of the antihydrogen ($\overline{H}$) in the final channel is constructed in the framework of time dependent perturbation theory.

The Differential cross section (DCS) is given by

$$\frac{d\sigma}{d\Omega}=\frac{k_f}{k_1k_2}[\frac{1}{4}|f+g|^2+\frac{3}{4}|f-g|^2] \quad (3)$$

where f is the direct amplitude and g is the exchange amplitude.

3. Results and Discussion

Laser assisted differential (MDCS) as well as total cross sections (MTCS) for the $\overline{H}$ formation in ground and excited states (2s, 2p) states are computed for multiphoton (MP) processes.

Figures 1-3, representing the 1s, 2s and 2p MDCS reveals that the cross section is significantly enhanced by a factor of 2.5 – 4.5 in the presence of the laser field.

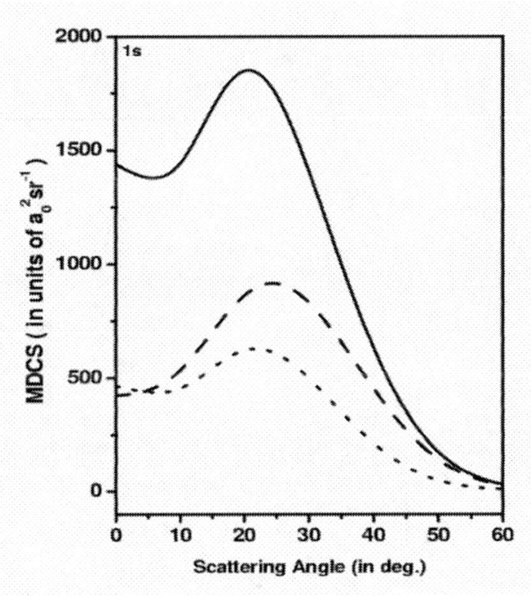

Figure 1. MP Differential cross sections (MDCS) for antihydrogen formation in the 1s state in the antiproton-positron-positron collisions for different values of E_1 and E_2. Solid line: $E_1=15,\ E_2=10\ eV$, dashed line: $E_1=10,\ E_2=15\ eV$, dotted line: $E_1=15,\ E_2=15\ eV$.

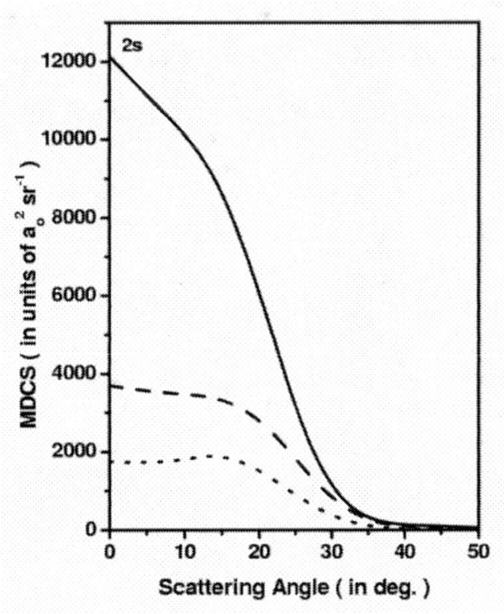

Figure 2. Same MDCS as in 2s state figure1 but for the 2s state.

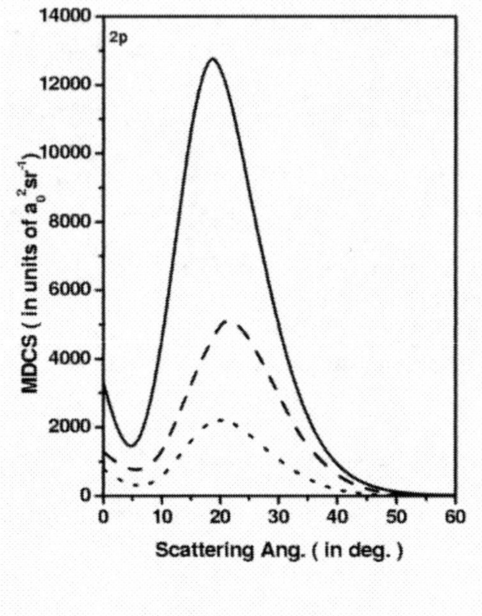

Figure 3. Same MDCS as in figure 1 but for the 2p state.

Fig. 4 displays the TCS against the active e^+ energy (E_1) at a fixed $E_2 = 15\ eV$ (spectator e^+). The inset of figure 4 displays the corresponding field free (FF) results. The general trend of the TCS is as follows: At low energies of E_1: $2p > 2s > 1s$ for both LA and FF cases, whereas at higher E_1, the behaviour is different in two cases, e.g., $2s > 2p > 1s$ for FF case and $2s > 1s > 2p$ for LA case.

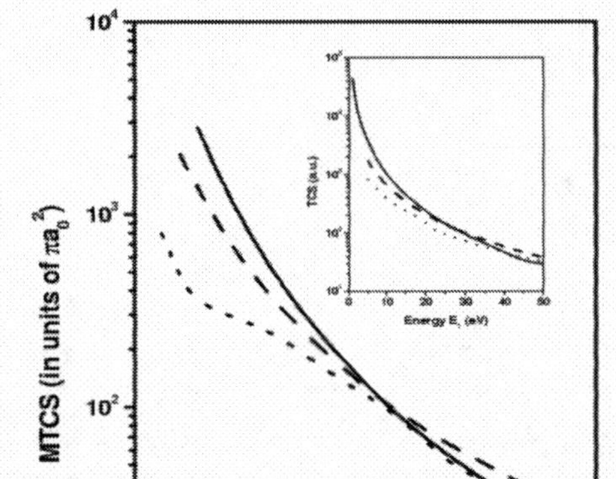

Figure 4. MP Partial total cross section (MTCS) for antihydrogen formation in 1s, 2s, 2p states with respect to E_1 for $E_2 = 15\ eV$. Inset shows the same TCS for the field free case. Solid line for 2p state, dashed line for 2s state, dotted line for 1s state.

Table 1. TCS for $\overline{H}$ formation in the ground (1s) and excited (2s, 2p) states along with different theoretical result

Energy (eV)	Mitroy et al (close coupling) [10]		Mitroy et al (Unitarised Born) [11]			Sinha et al (Charge transfer) [9]		Sinha et al FF TBR [8]			Present Results LA TBR		
States	1s	2s+2p	1s	2s	2p	FF(1s)	FA(1s)	1s	2s	2p	1s	2s	2p
13.6	1.923	8.44	1.46	0.66	3.81	0.92	8.32	373.07	705.4	530.1	383.6	743.9	1142.2
25.84	0.735	1.73	-	-	-	0.29	2.05	23.16	15.43	11.08	14.63	14.42	9.9

The following inferences could be drawn from the Table:

(1) The present TBR cross sections are >> (by orders of magnitude) than the other processes leading to $\overline{H}$ except at very high energies (~ 100 eV) where the radiative recombination (RR) process dominates over the TBR .

(2) At very low incident energies the present TBR dominates in higher excited states of $\overline{H}$ while at intermediate and high energies the ground state dominates.

4.Conclusions

No significant changes are noted in the qualitative behaviour of the LA cross sections regarding different quantum states or with respect to energies E_1 and E_2 while the quantitatively the LA TCS are enhanced by orders of magnitude as compared to FF ones. Among all the $\overline{H}$ formation processes, the TBR is the leading candidate and can be further enhanced if the $\overline{H}$ is in the excited states.

5.References

[1] Amoretti M et al 2002 *Nature (London)* **419** 456
[2] Amoretti M et al 2006 *Phys. Rev. Lett.* **97** 213401 (see also other refs. cited therein in this context)
[3] Gabrielse G et al 2002 *Phys. Rev. Lett.* **89** 213401
[4] Gabrielse G et al 2007 *Phys. Rev. Lett.* **98** 113002 (see also other refs. cited therein in this context)
[5] Pohl T, Sadeghpour H R and Gabrielse G 2006 *Phys. Rev. Lett.* **97** 143401
[6] Andresen G B et al 2007 Phys. Rev. Lett. **98** 023402
[7] Andresen G B et al 2008 *J. Phys. B: Atomic, Molecular and Optical Physics* **41** 011001
[8] Roy S, Ghosh Deb S and Sinha C 2008 *Phys. Rev. A* **78** 022706
[9] Chattopadhyay A and Sinha C 2006 *Phys. Rev. A* **74** 022501
[10] Mitroy J and Ryzhikh G 1997 *J. Phys. B: Atomic, Molecular and Optical Physics* **30** L371
[11] Mitroy J and Stelbovics A T 1994 *J. Phys. B: Atomic, Molecular and Optical Physics* **27** L79

Journal of Physics: Conference Series **185** (2009) 012044 doi:10.1088/1742-6596/185/1/012044

Positrons as Imaging Agents and Probes in Nanotechnology

Suzanne V Smith[1]

[1]Node Director, ARC COE Anti-matter Matter Studies, Australian Nuclear Science and Technology Organisation, PO Box 1, Menai, 2234 NSW, Australia

E-mail: suzanne.smith@ansto.gov.au

Abstract. Positron emission tomography (PET) tracks a positron emitting radiopharmaceutical injected into the body and generates a 3-dimensional image of its location. Introduced in the early 70s, it has now developed into a powerful medical diagnostic tool for routine clinical use as well as in drug development. Unrivalled as a highly sensitive, specific and non-invasive imaging tool, PET unfortunately lacks the resolution of Computer Tomography (CT) and Magnetic Resonance Imaging (MRI). As the resolution of PET depends significantly on the energy of the positron incorporated in the radiopharmaceutical and its interaction with its surrounding tissue, there is growing interest in expanding our understanding of how positrons interact at the atomic and molecular level. A better understanding of these interactions will contribute to improving the resolution of PET and assist in the design of better imaging agents. Positrons are also used in Positron Annihilation Lifetime Spectroscopy (PALS) to determine electron density and or presence and incidence of micro- and mesopores (0.1 to 10 nm) in materials. The control of porosity in engineered materials is crucial for applications such as controlled release or air and water resistant films. Equally important to the design of nano and microtechnologies, is our understanding of the microenvironments within these pores and on surfaces. Hence as radiopharmaceuticals are designed to track disease, nuclear probes (radioactive molecules) are synthesized to investigate the chemical properties within these pores. This article will give a brief overview of the present role of positrons in imaging as well as explore its potential to contribute in the engineering of new materials to the marketplace.

1. Positron Emission Tomography (PET) and Drug Development

Positron emission tomography (PET) was developed some 30 years ago, however it was not until the 1990s that it was deployed as a tool for imaging the physiology and pathology of the brain, and later the heart. It is now been widely used, in many countries for the detection and staging of a variety of malignant diseases. PET differs from other nuclear medicine procedures such as single photon emission tomography (SPECT), by its ability to detect coincidence signals, which allows the quantification of regional tissue radioactivity.[1-5] This is achieved by injecting a PET imaging agent or radiopharmaceutical incorporating a positron-emitting radioisotope such as ^{11}C, ^{13}N, ^{15}O or ^{18}F. These radioisotopes decay to emit a positron that loses sufficient energy to form positronium, and undergo annihilation to emit two 511 keV gamma rays (see figure 1). Those positrons that do not form positronium will undergo an alternate process called free electron annihilation. The probability of the former process is generally higher. The ability to detect these gamma rays in coincidence in two detectors at 180 degree is the key to locating the position of the radiopharmaceutical in the body.

[1]To whom any correspondence should be addressed.

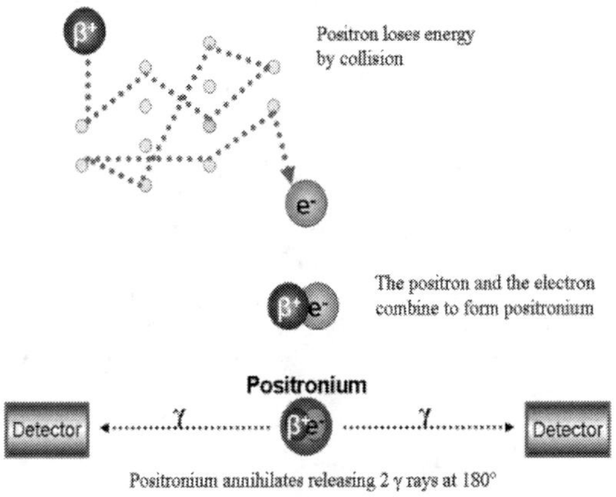

Figure 1 The positron annihilation process generates two gamma rays that are detected in coincidence by the PET camera.

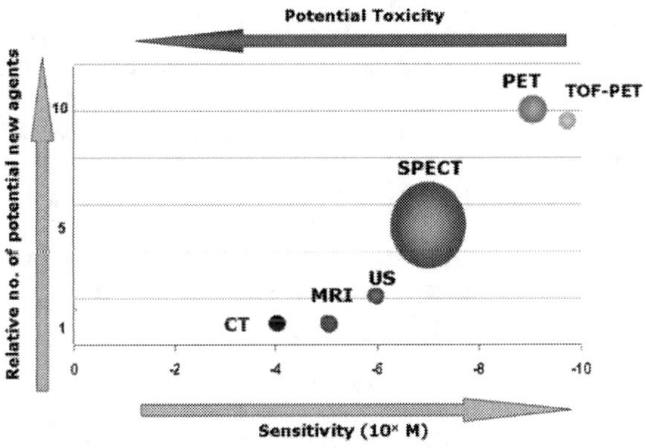

Figure 2 A comparison of spatial resolution (relative size of symbols) sensitivity and potential toxicity of non-invasive imaging modalities.

Compared with other non-invasive imaging tools, PET has greater sensitivity ($<10^{-8}$M) and more specificity for its targets, but it lacks the resolution of other non-invasive imaging agents, such as MRI and CT (see figure 2).[3] An advantage of PET, is the availability of an array of positron emitting radioisotopes, (with varying half lives and chemistries) which allow for the development of a wide range of PET radiopharmaceuticals for imaging many biological processes.

In more recent times PET has been used in the risk assessment of new pharmaceuticals. Here the drug is radiolabelled with the positron emitter and the resultant product or probe is used to investigate where the drug (e.g. peptide, protein, or nanoparticle) goes in the body. The strategy allows the pharmaceutical industry to monitor the uptake, distribution and pharmacokinetics of their drug *in vivo* (i.e. preclinical and humans trials). The ability to monitor the product movement non-invasively assists the company to make more accurate assessment of the product's therapeutic index and ultimately it's effectiveness in the long term. The current success rate of products through phase I clinical trial to the market is approximately 20%. When effectively implemented non-invasive imaging is estimated to generate cost savings of up to 50% for the commercialisation of a product.[6] In addition, non-invasive imaging can contribute to personalising treatments, by assessing the drug movement within an individual and providing information to ensure the dosage of drug intended is appropriate for that individual. This has the potential to reduce or highlight potential toxic effects.

2. Emerging Positron Radioisotopes

Many drug development companies have already implemented PET technology (using ^{11}C and ^{18}F radioisotopes) to fast track their research programmes. The short half lives and the requirement for specialised skills to incorporate these isotopes have limited the application of PET to small molecules (i.e. < 500 mol wt) with fast biological clearance rates. For larger molecules, (such as peptides, antibodies, DNA and nanoparticles) that have slower clearance rates, a PET isotope with longer half life is more appropriate. Copper 64, (^{64}Cu) is an emerging PET isotope that has a positron energy similar to ^{18}F, but a longer half life (7 fold) which makes it ideal for attaching to larger target agents (i.e. > 1000s mol wt). It can be produced in a reactor (using natural copper or enriched zinc targets) and in a cyclotron (using enriched nickel or zinc targets). [7-18] The first routine production for ^{64}Cu, as a by-product of the ^{67}Ga production (using enriched ^{68}Zn) using a 30 MeV was developed in

Australia by 1996.[11,19] In more recent times, it has been produced using low energy (15-12 MeV) cyclotrons and ^{64}Ni targets. [13]

Unlike ^{11}C and ^{18}F, ^{64}Cu can not be covalently attached or incorporated into the molecule and therefore needs to be attached through a bi-functional ligand. That is a ligand incorporating two parts; one for linking to the target agent or drug, and the other to complex the ^{64}Cu. An array of polyazamacrocycles, polyazacarboxylate macrocycles and hexaaza cage ligands have been designed for radiolabelling with ^{64}Cu. However, only the hexazacage, **sarar**[2] can complex ^{64}Cu quantitatively over a wide pH range (4-9) at room temperature which makes it ideal for easy production and use in "kit formulations" (e.g. mixing of two vials of reagent at clinical site).[19] Furthermore, the three-dimensional encapsulation of ^{64}Cu yields a kinetically inert complex that is ideal for the risk assessment of target agents. It ensures the PET signal is always associated with the target agent and therefore more accurate information regarding the biological pathway of the target agents can be obtained.

Table 1 Selected positron radioisotopes and their physical characteristics.

Isotope	Half-life	Positron Energy (average keV)	Gamma energy and proability (keV ; %)
^{82}Rb	72 sec	1535	776.5 ; 13
^{15}O	122.24 sec	735	Nil
^{11}C	20.39 min	385	Nil
^{18}F	109.77 min	249	Nil
^{68}Ga	67.63 min	836	1077 ; 3
^{64}Cu	12.7 hours	278.2	1345 ; 1.27
^{124}I	4.176 days	1540 (max)	603 ; 62.9 1691;10.88

While the range of positron radioisotopes (see Table 1) generates plenty of opportunity for developing radioactive mimics of drugs, the relatively poor resolution of PET hinders its wider deployment in patient management. The resolution of PET is highly dependent on the energy of the positron, as the annihilation process requires the positron to slow to energies below ~100 eV and form positronium or annihilate with free electrons. A comparison of Monte Carlo simulations of the path length of different positrons in various tissues illustrated in figure 3, shows how the resolution of the PET image can depend not only on the energy of the positron but also the type of tissue from which it annihilates.[20]

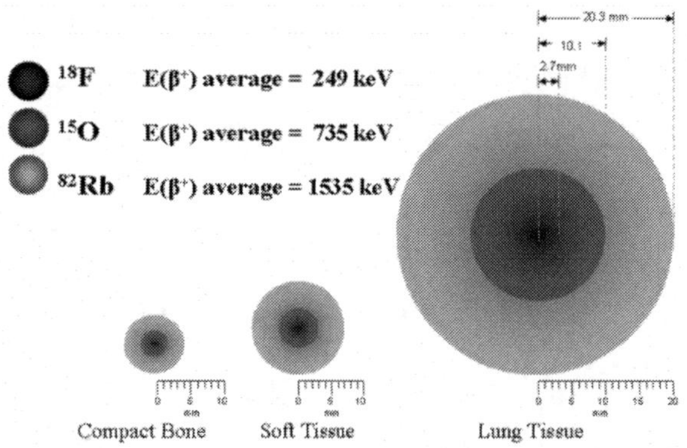

Figure 3 Monte Carlo calculations comparing the path length for different positrons in various tissues.[9]

For human PET cameras the resolution is of the order of 3 – 5 mm (for whole body) with ^{18}F radiopharmaceuticals. The potential resolution achievable for animal PET cameras is estimated to be 0.50 to 0.75 mm. However this will only be achievable with improvements in the following areas: sensitivity of detectors, count rate capability, statistical reconstructions, the development of high specific activity radiopharmaceuticals and the use of appropriate positron emitting radioisotopes. Understanding how positrons interact with molecules and materials is fundamental to developing strategies to improve the performance of PET and therefore its applications.

[2] See abbreviations for definitions

3. Positrons and Materials

Nature has developed exquisite biological systems that have the ability to sense, react, regulate, grow, regenerate and heal. Nature does this by designing systems (such as the cell) that are porous in nature and contain multiple microenvironments (e.g. nucleus, ribosome, and lysozomes) that can capture, conduct chemical reactions, release and adapt to their specific surroundings. The ability to design and synthesize materials (hard and soft matter) that can perform these types of operations and furthermore, progress their development into the marketplace is a key challenge for contemporary materials science. One fundamental property that appears to govern the design and performance of many new materials is their micro (< 2 nm) and mesoporosity (2 – 50 nm). Understanding the role of porosity in novel materials has implications in the design of catalysts, adsorbents, filtration and controlled release materials as well as anti-corrosive and water repellent films.

The sizes of the holes in materials can be readily determined using a number of routine techniques, such as transmission electron microscopy (TEM), Brunauer, Emmett and Teller (BET) surface analysis, and positron annihilation lifetime spectroscopy (PALS). TEM can give information at the atomic level however it does involve extensive sample preparation and many materials of interest are unstable under electron beam irradiation and high vacuum conditions. BET technique uses inert gases to probe the porosity and surface area of materials to give information on pore sizes from 3 – 100 nanometers. BET also requires high vacuum conditions and can give information about pore connectivity provided it is connected to surface pores. PALS can provide information on holes as small as an *atomic vacancy* (~0.2 nm) to those up to 10 nm in bulk materials as well as in films. The advantage of PALS is the minimal sample preparation, making it ideal for comparing changes in the properties of materials under going various processes.

PALS works by releasing a positron (from a ^{22}Na source; 10μCi) into desired material (~1 cm^3 position on either side of the source) that eventually annihilates with an electron in the pore or defect to emit gamma rays.[21] Hence the outgoing gamma rays carry information about the electronic density and pore architecture at the site of annihilation (see figure 4). Not only can the pore size and distribution be resolved, but information about the connectedness of pores can also be obtained. In figure 4 the extended decay profile for quartz indicates the presence of lower electron density or larger pores compared to the annealed nickel.

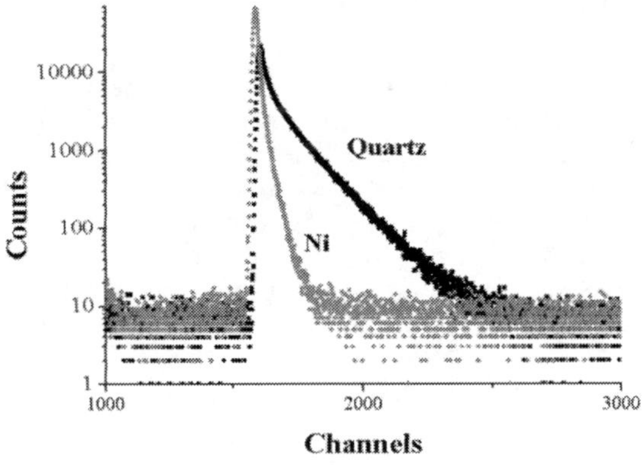

Figure 4 PALS spectra for Nickel and Quartz

These lifetime spectra are fitted to the continuous decay form:

$$S_{LT}(t) = \int \lambda\alpha(\lambda) \exp(\lambda t)d\lambda \quad (1)$$

where λ is an inverse of the positron lifetime ($=1/\tau$) and $\alpha(\lambda)$ is the probability density function (PDF) of the annihilation rate.[22]

The most common application of PALS has been in the semi-conductor industry; used to determine the presence and concentration of defects in these materials. More recent PALS has been used to study porosity in polymeric, ceramic, alloys and functionally gradient materials. Recently we have reported the use of PALS to study nanoporosity in complex biological materials such as wool and silk powders.[23]

4. Nanoporosity and Nuclear Probes

While understanding how the engineering of a material changes with the presence of micro and meso porosity, it is equally important to understand the chemistry within these microenvironments or

Journal of Physics: Conference Series **185** (2009) 012044 doi:10.1088/1742-6596/185/1/012044

nanopores. Particularly, when these materials are expected to react and respond to their changing environment by capturing a molecule from a solution for later release in another solvent system. Thus as one would design a drug for imaging disease in the body, nuclear probes are designed with different chemical characteristics (i.e. charge, size and hydrophobicity) to probe the physical and chemical properties of these nanopores.

Nuclear probes (radioactive target molecules) are highly sensitive (up to 10^{-5} ppb) and because they emit gammas, they can be used to study the interaction of materials with liquids (including high concentrations of electrolytes) and gases with negligible processing. Furthermore, studying of static or dynamic processes under experimentally relevant conditions can be adapted for high-throughput analysis. For example, three silk powders (Eri silk: Eri 1, Eri 2; Mulberry silk: Mulb; see figure 5), milled under various conditions, to different sizes (9 – 4.3 μm) and morphology; showed no change in microporosity (0.48 nm) or surface area when analysed by PALS or BET, respectively.[24] Exposing the silk powder to metal ions (i.e. $^{64/nat}Cu^{2+}$, $^{57/nat}Co^{2+}$ and $^{109/nat}Cd^{2+}$) at various pH showed changes in rate and therefore selectivity for these metal ions. Figure 5 shows a typical plot for $^{109/nat}Cd^{2+}$ absorption by these powders. These data suggest molecular structure of powders did not significantly alter compared to fibres on processing. However the processing the fibres into powders did increase the rate of diffusion of metal ions to binding sites.

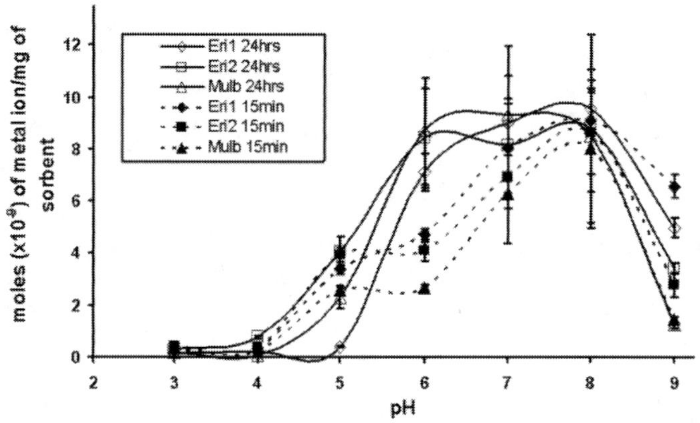

Figure 5 Absorption of $^{109/nat}Cd^{2+}$ to silk powder at various pH at 15minutes and 24 hours.[a]
[a] $[M^{2+}] = 10^{-4}$M; powder 10 mg; temp.23°C; vol, 1.0 mL; centrifuge; 5000 rpm

In addition, a range of nuclear probes (A – D; see figure 6) with overall positive and negative charges that are similar in size, show selective uptake by hollow silica shells over a range of pH. Negative (i.e. D=$[Co\text{-}dota]^{2-}$) and positively (A=$[Co\text{-}diamsar]^{2+}$; B=$[Co\text{-}sarar]^{2+}$; C=$[Co\text{-}sara\text{-}OH]^{2+}$) charged nuclear probes have optimum absorption into hollow silica particles at pH 3 and 7, respectively.[3] This indicates that there is an overall positive charge within the micropores of the silica particles that on incubation in buffer pH 7 could be made neutral or negative and therefore become responsive to the uptake of positively charge molecules as seen in figure 6. More interestingly, the subtle changes in molecular structures showed significant changes in uptake of these probes.

[3] See abbreviation for definitions

Journal of Physics: Conference Series **185** (2009) 012044 doi:10.1088/1742-6596/185/1/012044

Nuclear probes		**Percentage absorbed in silica shells**
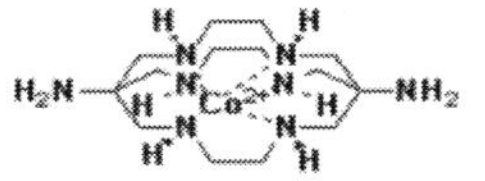	A	~71% at pH7
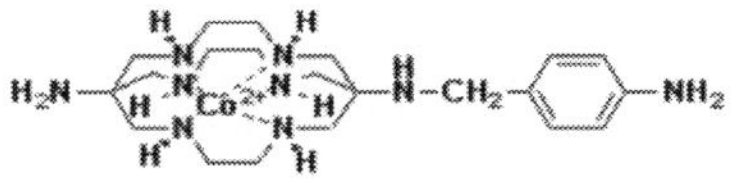	B	~ 83% at pH7
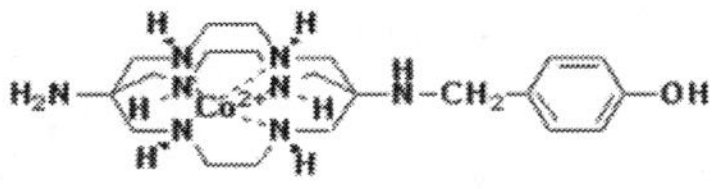	C	~97.5% at pH7
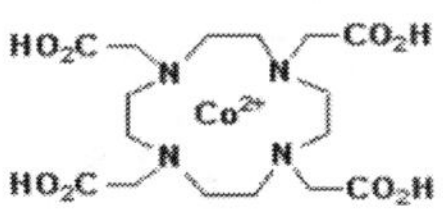	D	~20% at pH 7 and 50% at pH 3

Figure 6 Absorption of selected nuclear probes to silica particles at equilibrium.

5. Conclusion

A review of the strategies used to advance the development of pharmaceuticals to the marketplace offers valuable lessons for the commercialisation of nanotechnologies. In particular, the design of relevant structure/activity tests that can act as gateways or in the risk assessment of material early in the R and D program, are valuable long term strategies for the cost-effective development of materials (scheme 1). Positrons and highly sensitive and selective nuclear probes are essential tools for investigating the size, concentration and distribution of pores and defects in materials as well as their chemical properties. They can be readily applied within materials assessment programs and are adaptable to high-throughout analysis. Using nuclear probes, processes can be monitored within seconds, over days and up to months. Together with current materials characterisation tools they can increase our confidence in our ability to predict and accurately measure performance of materials. Nuclear probes can also be used to develop methods for the cost-effective scale up of synthesis and to assess or to forecast potential risk of these materials.

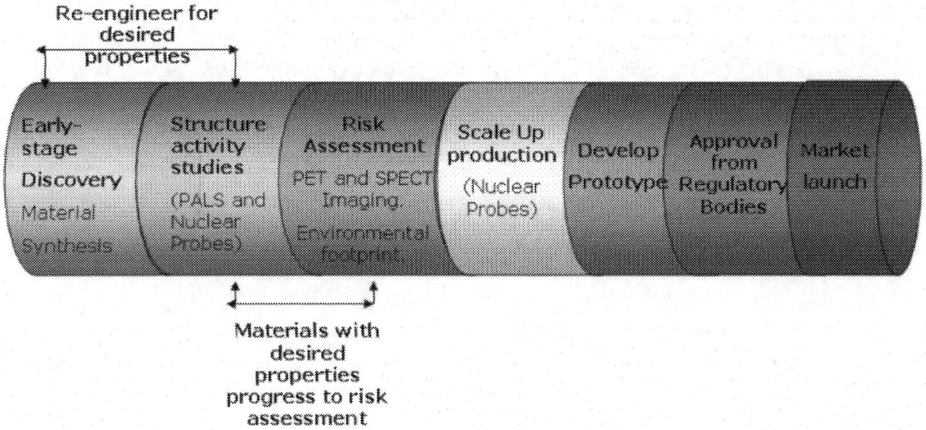

Scheme 1. Demonstrates the application of positrons and nuclear probes to the commercialisation path of new materials.

Acknowledgments
The author wishing to acknowledge financial support from Australian Research Council Center of Excellence in Antimatter Matter Studies and contribution from students post-doctorate and collaborators from Deakin University, University of Western Australia and Australian National University in Australia and University of Tsukuba in Japan. Excilia Ltd for supply of hollow silica particles.

References
[1] Buck A, Nekolla S, Ziegler S, Beer A, Krause B, Herrmann K, Scheidhauer K, Wester H and Rummeny E, Schwaiger M and Drzezga A 2008 Spect/Ct Journal Of Nuclear Medicine 49 19
[2] Harvey D 2006 Radiol. Today 7 20
[3] Smith S 2007 Expert Opinion in Drug Discov. 2(5) 72
[4] Ziegler S 2005 Proc. IEEE 93 70
[5] Phelps M 2002 Annu. Rev. Nucl. Part. Sci. 52 338
[6] Nunn A 2006 Invest Radiol 41 212; Dimasi J, Hansenr W and Grabowski H 2003 J. Health Econ. 185
[7] Mushtaq A, Karim H and Khan M 1990 J. Radioanal. Nucl. Chem 141 269.
[8] Szelecsenyi F, Blessing G and Qaim S 1993 Appl. Radiat. Isot 44 580.
[9] Boothe T 1991 Nucl. Instrum. Meth. Phys. Res. Sect. B: Beam Interact. Mater. Atoms B56-B57 1269.
[10] Hetherington E, Sorby P and Camakaris J 1986 Appl. Radiat. Isot. 37 1243.
[11] Smith S, Waters D and Di Bartolo N 1996 Radiochimica Acta 75 68.
[12] Mccarthy D, Shefer R, Klinkowstein R et al 1997 Nucl. Med. Biol. 24 43.
[13] Obata A, Kasamatsu S, Mccarthy D et al 2003 Nucl. Med. Biol. 30 539.
[14] Bonardi M, Groppi F, Birattari C et al 2003 J. Radioanal. Nucl. Chem. 257 241.
[15] Hilgers K, Stoll T, Skakun Y, Coenen H and Qaim S 2003 Appl. Radiat. Isot. 59 351.
[16] Tarkanyi F, Takacs S, Ditroi F et al 2004 Nucl. Instrum. Meth. Phys.Res. Sect. B: Beam Interact. Mater. Atoms 217 550.
[17] Bonardi M, Groppi F, Mainardi H et al 2005 J. Radioanal. Nucl. Chem. 264 105.
[18] Szelecsenyi F, Steyn GF, Kovacs Z et al 2005 Nucl. Instrum. Meth. Phys. Res. Sect. B: Beam Interact. Mater.Atoms 240 637.
[19] Smith S 2004 J. Inorg. Biochem. 98 1901
[20] Sanchez-Crespo A, Andreo P and Larsson S 2004 Eur. J. Nuc. Med. Mol. Imaging 31 51
[21] Rytsola J, Nissila J, Kokkonen J, Laakso A, Aavikko R and Saarinen K 2002 Appl. Sur. Sci. 194 263
[22] Saito H and Hyodo T 2003 Rad. Phys. Chem. 68, 431; Gregory R 1991 Nucl Instru Meth Phys Res Sect A, 303, 496.
[23] Guagliardo P, Uedono A, Sergeant T, Samarin S, Williams J and Smith SV 2008, Positron annihilation lifetime spectroscopy (PALS): a tool for exploring nanoporosity in biological materials. IC08, 14 - 18 December, Christchurch, New Zealand.
[24] Naik R, Rajkhowa R, Dharmaprakash MS, Hureau SA, Wang L, Wang X and Smith SV 2008, Metal absorption on eri and mulberry silk powders: a comparative study, IC08, 14 - 18 December, Christchurch, New Zealand.

Abbreviations
Because the IUPAC names for the ligands are long and complicated, those ligands described in this paper have been abbreviated as follows:
sarar = 1-*N*-(4-aminobenzyl)-3,6,10,13,16,19-hexaazabicyclo[6.6.6]eicosane-1,8-diamine
dota = (1,4,7,10-tetraazacyclododecane-*N*,*N*_,*N*_,*N*_-tetraacetic acid)
diamsar = 3,6,10,13,16,19-hexaazabicyclo[6.6.6]eicosane-1,8-diamine
hydroxybenzyldiamsar = 1-*N*-(4-hydroxybenzyl)-3,6,10,13,16,19-hexaazabicyclo[6.6.6]eicosane-1,8-diamine.
Complexes of these ligands will be denoted as M[(ligand)]*n*_, *e.g.* $[Cu(sar)]^{2+}$
CT Computer Tomography

Journal of Physics: Conference Series **185** (2009) 012044 doi:10.1088/1742-6596/185/1/012044

MRI Magnetic Resonance Imaging
US Ultrasound
SPECT Single Photon Emission Computer Tomography
PET Positron Emission Tomography
TOF-PET – Time of Flight Positron Emission Tomography

Zeeman splitting and Faraday rotation associated with Mn^+ $^7P_J \leftarrow {}^7S_3$ absorption: Photon-trap spectroscopy and quantum-theory analysis

Akira Terasaki[1], Takuya Majima[2,*] and Tamotsu Kondow[1]

[1] Cluster Research Laboratory, Toyota Technological Institute,
717-86 Futamata, Ichikawa, Chiba 272-0001, Japan
[2] East Tokyo Laboratory, Genesis Research Institute, Inc.,
717-86 Futamata, Ichikawa, Chiba 272-0001, Japan

E-mail: terasaki@clusterlab.jp

Abstract. A novel experimental method is presented for detection of a trace of optical absorption of mass-selected ions stored in an ion trap. Photon-trap spectroscopy (a generalized scheme of cavity ring-down spectroscopy) is the key technique, where a storage lifetime of photons in a high-Q cavity provides a ppm-level high sensitivity. Experiment was performed on the manganese ion, Mn^+, to record $^7P_{2,3,4} \leftarrow {}^7S_3$ absorption spectra at around 260 nm. Zeeman splitting and Faraday rotation of the polarization plane were observed as well in the presence of a magnetic field. The high-resolution spectra show complex structures, which are fully interpreted by the quantum theory incorporating hyperfine interaction due to the Mn nuclear spin (I = 5/2). The present technique provides a novel approach to electronic and magnetic structures of mass-selected atomic, molecular, and cluster ions.

1. Introduction

We have developed a novel methodology for absorption spectroscopy of ions in the gas phase [1,2]. The technique combines an ion trap with an optical cavity. The ion trap enables high-density storage of ions. The cavity allows us to perform photon-trap spectroscopy [1–4] (a generalized scheme of cavity ring-down spectroscopy [5]) for achieving extremely high sensitivity to detect extinction of light through the stored ions; a storage lifetime of photons in the cavity provides absorbance measurement free from intensity fluctuation of the light source. It is thus possible to perform absorption spectroscopy of ions without relying on laser-induced fluorescence (LIF) or photodissociation, which are often employed for molecules and clusters [6]. The advantages of the present method over other techniques targeting ions in a discharge [7–9] are its capabilities for mass selection and for application of a magnetic field. The latter provides an opportunity for magneto-optical spectroscopy as well. We demonstrated so far the first experiment on the atomic ion of manganese, Mn^+ [1], and then size-selected cluster ions of silver, Ag_9^+ [2].

In the Mn^+ experiment [1], we reported absorption and magneto-optical spectroscopy, where three absorption spectra due to $^7P_{2,3,4} \leftarrow {}^7S_3$ transitions were measured in the UV region along with their

* Present address: Department of Physics, Tokyo Metropolitan University, 1-1 Minami-Osawa, Hachioji, Tokyo 192-0397, Japan.

Zeeman splitting and Faraday rotation of the incident polarization plane. The stored Mn^+ ions were even spin-polarized with the aid of the cavity by incidence of a circularly polarized laser pulse [10]. In this paper, we describe the experimental method briefly, and present theoretical analysis of the spectra. In particular, the role of hyperfine interaction is illustrated in the spectrum simulation.

2. Photon-trap spectroscopy of trapped ions

2.1. Experimental procedure

The major part of the experimental setup is shown in figure 1. Briefly, ions were mass-selected by a quadrupole mass filter, guided by octopole ion guides (IG) and a quadrupole deflector (QD), and stored in a 40-cm linear ion trap filled with a buffer He gas at room temperature. Optionally, the buffer gas can be cooled down by liquid He. About 1×10^9 mass-selected ions were trapped; the ion density was about 3×10^7 ions/cm^3. The ion trap was located in a 10-cm bore of a 80-cm-long solenoidal superconductor magnet; the magnetic field was homogeneous over the ion trap volume within ±0.5%. The ions in the trap was further allowed to interact with photons trapped in a high-finesse optical cavity (a photon-trap cavity), which consisted of two mirrors (MR) with a reflectivity, R = 99.98%, and a radius of curvature, ROC = 6 m, separated by 1.6 m.

The round-trip absorbance was evaluated with a precision better than 1 ppm from the lifetime of photons in the cavity, as measured by the intensity-decay rate of the signal pulse. For the measurement of polarization rotation, the signal pulse was split into two orthogonal polarization components [11]; the intensity difference between the two components is proportional to sin 2θ, where θ is the angle of rotation. The angle, θ, was determined with a precision of about 1 μrad. per round trip.

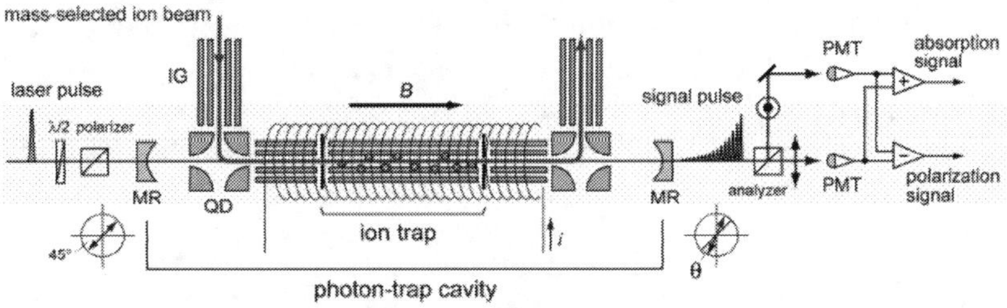

Figure 1. A schematic of the experimental setup. IGs: ion guides; QDs: quadrupole deflectors; MRs: cavity mirrors; PMTs: photomultiplier tubes.

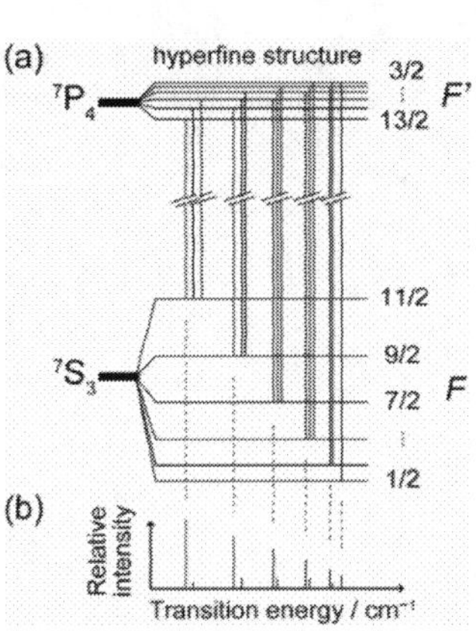

Figure 2. (a) Energy-level diagram of Mn^+. Vertical lines indicate allowed transitions. (b) Relative intensities of the transitions.

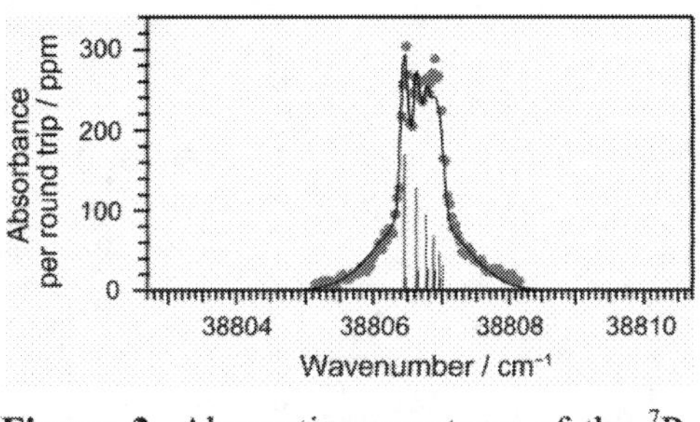

Figure 3. Absorption spectrum of the $^7P_4 \leftarrow {}^7S_3$ transition of Mn^+ without a magnetic field (B = 0 T). Solid circles: experimental data; bars: transition lines obtained theoretically; solid curves: spectrum simulation.

Journal of Physics: Conference Series **185** (2009) 012045 doi:10.1088/1742-6596/185/1/012045

2.2. Results

The present experiment was performed on atomic ion of manganese, Mn^+, which has a $^7S_3$ ground state. By the measurements of the photon-trap lifetime as a function of the laser frequency, three absorption bands associated with spin-allowed electronic transitions $^7P_J \leftarrow {}^7S_3$ ($J = 2$, 3, and 4) were observed at 38366.2, 38543.1, and 38806.7 cm^{-1}, respectively, as reported by emission spectroscopy [12,13]. Here we present the results of the $J = 4$ transition; relevant energy levels are illustrated in figure 2(a). An absorption spectrum measured is displayed in figure 3, which shows a hyperfine structure due to the nuclear spin of Mn ($I = 5/2$). The spectrum exhibited Zeeman splitting in the presence of a magnetic field of $B = 3$ T as displayed in figure 4; the two panels show spectrum simulations performed in different levels of calculation as described in Section 3. A spectrum of Faraday rotation was measured as well, which is shown in figure 5.

3. Analysis of absorption and magneto-optical spectra

The hyperfine structure of the absorption spectrum at $B = 0$ T was reconstructed from the transition energy (38806.689 cm^{-1}) and the hyperfine constants (A = 797.4 and −122.9 MHz for $^7S_3$ and $^7P_4$, respectively) reported by emission spectroscopy [14]. The bars in figure 3 (as well as figure 2(b)) indicate absorption lines satisfying the selection rule of $\Delta F = 0$ and ± 1, where F is the quantum number of the total angular momentum ($F = I + J$). The lengths of the bars are proportional to the relative intensity. These absorption lines correspond to the 15 transitions shown in figure 2(a). The profiles of these absorption lines were added together after convolution with an instrumental function, dominated by the laser bandwidth of about 0.13 cm^{-1} [1], to obtain a simulated spectrum; the solid curve in figure 3 thus obtained explains the experimental data very well.

The spectrum of Zeeman splitting was simulated by the procedure outlined by Tremblay et al. for alkali-metal atoms [15]. Briefly, the magnetic field causes mixing among the levels with a different quantum number, F, but with the same m_F via off-diagonal elements, $\langle F-1, m_F|H|F, m_F\rangle$, of the Hamiltonian, where H consists of unperturbed, hyperfine and Zeeman terms. The Hamiltonian was diagonalized to obtain eigenenergies and eigenfunctions both in the ground and in the excited states for calculation of transition energies and strengths.

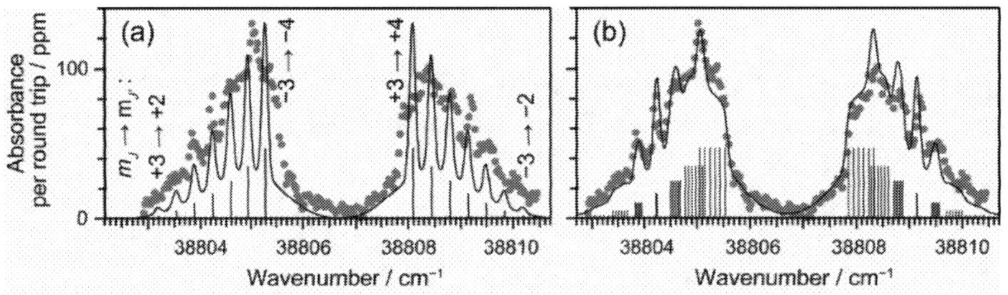

Figure 4. Zeeman splitting of the $^7P_4 \leftarrow {}^7S_3$ absorption spectrum of Mn^+ at $B = 3$ T. Solid circles: experimental data; bars: transition lines obtained theoretically; solid curves: spectrum simulation. (a) The simulation neglects hyperfine interaction; it does not account for the experiment perfectly. (b) The simulation incorporates hyperfine interaction, providing full explanation of the experimental data.

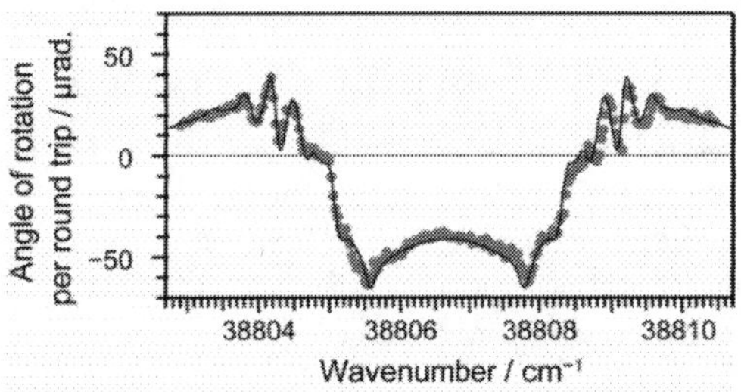

Figure 5. Spectrum of Faraday rotation associated with the $^7P_4 \leftarrow {}^7S_3$ transition of Mn^+ measured at $B = 3$ T. Solid circles: experimental data; solid curve: spectrum simulation.

The spectrum shows splitting primarily by m_J, as shown in figure 4(a), because the magnetic field is so strong as to be in the Paschen–Back regime. The selection rule, $\Delta m_J = \pm 1$, provides 7 lines for each component of blue- and red-shifts. However, the simulated spectrum does not agree with the experimental data; the peaks observed in the experiment do not show a regular interval, and even shoulder structures are discernible (for example, at 38805.3 and 38808.0 cm^{-1}). Actually, these 14 lines are subject to further splitting by hyperfine interaction; each line splits into 6 due to m_I (= −5/2, ..., +5/2) with a selection rule, $\Delta m_I = 0$. These 84 lines reproduced every detail of the experimental spectrum as shown in figure 4(b). The interval of the hyperfine splitting is dependent on the value of m_J, and thus a different magnitude of broadening is caused for the 14 lines.

The spectrum of the Faraday rotation was simulated as well. Each transition line of the Zeeman splitting was represented by a dispersion profile. They were added together for blue- and red-shifted components separately after convolution with the instrumental function. The difference of these two dispersion profiles is proportional to the magnitude of the polarization rotation, which is shown by the solid curve in figure 5. The experimental spectrum was fully reproduced by this simulation.

4. Summary

We have shown that absorption and magneto-optical spectra of mass-selected ions can be measured by our novel technique combining an optical cavity and an ion trap. The spectra of field-free absorption, Zeeman splitting and Faraday rotation measured for the atomic ion of manganese, Mn^+, were fully explained by the simulation based on the quantum theory of atomic transitions incorporating hyperfine interaction. Extension of the present magneto-optical spectroscopy toward cluster ions would provide clear evidence for specificity of their magnetism; Mn_2^+ and Mn_3^+ are considered to exhibit a ferromagnetic spin coupling in contrast to antiferromagnetic bulk manganese [6,16,17].

Acknowledgement

The present study is supported by the Special Cluster Research Project of Genesis Research Institute, Inc.

References

[1] Terasaki A, Majima T and Kondow T 2007 *J. Chem. Phys.* **127** 231101
[2] Terasaki A, Majima T, Kasai C and Kondow T 2009 *Eur. Phys. J.* D, in press
[3] Terasaki A, Kondow T and Egashira K 2005 *J. Opt. Soc. Am.* B **22** 675
[4] Egashira K, Terasaki A and Kondow T 2007 *J. Chem. Phys.* **126** 221102
[5] O'Keefe A and Deacon D A G 1988 *Rev. Sci. Instrum.* **59** 2544
[6] Terasaki A 2007 *J. Phys. Chem.* A **111** 7671
[7] Gudeman C S, Begemann M H, Pfaff J and Saykally R J 1983 *Phys. Rev. Lett.* **50** 727
[8] Stephenson S K and Saykally R J 2005 *Chem. Rev.* **105** 3220
[9] Kotterer M, Conceicao J and Maier J P 1996 *Chem. Phys. Lett.* **259** 233
[10] Majima T, Terasaki A and Kondow T 2008 *Phys. Rev.* A **77** 033417
[11] Engeln R, Berden G, van den Berg E and Meijer G 1997 *J. Chem Phys.* **107** 4458
[12] Kling R and Griesmann U 2000 *Astrophys. J.* **531** 1173
[13] Kling R, Schnabel R and Griesmann U 2001 *Astrophys. J. Suppl. Ser.* **134** 173
[14] Blackwell-Whitehead R J, Toner A, Hibbert A, Webb J and Ivarsson S 2005 *Mon. Not. R. Astron. Soc.* **364** 705
[15] Tremblay P, Michaud A, Levesque M, Thériault S, Breton M, Beaubien J and Cyr N 1990 *Phys. Rev.* A **42** 2766
[16] Terasaki A, Matsushita A, Tono K, Yadav R T, Briere T M and Kondow T 2001 *J. Chem. Phys.* **114** 9367
[17] Terasaki A, Briere T M, Kulawik M, Minemoto S, Tono K, Matsushita A and Kondow T 2003 *J. Chem. Phys.* **118** 2180

Journal of Physics: Conference Series **185** (2009) 012046 doi:10.1088/1742-6596/185/1/012046

Surface Processes of Highly Charged Ions

Masahide Tona and Shunsuke Ohtani

Institute for Laser Science, University of Electro-Communications, Chofu, Tokyo 182-8585, Japan

CREST, Japan Science and Technology Agency, Chofu, Tokyo 182-8585, Japan

E-mail: tona@ils.uec.ac.jp

Abstract. Striking features of radiation effects by a highly charged ion (HCI) impact onto a surface are shown. The HCI-impact effects depend strongly on the electronic property of the target material and are restricted within the nanometer-sized area on the surface. Further, a challenging experiment for application of HCI-based nano-processing is presented. One of the examples for the processing can be seen in the interaction of a HCI with a hydrogen-terminated Si surface. We try to produce a SiO nano-structure on the Si surface with a chemical reaction of oxygen molecules to the active region induced by individual HCI impacts.

1. Introduction

A highly charged ion (HCI) is one of the most active particles reacting violently with the matter, *e.g.*, electron, atom, molecule, cluster and solid surface. An electron beam ion trap (EBIT) is a useful device for investigating these reactions. An EBIT can not only trap efficiently HCIs with very high charge state but also extract the slow HCI-beam with low emittance. By using the Tokyo-EBIT facility [1] at the university of Electro-Communications, two kinds of research projects are going on, one is for atomic physics and the other is for surface science and technology. Representative research subjects in atomic physics are to reveal the Quantum electrodynamic (QED) and the relativistic effects, observed in resonant interaction between electrons and HCIs of heavy elements having few electrons [2, 3]. For the surface experiments, on the other hand, the mechanism of the interaction of slow HCIs with a surface is investigated by measuring secondary emission yields of various kinds, such as X-rays [4, 5, 6], visible photons [7] and ions [8], and by observing surface modification induced by individual HCI impacts [9, 10, 11].

In this article, we show a few limited results in the recent surface experiments. HCI-surface reaction is triggered by multiple electron transfer from the surface to the approaching HCI. The slow incident ion has sufficient time for capturing many electrons before arriving at the surface. The electronic system of the surface is, in turn, excited through this process. Consequently, dramatic radiation effects such as secondary ion and neutral emission, excitation of plasmons and excitons, and dislocation of surface atoms are induced. Since the charge transfer cross-section increases rapidly with incident ion charge q, these effects would be enhanced in a higher-q HCI impact. The mechanism of these phenomena is considered to be essentially different from that induced by energetic neutrals and singly charged ions where the kinetic energy of the primary particles plays a major role.

It is considered that the radiation effects are induced in close relation to the electronic property of a target material because of the strong coupling of the incident HCI with target

Journal of Physics: Conference Series **185** (2009) 012046 doi:10.1088/1742-6596/185/1/012046

electrons. The surface density of active electrons is thought to be one of the most important parameters for governing the degree of the effects. Accordingly, we investigate the HCI-induced surface modification for various kinds of target materials: gold (metal), silicon (semiconductor) and TiO_2 (oxide), using a scanning tunneling microscope (STM). The observation reveals that the HCI-surface reaction is restricted within the nanometer-sized area on the surface. This result suggests that the individual HCI impacts could be applied to nano-processing. Here, the preliminary result of the processing to produce a SiO nano-structure on a hydrogen-terminated silicon surface is presented.

2. Experimental

Iodine HCIs, I^{q+}, having $3\times q$ keV of kinetic energy were extracted from the Tokyo-EBIT and their charge states were selected by using a sector magnet. A collision chamber (base pressure: 2×10^{-8} Pa) was connected to an observation chamber equipped with a STM apparatus [12, 13]. The sample irradiated with HCIs was transported between the chambers in the vacuum. Before HCI irradiation, clean, well-defined surfaces were prepared by the ordinary methods: a repetition of annealing and ion-sputtering procedures was employed for Au(111) and TiO_2(110) surfaces, and a repetition of flashing procedures was for clean Si(111)-(7×7) and Si(100)-(2×1) surfaces. A hydrogen-terminated Si(100) -(2×1) surface was obtained with exposure of atomic hydrogen to a clean Si (100) surface.

3. Results and discussion

Figure 1 shows typical STM images of Au(111), Si(111), Si(100) and TiO_2(110) surfaces irradiated with I^{50+}. Each observation is performed in the high resolution so that the atomic surface structure can be seen. For the gold surface (metal target), no impact site is found although the irradiated area is carefully observed. Further, in a secondary ion mass spectrometry (SIMS) measurement for the surface, the yield of sputtered Au^+ is so weak that no clear peak of Au^+ appears in the SIMS spectrum. In a recent report for observation of dislocations on a gold surface bombarded with Xe^{q+} ($q < 45$) [14], it was concluded that nanometer sized

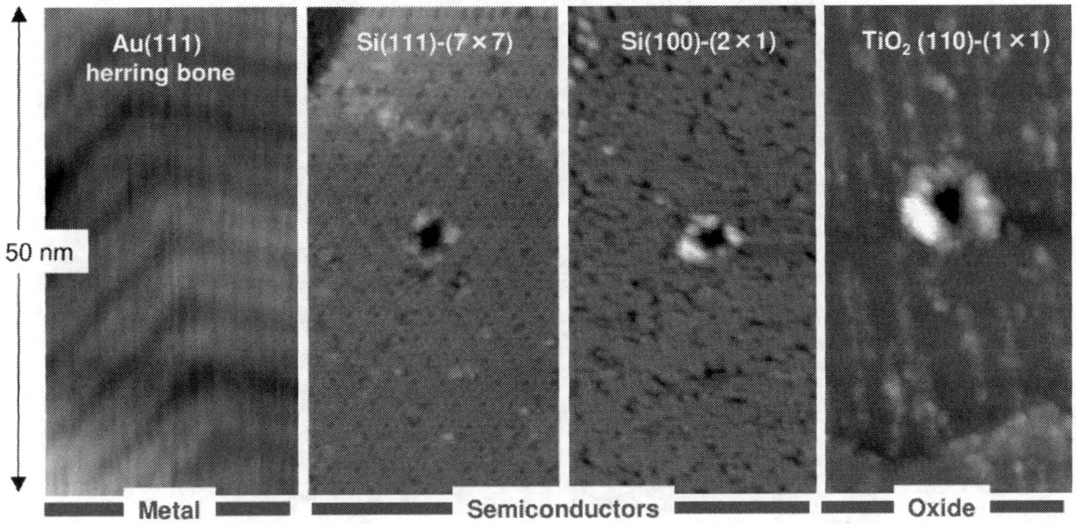

Figure 1. STM images of Au(111), Si(111), Si(100) and TiO_2(110) surfaces irradiated with I^{50+}.

structures were created by the nuclear stopping power of the primary ions rather than by the multiple electron transfer, that is, no q-dependence was observed. These results suggest that the radiation effect by individual HCI impacts is very week for the gold surface in which conduction electrons can rapidly compensate a depletion site created by the multiple electron transfer.

For the Si surfaces (semiconductor target), an HCI-impact site is seen as the deep crater structure at the center of each image, in addition to several small native defects due to missing surface atoms or adsorption spots of residual gas molecules in the chamber. The crater sizes are nearly the same for the both surfaces: the area of crater is approximately 6 nm^2 and the depth is at least 0.3 nm. In our previous work [10], it was revealed that the HCI-radiation effects were enhanced with q on a Si(111)-(7×7) surface by individual HCI impacts (I^{q+}, $q \geq 30$). For the TiO_2 surface (oxide target), a typical impact site shows the "caldera" structure having the relatively high outer rim of the crater. The maximum height reaches ~1 nm which is higher than that single atomic step of this surface (=~0.3 nm) and the depth is measured to be 1.5 nm at least. The STM observation shows the created impact site on the TiO_2 surface is much larger than that on the Si surface. Further, it was found from previous SIMS measurements for these surfaces [10, 11] that the secondary ion emission yield from TiO_2 is also larger than that from Si. From these results we could conclude that the degree of the HCI-radiation effect on the TiO_2 surface is higher than that of the Si surfaces.

As seen Fig. 1, nano-structures on the Si surfaces are created following that large amount of surface atoms are sputtered by individual HCI impacts. This result is very important for application of the HCI-beam to the nano-processing. One of the examples of the processing can be seen in the interaction of a HCI with a hydrogen-terminated Si (Si-H) surface. The Si-H surface is chemically stable because dangling bonds of Si atoms are fully terminated by hydrogen atoms. The nanometer sized crater structure might become an active region by removal of many hydrogen adsorbates and appearance of open valence bonds on the impact site. This implies that Si-O nano-structure can be created through selective reactions of oxygen gas with this active region.

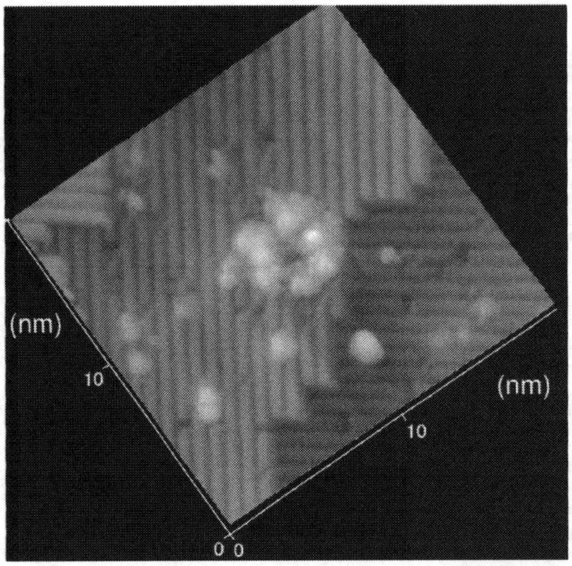

Figure 2. Nano-structure on a Si(100)-(2×1)-H surface by the I^{44+} bombardment. The HCI irradiation was performed with O_2 gas exposure.

Figure 2 shows a nano-structure on a Si(100)-(2×1)-H surface by the I^{44+} bombardment. HCI irradiation was performed under the condition that partial pressure of O_2 gas was 2×10^{-6} Pa and irradiation time was 4 hours. The surface atomic structure of dimer lines can be seen in this image, which is kept although O_2 gas was exposed more than several hundreds Langmuir to the surface. On the other hand, this structure cannot be seen on the H-removal region at the impact site, which is instantly contaminated by O_2 gas with a high sticking probability [15]. The impact site with several nanometers in diameter would contain Si-O_x bondings formed by the surface chemical reaction. The same nano-structre can be produce by using STM where a pulse current is injected from the STM tip into a nano-portion to sputter hydrogens [16]. The HCI-processing has an advantage that the nano-structure can be efficiently produced on the semiconductor substrate with high density ($> 10^{11}$ structures/cm^2). Physical and chemical properties of this nano-structure are being investigated with electron energy loss spectroscopy, FTIR spectroscopy and scanning tunneling spectroscopy.

Acknowledgments

This work has been supported by the CREST program, "Creation of Ultrafast, Ultralow Power, Super-performance Nanodevices and Systems" in the Japan Science and Technology Agency, and has been performed under the activity of the 21st Century Center of Excellence (COE) Program, "Innovation in Coherent Optical Science" at the University of Electro-Communications.

References

[1] H. Watanabe, J. Asada, F. Currell, T. Fukami, T. Hirayama, K. Motohashi, N. Nakamura, E. Nojikawa, S. Ohtani, K. Okazaki, M. Sakurai, H. Shimizu, N. Tada, and S Tsurubuchi: J. Phys. Soc. Jpn. **66** (1997) 3795.

[2] N. Nakamura, A. P. Kavanagh, H. Watanabe, H. A. Sakaue, Y. Li, D. Kato, F. J. Currell, and S. Ohtani: J. Phys.: Conf. Ser. **88** (2007) 012066.

[3] N. Nakamura, A. P. Kavanagh, H. Watanabe, H. A. Sakaue, Y. Li, D. Kato, F. J. Currell, and S. Ohtani: Phys. Rev. Lett. **100** (2008) 073203.

[4] H. Watanabe, S. Takahashi, M. Tona, N. Yoshiyasu, N. Nakamura, M. Sakurai, C. Yamada, and S. Ohtani: Phys. Rev. A **74** (2006) 042901.

[5] H. Watanabe, J. Sun, M. Tona, N. Nakamura, M. Sakurai, C. Yamada, N. Yoshiyasu, and S. Ohtani: Phys. Rev. A **75** (2007) 062901.

[6] J. Sun, H. Watanabe, M. Tona, T. Watanabe, N. Nakamura, C. Yamada, and S. Ohtani: Phys. Rev. A **77** (2008) 032901.

[7] M. Tona, T. Abe, H. Watanabe, J. Sun, N. Nakamura, C. Yamada, M. Kotani, and S. Ohtani: Phys. Rev. A **77** (2008) 052902.

[8] M. Tona, K. Nagata, S. Takahashi, N. Nakamura, N. Yoshiyasu, M. Sakurai, C. Yamada, and S: Ohtani: Surf. Sci. **600** (2006) 124.

[9] Y. Baba, K. Nagata, S. Takahashi, N. Nakamura, N. Yoshiyasu, M. Sakurai, C. Yamada, and S. Ohtani, and M. Tona: Surf. Sci. **599** (2005) 248.

[10] M. Tona, H. Watanabe, S. Takahashi, N. Nakamura, N. Yoshiyasu, M. Sakurai, T. Terui, S. Mashiko, C. Yamada, and S. Ohtani: Surf. Sci. **601** (2007) 723.

[11] M. Tona, Y. Fujita, C. Yamada, and S. Ohtani: Phys. Rev. B **77** (2008) 155

[12] H. Shimizu, F. J. Currell, S. Ohtani, E. J. Sokell, C. Yamada, T. Hirayama, and M. Sakurai: Rev. Sci. Instrum. **71** (2000) 681.

[13] M. Tona and S. Takahashi: J. Phys. Conf. Ser. **2** (2004) 57.

[14] J. M. Pomeroy, A. C. Perrella, H. Grube, and J. D. Gillaspy: Phys. Rev. B **75** 241409(R) (2007).

[15] M. Tona, H. Watanabe, S. Takahashi, N. Nakamura, N. Yoshiyasu, M. Sakurai, T. Terui, S. Mashiko, C. Yamada, and S. Ohtani:Nucl. Instrum. and Meth. B **256** (2007) 543.

[16] T. Hallam, F. J. Rueβ, N. J. Curson, K. E. J. Goh, L. Oberbeck, M. Y. Simmons, and R. G. Clark: Appl. Phys. Lett. **86** (2005) 143116.

A new time-dependent scattering theory and its application to the capture of antiprotons by atoms

X. M. Tong[1,2], K. Hino[2,3], and N. Toshima[1]

[1] Doctoral Program in Materials Science, Graduate School of Pure and Applied Sciences, University of Tsukuba, 1-1-1 Tennodai, Tsukuba, Ibaraki 305-8573, Japan
[2] Center for Computational Sciences, University of Tsukuba,1-1-1 Tennodai, Tsukuba, Ibaraki 305-8577, Japan
[3] Doctoral Program in Frontier Science, Graduate School of Pure and Applied Sciences, University of Tsukuba, 1-1-1 Tennodai, Tsukuba, Ibaraki 305-8573, Japan

E-mail: tong@ims.tsukuba.ac.jp

Abstract. We presented a theoretical method to study the capture of the antiprotrons by atoms solving a Chew-Goldberger-type integral equation directly. The scattering boundary conditions are automatically satisfied by adiabatically switching on the interaction between the antiprotons and targets. Hence the outgoing wave function is obtained without the tedious procedure of adjusting the total wave function in the asymptotic region. All the dynamical information can be derived from the scattering wave function obtained on pseudo-spectral grids numerically. Using this method, we obtained the state-specified capture cross sections when antiprotons collide with helium atoms. Differing from the capture processes of antiprotons by hydrogen atoms, the anomalous bumpy structures are revealed in the angular momentum dependent capture cross sections by helium atoms. Further analysis shows that the bumps arise from the partial channel closing due to the removal of the energy degeneracy in the antiprotonic helium atoms.

1. Introduction

Coulomb three-body rearrangement collision is a prototype of general ion-atom collisions and electron-atom collisions. For heavy negative charged particles captured by atoms, the number of capture channels increases dramatically due to the huge mass difference between electron and negative charged particles. Thus, the computational time required for the nonperturbative approaches, e.g., the close-coupling method, inevitably increases dramatically as discussed by Esry and Sadeghpour [1]. It is very difficult to solve the scattering equations without reducing the antiproton mass artificially. Protonium formation cross sections were calculated by scaling down the proton and antiproton masses to 20 a.u. Hesse *et al.* [2] used a diabatization technique to calculate the protonium formation but they were also obliged to use scaled proton and antiproton masses of 100 a.u. Their formation cross sections are peaked at $n = 7$, which is much lower than the estimated value of $n = 30$ for the protonium with the real mass, 1836 a.u. Cohen [3] calculated the protonium formation cross section by the classical trajectory Monte Carlo (CTMC) method. The applicability of the classical mechanics at such low energies is not clear, especially for the motion of electron. Besides ambiguity occurs in extracting the discrete quantum numbers n and ℓ from the classical continuous quantities.

Sakimoto [4] studied the formation process using a time-dependent wavepacket method but he did not obtain the state-specified formation cross sections due to the numerical difficulty. Ovchinnikov and Macek [5] applied the advanced adiabatic approach to calculate the n-dependent formation cross sections. Although their sophisticated method may be accurate qualitatively for the relevant process, it is not easy to estimate the quantitative reliability for the minute information such as the (n, ℓ)-distribution. Yamanaka and Ichimura [6] studied the protonium formation by solving Faddeev equations involved the ejected electron in the s-wave states only. To the best of our knowledge, reliable state-specified formation cross sections have not been obtained yet from nonperturbative quantal calculations apart from our recent works [7, 8, 9].

As mentioned above, the traditional coupled-channel method for the time-independent differential Schrödinger equation is not practically applicable to the process. To solve the problem differently, we started from the Chew-Goldberger integral equation of the scattering equation. The advantage of this method is that we transfer the complicated boundary conditions into a simple initial condition and convert the time-independent scattering theory into the time-dependent scattering theory. The theoretical details will be presented in the following section, followed by the applications of the theory to the captures of antiprotons by helium atoms.

2. Theoretical Method

2.1. General time-dependent formalism

The time-independent Schrödinger equation of a Coulomb three-body system in the center-of-mass frame is written as (atomic units $\hbar = m_e = e = 1$ are used hereafter unless otherwise stated)

$$H|\Psi> = E|\Psi>, \tag{1}$$

with

$$H = -\frac{1}{2\mu_e}\nabla_{\mathbf{r}}^2 - \frac{1}{2\mu_{\bar{p}}}\nabla_{\mathbf{R}}^2 + V_{n,e}(\mathbf{r}) + V_{n,\bar{p}}(\mathbf{R}) + V_{\bar{p},e}(\mathbf{r},\mathbf{R}) \tag{2}$$

Here $E = E_c + E_{1s}$ is the total energy of the collision system in the center-of-mass frame with E_c the incident energy of the projectile and E_{1s} the ground state energy of the target atom. $\mathbf{r}$ and $\mathbf{R}$ are respectively the distances from the nucleus to the electron and from the center-of-mass of the atom to the projectile. $V_{n,e}(\mathbf{r})$, $V_{n,\bar{p}}(\mathbf{R})$, and $V_{\bar{p},e}(\mathbf{r},\mathbf{R})$ are the interactions of the nucleus and the electron, the nucleus and the antiproton, and the electron and the antiproton, respectively. μ_e is the reduced mass of the nucleus and the electron and $\mu_{\bar{p}} = M_{\bar{p}}(m_e + M_n)/(M_n + m_e + M_{\bar{p}})$. M_n, m_e and $M_{\bar{p}}$ are the masses of nucleus, electron and antiproton, respectively. The solutions of Eq. (1) satisfying proper boundary conditions can be expressed as an integral expression;

$$\Psi^+(t) = \Psi_i - i\int_{-\infty}^{t} e^{-i(H-E)(t-t')} f(t')V_i\Psi_i dt'. \tag{3}$$

Here, V_i is the interaction between the initial state and the complete state. The time-dependent wave function $\Psi^+(t)$ goes back to the initial state as $t \to -\infty$, and it also satisfies Eq. (1) at $t = 0$ if $f(-\infty) = 0$, $f(0) = 1$ and $f(t)$ switches on adiabatically [10], i.e., $\frac{d}{dt}f(t) \to 0$. A choice of $f(t) = \exp(-\eta|t|)$ gives Chew-Goldberger formal solution at $t = 0$ as

$$\Psi^+ = \Psi_i + \frac{1}{E - H + i\eta}V_i\Psi_i, \tag{4}$$

where η is a positive infinitesimal number. We calculated the time-dependent wavefunction by integrating Eq. (3) instead of solving Eq. (1) or Eq. (4). $f(t)$ can be any function so long as it is switched on adiabatically. For instance, $f(t) = \exp(-t^2/\tau^2)$ can be used for $\tau \to \infty$. Once obtaining $\Psi^+(t = 0)$ by propagating Eq.(3), we can derive all the dynamical information of the collision system.

Journal of Physics: Conference Series **185** (2009) 012047 doi:10.1088/1742-6596/185/1/012047

2.2. Numerical procedures

Coulomb three-body systems cover a variety of physical processes. The characteristic of each system depends strongly on the combination of the masses and the charges of the constituting particles. We focused on dealing with heavy negatively-charged particles, antiprotons, colliding with helium atoms as an example to show how to propagate Eq. (3) effectively and accurately. The initial wave function Ψ_i of the system is the product of the ground state wave function of the target atom and the wave function of the projectile, which is expressed as

$$\Psi_i(\mathbf{r},\mathbf{R}) \;=\; \psi_{1s}(\mathbf{r})\psi_c(\mathbf{k}_0,\mathbf{R}). \tag{5}$$

Here $\mathbf{k}_0$ is the initial momentum of the projectile in the center-of-mass frame, $\psi_{1s}(\mathbf{r})$ the ground state wave function of the Hamiltonian $H(r) = -\frac{1}{2\mu_e}\nabla^2_{\mathbf{r}}+V_{n,e}(\mathbf{r})$ and $\psi_c(\mathbf{k}_0,\mathbf{R})$ the wave function of the Hamiltonian $H = -\frac{1}{2\mu_{\bar{p}}}\nabla^2_{\mathbf{R}} + V_{n,\bar{p}}(\mathbf{R}) - \frac{1}{R}$ with the incoming wave boundary condition. The Hamiltonian of the initial channel and the corresponding interaction are

$$H_i \;=\; -\frac{1}{2\mu_e}\nabla^2_{\mathbf{r}} + V_{n,e}(\mathbf{r}) - \frac{1}{2\mu_{\bar{p}}}\nabla^2_{\mathbf{R}} + V_{n,\bar{p}}(\mathbf{R}) + \frac{1}{R}, \tag{6}$$

$$V_i \;=\; V_{\bar{p},e}(\mathbf{r},\mathbf{R}) - \frac{1}{R}. \tag{7}$$

Because the total angular momentum L, its component M onto the incident beam direction, and the parity are good quantum numbers, we expand the time-dependent wave function as

$$\Psi(t) \;=\; \sum_{LM}\Psi_{LM}(\mathbf{r},\mathbf{R},t) = \sum_{LM}\sum_{\alpha} F^{\alpha}(r,R,t)\Omega^{\alpha}_{LM}(l_e,l), \tag{8}$$

with

$$\Omega^{\alpha}_{LM} = \sum_{m_e,m} < l_e m_e l m | LM > Y_{l_e m_e}(\hat{r})Y_{lm}(\hat{R}). \tag{9}$$

Here α is a index to represent the angular momentum (l_e) of the electron , the angular momentum (l) of the antiproton, and the parity collectively. Each $\Psi_{LM}(\mathbf{r},\mathbf{R},t)$ can be propagated independently. The radial Hamiltonian for a given LM is written as

$$\begin{aligned} < \Omega^{\alpha}_{LM}|H|\Omega^{\alpha'}_{LM} > \;=\;& \left(-\frac{1}{2\mu_{\bar{p}}}\frac{\partial^2}{\partial R^2} + \frac{l(l+1)}{2\mu_{\bar{p}}R^2} + V_{n,\bar{p}}(\mathbf{R})\right)\delta_{\alpha,\alpha'} \\ &+\left(-\frac{1}{2\mu_e}\frac{\partial^2}{\partial r^2} + \frac{l_e(l_e+1)}{2\mu_e r^2} + V_{n,e}(\mathbf{r})\right)\delta_{\alpha,\alpha'} \\ &+ < \Omega^{\alpha}_{LM}|V_{\bar{p},e}(\mathbf{r},\mathbf{R})|\Omega^{\alpha'}_{LM} > \\ =\;& \mathbf{H}^{\alpha}(R) + \mathbf{H}^{\alpha}(r) + \mathbf{V}^{\alpha,\alpha'}(r,R). \end{aligned} \tag{10}$$

After the partition, the time propagation from time t to $t+\Delta t$ was performed by the split-operator method in the energy representation [11, 12] as

$$U(t+\Delta t,t) \;=\; e^{-i\mathbf{V}(r,R)\Delta t/2}e^{-i\mathbf{H}(R)\Delta t}e^{-i\mathbf{H}(r)\Delta t}e^{-i\mathbf{V}(r,R)\Delta t/2} + O(\Delta t^3).$$

We first propagated the wave function a half-time step in $\mathbf{V}(r,R)$, and then a full-time step in $\mathbf{H}(r)$ and $\mathbf{H}(R)$, and finally another half-time step in $\mathbf{V}(r,R)$. The advantage of this time-propagation scheme is that each operator is diagonal in the other two dimensions. $\mathbf{V}(r,R)$ couples only different channels $\alpha \neq \alpha'$ and $\mathbf{H}(R)$ and $\mathbf{H}(r)$ are, on the other hand, diagonal for (α,α'). These characters make the time propagation computationally efficient.

Journal of Physics: Conference Series **185** (2009) 012047 doi:10.1088/1742-6596/185/1/012047

For three-body rearrangement collisions, we have to deal with continuum wave functions. In the direct integration method on grid points, the coordinate space is confined in a finite box. We have to filter out the outgoing waves at the boundary in order to eliminate the unphysical reflection. For this purpose, we added an optical potential near the boundary to absorb the outgoing flux. The optical potential we used is of the form of

$$V_{opt}(\epsilon, r) = \begin{cases} 0 & \text{for } r < r_c \\ -iV_0\left(\frac{r-r_c}{r_{max}-r_c}\right) & \text{for } r \geq r_c \end{cases}, \tag{11}$$

with V_0 in the form suggested in Ref. [13]. Here ϵ is the energy of the basis wave function and r_{max} and r_c are the position of the boundary in the radial direction and the start of the filter, respectively. μ is the reduced mass associated with coordinate r. The optical potential added to Hamiltonian $\mathbf{H}(r)$ is

$$V_{opt}(r) = \sum_i V_{opt}(\epsilon_i, r)|\psi_i >< \psi_i| \quad \text{for} \quad \epsilon_i \geq 0, \tag{12}$$

where $\{\epsilon_i\}$ and $\{\psi_i\}$ are the eigen energies and eigen wave functions of $\mathbf{H}(r)$. $V_{opt}(r)$ depends on the grid structure but it does not depend on the collision energy. It is a universal optical potential which can be used to absorb the ejected electron in a broad energy range. Similarly we add an optical potential to $\mathbf{H}(R)$. After a time propagation from $t = -\infty$ to $t = 0$, in the interaction region, we obtain the scattering wave function $\Psi^+(t = 0)$, which satisfies the proper boundary conditions. The final wave function corresponding to the capture to the state of a principal quantum number n and an angular momentum l is expressed as

$$\Psi_f^\alpha = j_{l_e}(k_{nl}r)\psi_{nl}(R)\Omega_{LM}^\alpha \tag{13}$$

where k_{nl} is the ejected electron momentum associated with the formation of an exotic atom in the nl state. The T-matrix element is given by

$$T_{nl}^\alpha =< \Psi_f^\alpha|V_f|\Psi_{LM}^+ >, \tag{14}$$

and the state specified capture cross section is readily obtained as

$$\sigma_{nl}(L) = \frac{4\mu_{\bar{p}}\,\mu_e\,k_{nl}}{k_0}\sum_\alpha |T_{nl}^\alpha|^2. \tag{15}$$

3. Results and Discussion

Based on the above theoretical method, we can investigate the dynamical processes involved in a Coulomb three-body problem or many-body problem which can be reduced to a three-body problem approximately. Following, we will present our studies on the antiproton captured by helium atoms. Frozen one of the two electrons in the collision system, the electron-core interaction $V_{n,e}(\mathbf{r})$ is approximated as $V_{eff}(r)$, a six-parameter model potential [14] and the antiproton-core interaction $V_{n,\bar{p}}(\mathbf{R})$ is approximated as $U(R)$, the effective potential of the antiproton moving in the combined field of the helium nucleus and the passive electron in the ground state and it can be calculated from the following equation adiabatically

$$H_a\Psi(\mathbf{r}_b; R) = U(R)\Psi(\mathbf{r}_b; R).$$

The electron-antiproton interaction is a pure Coulomb interaction. With the above approximations, we solve the time-dependent Schrödinger equation (Eq. (7)) as we did for antiproton captured by hydrogen atoms and obtain the state-specified capture cross-sections as

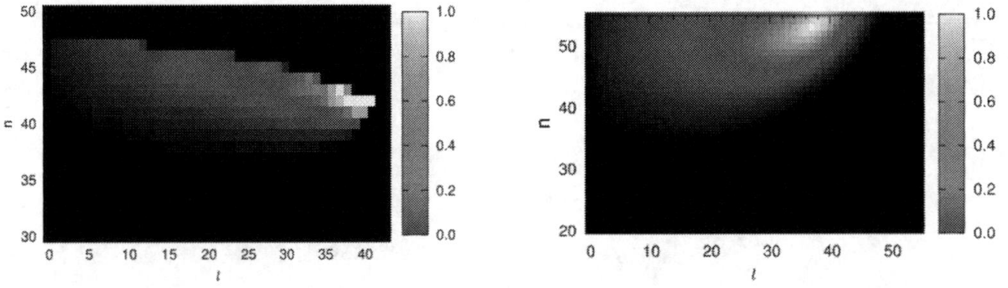

Figure 1. The state-specified capture cross sections normalized to the total capture cross section of 10.0 eV incident energy for antiprotons captured by helium atoms (left panel) and hydrogen atoms (right panel).

shown in Fig. 1. We also plot the state-specified capture cross-sections by hydrogen atoms for comparison. For hydrogen atoms, the capture cross-section changes smoothly as a function of the (n, ℓ). While for helium atoms, the capture cross-section peaks at higher n for lower ℓ. At the same incident energy, the antiproton is captured to a higher n states for hydrogen atoms than that of helium atoms. This can be understood since the effective charge of helium atoms is larger than that of hydrogen atoms and the antiproton can only be captured to the lower n states due to the conservation of energy. Although the state-specified cross sections contain all the necessary information to analyze the experiment, it is difficult to see the details from the 2-dimensional color plot. Now we sum over all n for a given ℓ to study the ℓ-dependent capture cross-section as shown on the left panel in Fig. 2.

For the capture by hydrogen atoms, the capture cross sections increase smoothly as ℓ increases, reach a peak at $\ell = 35$ and then drop quickly as ℓ increases further. For the capture by helium atoms, the capture cross sections first increase in an irregular manner, reach the maximum at $\ell = 38$, and then, drop sharply. This global feature is similar to the capture of antiprotons by hydrogen atoms. Unlike the hydrogen case, however, the present cross sections show the anomalous bumpy structures. It is notable that the present calculations are more difficult than those of the hydrogen case due to the larger reduced mass μ_p and the larger effective charge of helium. To check the numerical accuracy, the simulation conditions, such as the number of grids, the size of the space surrounding the whole system, and so on, have been varied. As a result, it is convinced that the anomalies manifested on the left panel of Fig. 2 are insensitive to the changes of these parameters. Thus, the anomalies are ascribable to a certain

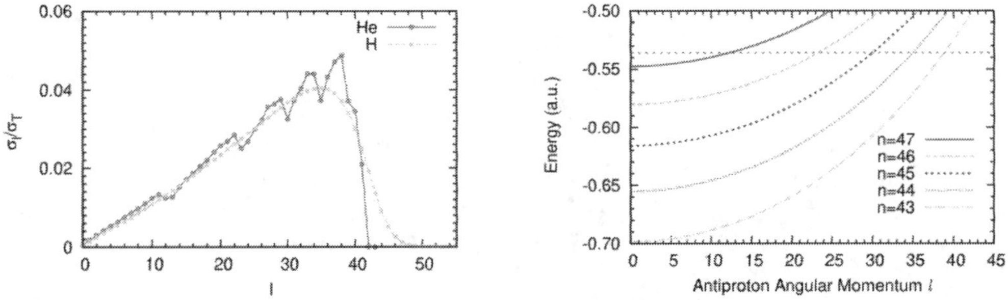

Figure 2. The ℓ-dependent capture cross sections of antiprotons by helium (solid line) and hydrogen atoms (dashed line) (left panel) and the energy levels of antiprotonic helium (right panel). The horizontal dashed line on the left panel is the total energy for 10 eV incident energy.

physics, not due to the numerical inaccuracy.

Although the capture process of the antiprotons by hydrogen atoms and helium atoms are similar within the present approximation, the energy structures of protonium ($\bar{p}p$) and antiprotonic helium ($\bar{p}He^+$) are different. Replacing the Coulomb potential in the former by the effective potential in the latter, the energy degeneracy assigned by n and ℓ is removed, where n and ℓ are the principle and angular momentum quantum numbers, respectively. As shown on the right panel of Fig. 2, the energy curves for a given n and different ℓ cross over with the total energy (dashed straight line). The cross over positions are in good agreement with the dip positions in the ℓ-dependent capture cross sections. Let us focus on $n = 47$ as an example to show the origin of the dips. With the 10 eV incident energy, the antiprotons can be captured to $n = 47$ with $\ell \leq 11$ due to the energy conservation. This means that this channel is opened for the capture to $\ell \leq 11$ states and is closed for the capture to $\ell \geq 12$ states. This fact would allow us to speculate that the manifestation of the anomaly around ℓ =12 is attributed to the channel closing of the states with $l \geq 12$. Here, the channel closing means that for a given n the capture process is forbidden when the ℓ is larger than a critical value. In a similar way, it is shown that the next dip corresponds to the closing of channel $n = 46$ at ℓ =23, and so on. In general, the partial inelastic scattering cross sections exhibit cusps when the incident energies cross a new threshold since the cross sections differ significantly for the energies below and above the threshold. Here we found cusps from a different aspect, i.e., the cross sections as a function of the ℓ for a given incident energy and n. The similar phenomena are also observed at other incident energies with the change of the dip positions and hence this anomaly is considered as a general phenomenon.

To summarize, we presented a new time-dependent scattering theory, which can be used to study general Coulomb three-body rearrangement processes. The key procedure is that we transferred the very complicated boundary conditions into a simple initial condition. Using this method, we obtained the state-specified capture cross sections when antiprotons collide with hydrogen atoms or helium atoms. Differing from the capture processes of antiprotons by hydrogen atoms, the anomalous bumpy structures are revealed in the total angular momentum dependent capture cross sections by helium. Further analysis shows that the bumps arise from the partial channel closing due to the removal of the energy degeneracy in the antiprotonic helium atom.

Acknowledgments
This work was supported by Grant-in-Aids for Scientific Research (C) from the Japan Society for the Promotion of Science.

References
[1] Esry B D and Sadeghpour H R 2003 *Phys. Rev. A* **67** 012704.
[2] Hesse M, Le A T and Lin C D 2004 *Phys. Rev. A* **69** 052712
[3] Cohen J S 1987 *Phys. Rev. A* **36** 2024–2030
[4] Sakimoto K 2002 *J. Phys. B* **35** 997–1014
[5] Ovchinnikov S Y and Macek J H 2005 *Phys. Rev. A* **71** 052717
[6] Yamanaka N and Ichimura A 2006 *Phys. Rev. A* **74** 012503
[7] Tong X M, Hino K and Toshima N 2006 *Phys. Rev. Lett.* **97** 243202
[8] Tong X M, Shirahama T, Hino K and Toshima N 2007 *Phys. Rev. A* **75** 052711
[9] Tong X M, Hino K and Toshima N 2008 *Phys. Rev. Lett.* **101** 163201
[10] Avron J E and Elgart A 1998 *Phys. Rev. A* **58**(6) 4300–4306
[11] Tong X M and Chu S I 1997 *Chem. Phys.* **217** 119–130
[12] Telnov D A and Chu S I 1999 *Phys. Rev. A* **59** 2864
[13] Perie J and Jolicard G 1993 *J. Phys. B* **26** 4491–4502
[14] Tong X M and Lin C D 2005 *J. Phys. B:* **38** 2593–2600

Journal of Physics: Conference Series **185** (2009) 012048 doi:10.1088/1742-6596/185/1/012048

Space distribution of the rescattering electron wavepacket in the laser-atom interactions

X. M. Tong[1,2], K. Sasaki[1], K. Hino[2,3] and N. Toshima[1]

[1] Doctoral Program in Materials Science, Graduate School of Pure and Applied Sciences, University of Tsukuba, 1-1-1 Tennodai, Tsukuba, Ibaraki 305-8573, Japan

[2] Center for Computational Sciences, University of Tsukuba,1-1-1 Tennodai, Tsukuba, Ibaraki 305-8577, Japan

[3] Doctoral Program in Frontier Science, Graduate School of Pure and Applied Sciences, University of Tsukuba, 1-1-1 Tennodai, Tsukuba, Ibaraki 305-8573, Japan

E-mail: tong@ims.tsukuba.ac.jp

Abstract. We investigated the space distribution of the rescattering electron wavepacket created in the laser-atom interaction by a full quantum simulation and a semiclassical calculation. Both the quantum simulation and the semiclassical calculation showed that the rescattering electron beam current intensity can reach the order of 10^9 A/cm^2, much intense than the conventional electron beam. Different from the convention electron beam, the rescattering electron beam is of a non-uniform distribution both in energy and space. The simulated information is important for analyzing the molecular structure in the rescattering imaging experiments.

The laser-material interaction is a hot research topics owing to the rapid advance of the laser technology. Most of the observed phenomena can be explained by the rescattering model proposed by Corkum [1]. The electron is firstly released by the laser through tunneling ionization when the laser field reaches its peak and the released electron is bounced back when the laser field changes the polarization direction. The collision of the bounced back electron with the parent core is similar to an electron beam incidents on the target. In principle, all the dynamics, which can be investigated by the conventional electron beam, can also be investigated with the rescattering electron beam. More attractively, the rescattering electron beam is a coherence beam and the beam intensity is much higher than the conventional one. Thus, the rescattering electron beam provides a potential way to image the molecular structure [2], like the conventional (e,2e) experiment [3]. For the conventional electron beam, the beam intensity, beam energy and so on can be controlled in the experiment. For the rescattering electron beam, the full information [4] about the rescattering beam, namely, the energy distribution and the space distribution, is unknown. Without such knowledge, the beam quality cannot be controlled experimentally. Since the rescattering is an intermediate process, the information cannot be obtained directly from the experimental measurement. It has to rely on the numerical simulation.

In principle, all the dynamic information can be obtained by solving the time-dependent Schrödinger equation. In practice, directly getting the rescattering information is not easy because (1) the wavefunction corresponds to the tunneling ionized electron is just a small factional part of the total wavefunction; (2) the rescattering electron wavepackets created at different laser cycles mix up and so it is difficult to decompose the n-th returns clearly. Of course, the rescattering information can be obtained by a semiclassical simulation [5] but the reliability of the classical results is not clear. In our previous work [6], we proposed a numerical procedure to obtain all the rescattering information directly by solving the time-integral equation instead of the differential Schrödinger equation and we extracted all the rescattering information. So far most of the works focused on the returning time and energy distribution, there is no report on the space distribution of the scattering electron, which is another crucial information for the molecular structure imaging study using rescattering electron beam. In this paper, we study the space distribution of the rescattering electron beam in a full quantum, nonperturbative method.

Similar to our previous work [6], we rewrote the time-dependent Schrödinger equation into an integral equation and factored out the non-active part which corresponds to the initial state without interacting with the laser field. The time-dependent wavefunction is formally written as (atomic units $\hbar = m = e = 1$ are used)

$$\Psi(t) = -i \int_{t_0}^{t_1} e^{-i\int_\tau^t [H_0+V(t')]dt'} V(\tau) e^{-iH_0\tau} \Phi_0 d\tau. \quad (1)$$

Here Φ_0 is the initial state, H_0 is the laser-field free Hamiltonian (we took the hydrogen atom as an example) and $V(t) = -zE_0 \cos\omega t$ is the laser-electron interaction. $\Psi(t)$ is interpreted as the time dependent wavefunction due to the interaction of the atom with laser field during the time interval from t_0 to t_1, with E_0 the laser field peak strength and ω the laser frequency. Note that the exponential operator is an time-ordered operator. The time-propagation is performed by the split-operator method with a generalized pseudospectral grid in the energy representation [7, 8]. The laser-atom interaction V appears twice in the above equation. The intuitive physical picture is that the atom is ionized due to the laser-atom interaction $V(\tau)$ and the ejected electron is propagated in the atomic field combined with the laser-atom interaction $V(t')$. At time t, we calculated the component of a given energy E and partial wave l of the laser-field free wavefunction in the inner region (collision region) as $C_l(E,t)$. With the coefficients $\{ C_l(E,t)\}$, we obtained the energy distribution of the rescattering electron wavepacket as

$$\frac{dP(E,t)}{dE} = \sum_l |C_l(E,t)|^2. \quad (2)$$

Because in the quantum mechanics, the impact parameter b is not well defined, we approximate it as $b^2 = l(l+1)/(2E)$ and define the rescattering beam intensity as

$$I(b,t) = \frac{dP(b,t)}{2\pi b db} = \int \sum_l \frac{\sqrt{2E}}{\pi} |C_l(E,t)|^2 \delta\left(b^2 - \frac{l(l+1)}{2E}\right) dE. \quad (3)$$

In the present simulation, we chose the laser peak intensity 10^{14} W/cm^2 and wavelength 800 nm (the parameters most commonly used in the experiments). The tunneling ionization happens within half cycle from $\omega t_0 = -\pi/2$ to $\omega t_1 = \pi/2$. We traced the electron wavepacket when the ionization process is over. The upper panel in Fig. 1 shows the rescattering electron beam intensity $I(b,t)$. To compare with the conventional electron beam intensity, we converted the rescattering electron beam intensity from atomic unit to A/cm^2. The rescattering electron revisits the parent core at the returning times well predicted by the simple classical model [1] and the electron beam intensity reaches as high as 10^9 A/cm^2, which is several orders intense than the electron beam intensity in the electron beam ion trap [9], which is of the order of 10^3 A/cm^2. Note that the beam intensity obtained in the present work is about two orders smaller than the one predicted from the simple model [10]. The space distribution is not smooth but contains several strips within a give return time. These strips come from the discretized angular momenta. The most left one on the upper panel in Fig. 1 corresponds to the first

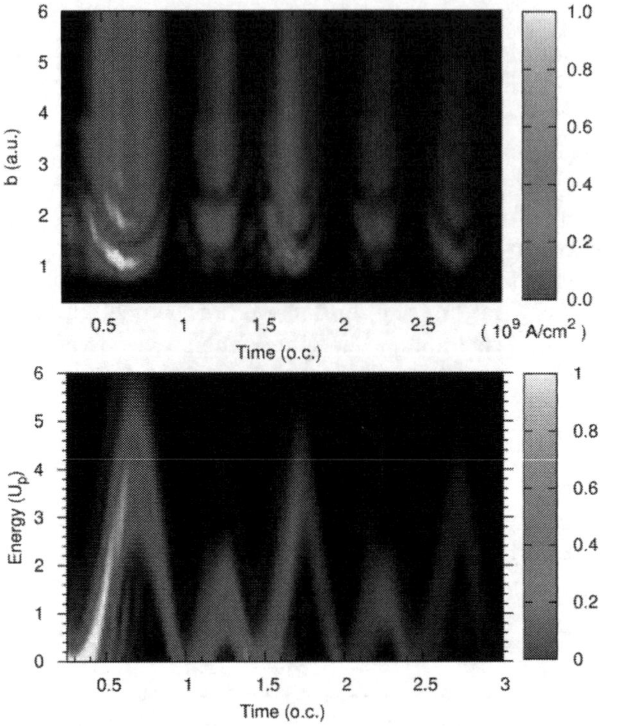

Figure 1. The rescattering electron beam intensity as a function of the rescattering time and the impact parameter b from the quantum calculation (upper panel) and the rescattering energy and time distributions. The color coding represents the relative value of $\frac{dP(E,t)}{dE}$ (lower panel).

return, then the second one, the third one and so on. The space distribution is rather smooth, or uniformly distributed for the first return. At the third return, the electron is concentrated to the inner region, which corresponds to the close collision with the parent core. We attributed this to the Coulomb focusing effect. To explain the Coulomb focusing effect, we showed the time-dependent energy distribution of the rescattering electron on the lower panel in Fig 1. The rescattering energy reaches highest for the first return. The Coulomb focusing effect is less important for the high energy electron since it spends a less time in the inner region. When the electron returns to parent core again (the second return), the electron energy is much lower than the first return energy and the returning electron beam is focused due to the electron parent core Coulomb interaction. The focused electron moves along the laser polarization direction in a narrow region on the perpendicular plane to the polarization direction. When the focused electron beam is bounced back by the laser field with a relatively high energy in the third return, the Coulomb force does not affect the motion of the returning electron significantly. But the electron has already been focused in the small region on the plane perpendicular to the polarization direction during the second return. Thus the Coulomb focusing effect in the second return results in a high intensity electron beam at the third return.

With the understanding of the detail of the Coulomb focusing, let us revisit the rescattering induced double ionization experiments [10, 11, 12] of H_2 molecules. The strong third return peak was explained due to (a) the vibration wavepacket moves to the large R, the nuclei separation, results in a small ionization potential; (b) the third return energy is lower than the first return energy and closer to the ionization threshold or the peak position of the electron impact ionization cross section [13, 14]. From the present study, we know the Coulomb focusing

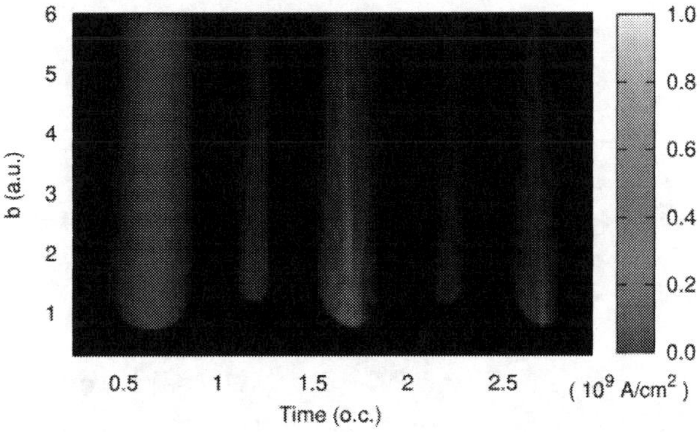

Figure 2. The rescattering electron intensity as a function of the rescattering time and the impact parameter b from the semi-classical simulation.

at the second return also contribute to the enhancement of the kinetic energy release at the third return. In the previous semi-classical simulations [13, 14], such an effect was taken into account in the simulation without analysis in detail.

To cross check the reliability of the quantum calculation, we also performed a semi-classical simulation [5]. In the simulation we monitored the impact parameter b-dependent rescattering electron beam intensity as shown in Fig. 2. The electron beam intensity distribution is rather smooth because of the nature of the classical behavior. The obtained results are consistent with the quantum simulated ones both in the returning times and the beam intensity. The Coulomb focusing effect on the third return is also clearly seen. The major physical conclusions, like the smooth space distribution for the first return, are consistent in both calculations.

To summarize, we studied the rescattering electron space distribution when it revisits the parent core. The rescattering electron beam intensity can reach as high as 10^9 A/cm^2, which is much higher than the convention electron beams. With the understanding of the rescattering electron beam, it may help us to analyze the molecular electronic structure, analog to the conventional (e,2e) experiment.

Acknowledgments
This work was supported by Grant-in-Aids for Scientific Research (C) from the Japan Society for the Promotion of Science.

References
[1] Corkum P B 1993 *Phys. Rev. Lett.* **71** 1994–1997
[2] Itatani J, *et al.*, 2004 *Nature* **432** 867–871
[3] Brion C E, Cooper G, Zheng Y, Litvinyuk I V and McCarthy I E 2001 *Chem. Phys.* **270** 13–30
[4] Spanner M, Smirnova O, Corkum P B and Ivanov M Y 2004 *J. Phys. B* **37** L243
[5] Tong X M and Lin C D 2007 *Phys. Rev. Lett.* **98** 123002
[6] Tong X M, Watahiki S, Hino K and Toshima N 2007 *Phys. Rev. Lett.* **99** 093001
[7] Tong X M and Chu S I 1997 *Chem. Phys.* **217** 119–130
[8] Tong X M, Hino K and Toshima N 2006 *Phys. Rev. A* **74** 031405
[9] Nakamura N,*et al.*, 1997 *Physica Scripta* **T73** 362–364
[10] Niikura H,*et al.*, 2002 *Nature* **417** 917–922
[11] Alnaser A S, *et al.*, 2003 *Phys. Rev. Lett.* **91** 163002
[12] Alnaser A S, *et al.*, 2004 *Phys. Rev. Lett.* **93** 183202
[13] Tong X M, Zhao Z X and Lin C D 2003 *Phys. Rev. A* **68** 043412
[14] Tong X M, Zhao Z X and Lin C D 2003 *Phys. Rev. Lett.* **91** 233203

Journal of Physics: Conference Series **185** (2009) 012049 doi:10.1088/1742-6596/185/1/012049

Atomic collision and spectroscopy experiments with ultra-low-energy antiprotons

Hiroyuki A. Torii[1], Yugo Nagata[2], Hiroshi Toyoda[1], Hiroshi Imao[2], Naofumi Kuroda[1], Victor L. Varentsov[3], Yasunori Yamazaki[1,2], on behalf of ASACUSA collaboration[4]

[1] Institute of Physics, University of Tokyo, 3-8-1 Komaba, Meguro-ku, Tokyo 153-8902, Japan
[2] Atomic Lab., RIKEN, 2-1 Hirosawa, Wako-shi, Saitama 351-0198, Japan
[3] Gesellschaft für Schwerionenforschung (GSI), Darmstadt, Germany
[4] CERN, Genève 23, CH-1211, Switzerland

E-mail: torii@radphys4.c.u-tokyo.ac.jp

Abstract. Antiproton, the antiparticle of proton, is a unique projectile in the study of atomic collision physics, which can be treated theoretically either as a 'negative proton' or a 'heavy electron'. Atomic capture of an antiproton will result in formation of a highly excited exotic atom. Antiprotonic helium atom has been studied intensively by means of precision laser spectroscopy, which has led to a stringent determination of antiproton mass and charge to a level of ppb. Comparison of these values with those of proton gives one of the best tests of CPT invariance, the most fundamental symmetry in physics. However, the dynamic processes of antiproton capture remain unclarified. With an aim to produce an antiproton beam at atomic-physics energies for 'pure' collision experiments, we have so far developed techniques to decelerate, cool and confine antiprotons in vacuo, using a sequential combination of the Antiproton Decelerator (AD) at CERN, a Radio-Frequency Quadrupole Decelerator (RFQD), and an electromagnetic trap. Our recent success in stable extraction of monoenergetic ultra-slow antiprotons, about 3×10^5 in number available every 5 minutes, has opened up the possibility to study ionization and atomic capture processes between an antiproton and an atom under the single collision condition. Our design and strategy of the cross-beam experiments are presented, together with technical challenges in the detection system to identify the rare events with a reaction rate of 10^{-4}.

1. Introduction

Antiproton is a unique probe for atomic physics researches. Having the same mass as the proton but with opposite charge, It behaves as a "heavy electron" or as a "negative nucleus", and gives a new test ground for studies of atomic collision dynamics. In fact, it is a "theoreticians' ideal projectile" because lack of electron capture processes avoids complication in theoretical treatments. In addition to various CTMC (Classical Trajectory Monte Carlo) calculations hitherto available [1], recently semi- and full-quantum treatments have become possible [2, 3]. Yet so far no experiment was possible at energies less than around 5 keV [4].

Experimentally, antiprotons can be produced at accelerator facilities, but only at a huge energy in the GeV range via the reaction $p + p \to p + p + p + \overline{p}$. The Antiproton Decelerator (AD) at CERN, Geneva, is a unique facility which provides cooled low-energy antiprotons with MeV energies. There, the ASACUSA collaboration has been doing researches on "Atomic Spectroscopy And Collisions Using Slow Antiprotons" [5].

One branch of our activities is the spectroscopy of metastable antiprotonic helium atom [6]. This three-body atom consisting of a helium nucleus, an electron and an antiproton is a peculiar system, which can be regarded as a mixture of matter and antimatter, and yet shows longevity of a few microseconds at large quantum numbers of $n \sim 38$ and $l \lesssim n$. Laser and microwave spectroscopy has not only revealed the level structure of this exotic atom [7, 8], but also assisted development of calculation techniques for the Coulombic three-body system, in the precision competition between the experiment and the theory [9]. Recent high precision in our laser spectroscopy has reached the level of 10^{-9} in determination of transition frequencies, and comparison with the theoretical QED calculations now gives one of the most precise values for the antiproton-to-electron mass ratio as well as the antiproton-to-proton mass ratio [10], contributing to the precise determination of the fundamental constant and to the test of CPT invariance theorem.

For the study of dynamic processes of atomic ionization and capture, an antiproton–atom cross-beam experiment would require an ultra-slow antiproton beam at a well-defined energy in the eV to keV range. We the MUSASHI group of the ASACUSA collaboration have developed a system consisting of a Radio-Frequency Quadrupole Decelerator (RFQD), an electromagnetic trap and an extraction beam line at the AD facility, and have recently succeeded in production of ultra-low-energy monoenergetic antiproton beams. These beams are crossed with an atomic gas-jet target, to study ionization and atomic capture processes of an antiproton projectile against various atoms "A", $\overline{p} + \mathrm{A} \to \overline{p} + \mathrm{A}^+ + e^-$ and $\overline{p} + \mathrm{A} \to (\overline{p}\mathrm{A}^+)^0 + e^-$ for the first time under the single-collision condition at very low energies. This study will also reveal the properties of exotic antiprotonic atoms formed in vacuo.

2. Production of ultra-low-energy antiproton beams

The antiproton beam at 5 MeV from the AD at CERN needs further deceleration for them to be captured in vacuo electrostatically. We developed an RFQ Decelerator in which RF waves were applied to the cavities in such a way as to decelerate microbunches of the antiproton beam at 5.3 MeV down to 110 keV with an efficiency of 30% [11]. The output beam from the RFQD was injected into an eletromagnetic trap in a strong magnetic field of 2.5 T produced by a superconducting solenoid. The beam was focused to a diameter of 3–4 mm on a thin Mylar foil used to isolate ultrahigh vacuum of 10^{-12} mbar inside the trap.

Antiprotons were then captured and confined in the trap, a Multi-Ring Trap (MRT) [12] consisting of 14 cylindrical electrodes, where they were sympathetically cooled by a plasma of typically 3×10^8 electrons preloaded in the harmonic potential of 50 V depth. The total number of trapped antiprotons can be estimated from the signal counts of antiproton annihilation detected by scintillation track detectors installed beside the trap, with their detection efficiency of $\epsilon \sim 5\%$ taken into account. With this trap, we successfully confined 1.2×10^6 cooled antiprotons until the end of our trap cycle of 1–5 minutes [13]. Typically 1 million antiprotons were trapped for each AD shot. We also demonstrated accumulation of antiprotons for several AD shots. This technique of "stacking" also worked fine, and we trapped 1.2×10^7 antiprotons simultaneously for stacking of 12 AD shots, the largest number of antiprotons ever accumulated.

After quick kick-out of electrons from the trap, the antiprotons were eventually released as the potential was gradually shallowed, and were extracted as an ultra-slow continuous beam. Since the antiprotons tend to expand in radial direction when they follow the strongly diverging magnetic field line, it was essential that the antiproton cloud be well compressed radially in the trap. To actively compress the antiproton cloud, it was given torque by a rotating electric field, applied to one of the ring electrodes azimuthally segmented into four parts. The antiproton cloud was thus compressed from a radius of 3.4 mm to 0.25 mm in a period of 200 s [14].

The extraction beam line was designed to transport antiproton beams over a length of 3 m, at variable energies ranging from 10 to 1000 eV. The antiproton beams were refocused three times by sets of Einzel lenses at the position of apertures. These variable apertures of diameter 4–10 mm allowed differential pumping of 6 orders of magnitude along the beam line [15], which was necessary to keep the trap region at an extremely high vacuum better than 10^{-12} mbar so as to avoid antiproton annihilation [11, 16].

The MRT, the superconducting solenoid and the transport line for the ultra-slow beam are jointly known as "MUSASHI", or the Monoenergetic Ultra-Slow Antiproton Source for High-precision Investigations. MUSASHI opens a new research field ranging from atomic physics to nuclear physics including our near-future project of antihydrogen synthesis in a cusp trap [17].

3. Experimental designing of collision experiments

With this unique beam, atomic formation and ionization processes by very-low-energy antiprotons can now be studied under single collision conditions. Since the number of available antiprotons is very much limited, the reaction probability must be maximized for their best use. The number of antiprotonic atoms formed is given by the formula $N_{\overline{\mathrm{p}}\mathrm{A}^+} = \sigma\, n_{\mathrm{A}}\, L\, N_{\overline{\mathrm{p}}}$, where σ is the formation cross section, n_{A} the number density of the atomic target, L the interaction length, and $N_{\overline{\mathrm{p}}}$ is the number of antiprotons.

Helium atom was chosen for our first target, for technical simplicity and because of the fact that the lifetime of antiprotonic helium atoms are already known from our spectroscopic measurements [18]. Our apparatus has 5 chamber stages in order to achieve good jet collimation and efficient pumping. A supersonic gas jet emerging at a stagnation pressure of 25 bar from a 0.1-mm-diameter nozzle was skimmed by two stages of conically shaped skimmers [19]. The gas jet was collimated to form a target of about 1 cm diameter at the cross point in the main chamber, before being collected at the dump stages. The vacuum in the main chamber was kept to a level of less than 10^{-6} mbar. With this setup, we have achieved a gas target density of $n_{\mathrm{A}} = 3 \times 10^{12}$ atoms/cm^3. Suppose a beam with 3×10^5 antiprotons is available every shot per several minutes, we will obtain 10–100 antiprotonic atoms for a reaction cross section of the order of 10^{-16}–10^{-17} cm^2 [1]. The reaction probability is of the order of 10^{-4}. Rigorous particle identification is necessary in order to assure enough statistical significance for the reaction events to be detected and distinguished from background events.

Antiprotons are extracted as a beam in a slow-extraction mode. This continuous beam allows event-by-event data acquisition associated with each single antiproton extracted. Figure 1 shows

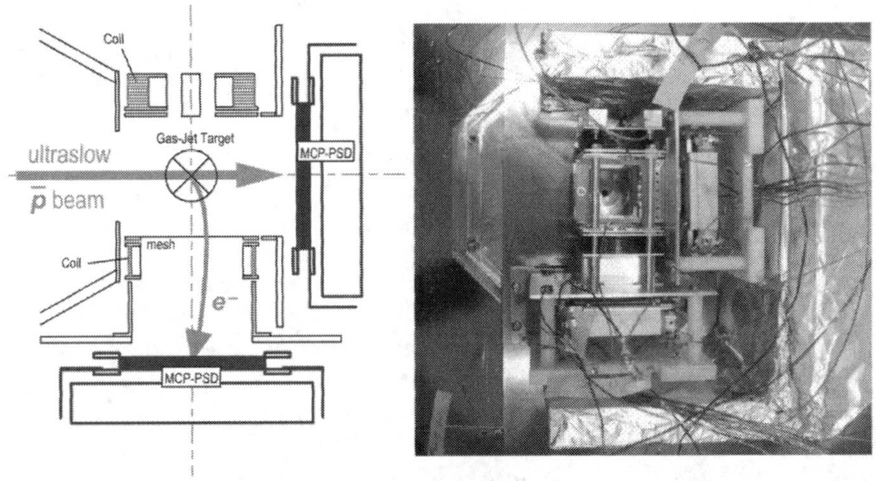

Figure 1. Schematic setup (left) and a photo image (right) of the modified version of detectors surrounding the collision point of the antiproton beam and the gas-jet target.

a schematic layout of the detectors. Two MCP-PSDs were placed inside the gas-jet chamber, one downstream of the cross point along the antiproton beam path for detection of antiprotons hitting it, and another one sideways of the antiproton beam, to selectively detect electrons. They were surrounded by a box of scintillator plates for identification of antiproton annihilation by detecting passage of its annihilation products such as pi mesons. A set of electrodes and coils were placed near the collision point to guide electromagnetically the electrons perpendicular to the antiproton beam. The TOF is shorter for electrons than for antiprotons due to the difference in their velocity. Thus a reaction event can be recognized by an electron signal (i.e. side-MCP hit and no scintillator hit) followed by an antiproton signal (i.e. coincidence between downstream-MCP hit and scintillator hit) with an appropriate TOF interval.

4. Summary and Outlook

We have succeeded in production of ultra-low-energy antiproton beams, using a combination of the AD facility, the RFQ Decelerator, the electromagnetic trap and the extraction beam line. This unique beam opens a new research field of atomic physics. We are planning in the future to make measurements in the 100–1000 eV region to reveal the ionization cross section curve which is totally unknown. Even theoretically, no reliable work has been done. A realistic calculation must include diabatic effects, whose methodology is not yet established. Atomic capture process will be studied at lower eneries of 10–30 eV. Capture of an antiproton by an atom results in formation of an antiprotonic atom. The cross-beam experiment in vacuo is expected to populate higher-lying states at $n > 40$ with a much longer lifetime of 10–100 μs, in contrast to 3 μs in the case of formation at $n \approx 38$ in media. Our new measurements are thus expected to clarify the collision dynamics under well-defined single-collision conditions at fixed energies.

Acknowledgments

We are grateful to the AD and PS stuff as well as the RFQD support team in the AB department of CERN for their tireless effort in providing us with the antiproton beam. This work was supported by the Grant-in-Aid for Creative Scientific Research (10P0101 and 19002004) of the Japanese Ministry of Education, Culture, Sports, Science and Technology (MonbuKagaku-shō), Special Research Projects for Basic Science of RIKEN, and the Hungarian National Science Foundation (OTKA T033079).

References

[1] J. S. Cohen, 2004 *Rep. Prog. Phys.* **67** 1769; 2005 *AIP Conf. Proc.* **793** 98.
[2] K. Sakimoto, 2006 *Phys. Rev. A* **74** 022709; 2006 *Phys. Rev. A* **74** 042711.
[3] X. M. Tong, K. Hino and N. Toshima, 2006 *Phys. Rev. Lett.* **97** 243202; X. M. Tong, T. Shirahama, K. Hino and N. Toshima, 2007 *Phys. Rev. A* **75** 052711.
[4] H. Knudsen *et al.*, 2008 *Phys. Rev. Lett.* **101** 043201.
[5] ASACUSA collaboration web page, http://cern.ch/ASACUSA
[6] T. Yamazaki *et al.*, 2002 *Phys. Rep.* **366** 183.
[7] N. Morita *et al.*, 1994 *Phys. Rev. Lett.* **72** 1180
[8] E. Widmann *et al.*, 2002 *Phys. Rev. Lett.* **89** 243402.
[9] H. A. Torii *et al.*, 1999 *Phys. Rev.* A **59** 223.
[10] M. Hori, A. Dax *et al.*, 2006 *Phys. Rev. Lett.* **96** 243401.
[11] H. A. Torii *et al.*, 2005 *AIP Conf. Proc.* **793** 293.
[12] A. Mohri, H. Higaki *et al.*, 1998 *Jpn. J. Appl. Phys.* **37** 664.
[13] N. Kuroda, H. A. Torii, K. Yoshiki Franzen *et al.*, 2005 *Phys. Rev. Lett.* **94** 023401.
[14] N. Kuroda, H. A. Torii, M. Shibata *et al.*, 2008 *Phys. Rev. Lett.* **100** 203402.
[15] K. Yoshiki Franzen, N. Kuroda, H. A. Torii *et al.*, 2003 *Rev. Sci. Instrum.* **74** 3305.
[16] H. A. Torii *et al.*, 2007 *J. Phys. Conf. Ser.* **80** 012022.
[17] A. Mohri and Y. Yamazaki, 2003 *Europhys. Lett.* **63** 207.
[18] M. Hori *et al.*, 2004 *Phys. Rev.* A **70** 012504.
[19] V. L. Varentsov *et al.*, 2005 *AIP Conf. Proc.* **793** 328.

The 8th Asian International Seminar on Atomic and Molecular Physics IOP Publishing
Journal of Physics: Conference Series **185** (2009) 012050 doi:10.1088/1742-6596/185/1/012050

Photoelectron interference fringes by super intense x-ray laser pulses

Koudai Toyota[1], **Oleg I. Tolstikhin**[2], **Toru Morishita**[1,3] **and Shinichi Watanbe**[1]

[1] Department of Applied Physics and Chemistry, University of Electro-Communications, 1-5-1, Chofu-ga-oka, Chofu-shi, Tokyo, Japan
[2] Russian Research Center "Kurchatov Institute," Kurchatov Square 1, Moscow 123182, Russia
[3] PRESTO, Japan Science and Technology Agency, Kawaguchi, Saitama 332-0012, Japan

E-mail: toyota@power1.pc.uec.ac.jp

Abstract. The photoelectron spectra of H^- produced by circularly polarized strong high-frequency laser pulses are theoretically studied. An oscillating substructure in the above-threshold ionization (ATI) peaks is observed, which extends the validity of the earlier findings in the 1D calculations [K. Toyota et al., Phys. Rev. A 76, 043418 (2007)] and 3D calculations for linear polarization [O. I. Tolstikhin, Phys. Rev. A 77, 032712 (2008)]. Its origin is due to an interference between a pair of photoelectron wave packets created in the rising and falling part of the pulse, which appears clearly in the stabilization regime.

1. Introduction

Recent experimental developments in free-electron lasers [1] provide us with coherent light sources in the X-ray range. Such new light sources experimentally opened up the so-called *high-frequency regime*, and revealed interesting dynamics of atoms [2, 3]. The experimentalists may further aspire to develop an ultra intense laser pulse to reach the *stabilization regime* in future. The stabilization is a phenomenon in which the total ionization yield decreases once the laser intensity exceeds a certain critical value. It was first predicted by theory (see [4, 5] for the history and detailed discussion of the stabilization).

In this paper, we study the photoelectron spectra of H^- in the stabilization regime, where oscillating substructures appear in the ATI peaks, an interest feature first found in a 1D case [6]. It is shown that this is due to the interference of a pair of photoelectron wave packets produced at different moments in time, and that the stabilization plays a key role . Subsequently, the effect was confirmed by 3D calculations for linear polarization [7]. Here, we follow these preceding papers but consider circular polarization. We interpret the effect in terms of an adiabatic version of the high-frequency Floquet theory (HFFT) [8]. Although this analysis is similar to that in [6], the generalization to the 3D case is not straightforward, and it is not evident *a priori* that such a simple approximation applies to the 3D case as well.

2. Theoretical approach

We consider a negative hydrogen ion H^- interacting with a laser pulse. The ion is treated in the single-active-electron approximation. The time-dependent Schrödinger equation (TDSE) in the Kramers-Henneberger (KH) frame reads (the atomic units are used in this paper),

$$i\frac{\partial}{\partial t}\psi(\vec{r},t) = \left(-\frac{1}{2}\Delta + V(|\vec{r}+\boldsymbol{\alpha}(t)|)\right)\psi(\vec{r},t), \tag{1}$$

where $V(r)$ is the atomic potential and $\boldsymbol{\alpha}(t)$ is the classical trajectory of the free-electron in a laser electric field $\mathbf{F}(t)$. The potential is modeled by [7] $V(r) = -V_0 \exp(-r^2/r_0^2)$ with $V_0 = 0.383\,108\,7$ and $r_0 = 2.5026$. This potential supports only one bound state with energy $E_0 = -0.027\,751\,0$. The field $\mathbf{F}(t)$ is assumed to vanish beyond the time interval $0 \le t \le T$. The classical trajectories satisfies $\ddot{\boldsymbol{\alpha}}(t) = -\mathbf{F}(t)$ with initial conditions $\dot{\boldsymbol{\alpha}}(0) = \boldsymbol{\alpha}(0) = \mathbf{0}$. In the KH frame, the effect of the laser field is combined into the atomic potential. This transformed potential is called the KH potential. The interaction between the electron and the laser field is thus described by a quiver motion of the KH potential along the classical trajectory. We solve Eq. (1) using the procedure described in [7].

3. Photoelectron spectra in the stabilization regime

3.1. Laser pulse

For circularly polarized pulse, we assume that the polarization vector lies in the xz plane. The components of the electric field $\mathbf{F}(t)$ are represented by $(0 \le t \le T)$

$$\mathbf{F}(t) = (\varepsilon F_0 f(t)\sin\omega t, 0, F_0 f(t)\cos\omega t) \tag{2}$$

where ε, F_0, ω, and T are the ellipticity, amplitude, frequency, and duration of the pulse, respectively. The pulse envelope $f(t)$ is defined so that the classical trajectory satisfies $\boldsymbol{\alpha}(T) = \dot{\boldsymbol{\alpha}}(T) = 0$.

3.2. Numerical results

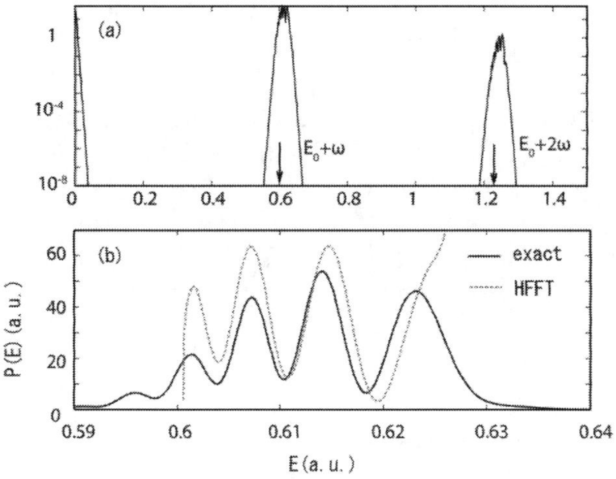

Figure 1. The photoelectron spectrum $P(E)$ of H^- produced by the high-frequency strong laser pulse with the parameters $\varepsilon = 1$, $F_0 = 1.2$, $\omega = \pi/5$, and $T = 2000$. (a) The overall view of the spectrum. The multiphoton absorption energies $E_0 + n\omega$ ($n = 1, 2$) are shown by arrows. (b) The one-photon absorption peak at $E \approx 0.601$ on a magnified scale. The solid and broken lines represent the exact TDSE result and the reconstruction by the high-frequency Floquet theory (HFFT), respectively.

The photoelectron spectrum in the stabilization regime is illustrated in Fig. 1(a). The laser parameters are $F_0 = 1.2$, $T = 2000$, $\omega = \pi/5$. An oscillatory substructure in the ATI peaks is clearly seen in Fig. 1(b). In the rest of the paper, we discuss this substructure and clarify the underlying interference mechanism focusing on the first ATI peak. In doing so, we follow a train of thought similar to that in [6, 7], but deal with circular polarizations.

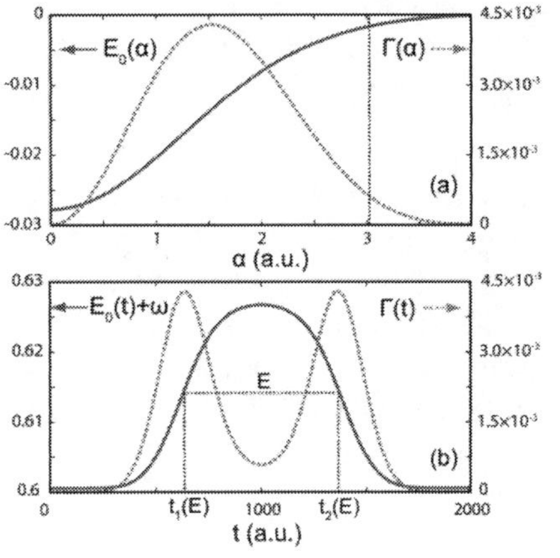

Figure 2. (a) The energy (solid line) and one-photon decay width (dashed line) of the initial dressed state as functions of the excursion amplitude $\alpha = F_0/\omega^2$ for $\omega = \pi/5$. (b) Same as in (a), but as functions of time recalculated using $\alpha \to \alpha(t) = \frac{F_0}{\omega^2} f(t)$ for the laser pulse with $F_0 = 1.2$, $\omega = \pi/5$, $T = 2000$. The maximum value of $\alpha(t)$ for this pulse is shown by the dotted vertical line in the upper panel.

3.3. Analysis in terms of the high-frequency Floquet theory

Here we develop an approximate picture of the dynamics and reconstruct the first ATI peak using an adiabatic version of the HFFT [8]. The present analysis generalizes that of [6] to the 3D case. It is convenient to rotate the coordinate axes in such a way that the polarization plane coincides with the xy plane, thus the laser pulse propagates along the z axis.

Let us consider a monochromatic laser field, i.e., temporarily omit the envelope factor $f(t)$ in Eq. (2). The classical trajectory in this case is given by $\boldsymbol{\alpha}(t) = (\alpha\cos\omega t, \alpha\sin\omega t, 0)$, where $\alpha = F_0/\omega^2$ is the excursion amplitude. The KH potential can be expanded into a Fourier series,

$$V(|\mathbf{r} + \boldsymbol{\alpha}(t)|) = \sum_{n=-\infty}^{\infty} V_n(r, \theta; \alpha) e^{in(\varphi - \omega t)}, \tag{3}$$

where θ and φ are the polar angles defining the direction of $\mathbf{r}$. In the zeroth order of the HFFT [8], the system is described by the stationary 'dressed' Hamiltonian

$$H_{\mathrm{HFFT}}(\alpha) = -\frac{1}{2}\Delta + V_0(r, \theta; \alpha). \tag{4}$$

The eigenfunctions of $H_{\mathrm{HFFT}}(\alpha)$ will be called the dressed states. Let $\psi_0(r, \theta; \alpha)$ and $E_0(\alpha)$ denote the eigenfunction and eigenvalue, respectively, of the initial bound dressed state which coincides with the ground state of the unperturbed atom for $\alpha = 0$. Let $\psi(\mathbf{r}, \mathbf{k}; \alpha)$ be the scattering dressed state corresponding to the momentum $\mathbf{k} = (k, \Omega)$ and energy $k^2/2$, normalized to unit amplitude of the incoming plane wave. Then, in the first order of the HFFT [8], the partial width of the initial state associated with the absorption of one photon is given by

$$\Gamma(\alpha) = \frac{k(\alpha)}{(2\pi)^2} \int |A(k(\alpha), \Omega; \alpha)|^2 \, d\Omega, \tag{5}$$

where $A(k, \Omega; \alpha)$ is the transition amplitude, $A(k, \Omega; \alpha) = \int \psi^*(\mathbf{r}, \mathbf{k}; \alpha)\, V_1(r, \theta; \alpha) e^{i\varphi}\, \psi_0(r, \theta; \alpha)\, d\mathbf{r}$, and $k(\alpha)$ is the momentum of the photoelectron, $k(\alpha) = \sqrt{2(E_0(\alpha) + \omega)}$. The width $\Gamma(\alpha)$ calculated for $\omega = \pi/5$ is shown in Fig. 2(a). It first grows with α, but then decays after α passes the critical value $\alpha_c \approx 1.5$. Such a behavior of $\Gamma(\alpha)$, which is a signature of stabilization, is the key to understand the ionization dynamics.

The remaining part of the analysis parallels that in [6]. A part of the photoelectron spectrum near the first ATI peak ($E_0 + \omega \approx 0.601$) calculated for this pulse is shown in Fig. 1(b). This pulse is not monochromatic. However, its envelope varies slowly, the pulse contains $n_{\mathrm{oc}} = 200$ optical cycles. So, one could expect that the picture suggested by the HFFT is followed adiabatically. The adiabatic approximation is implemented by the substitution $\alpha \to \alpha(t) = \frac{F_0}{\omega^2} f(t)$. The maximum value of $\alpha(t)$ for the present parameters is $F_0/\omega^2 = 3.04$; it is shown by the vertical dotted line in Fig. 2(a). The behavior of $E_0(t)$ and $\Gamma(t)$, now as functions of t recalculated using the adiabatic approximation, is shown in Fig. 2(b). The decay rate $\Gamma(t)$ indeed has two humps. The probability to stay in the initial state until the moment t is, in this approximation, $P_0(t) \approx \exp\left[-\int_0^t \Gamma(t')\,dt'\right]$. This gives $P_0(T) \approx 0.063$, which is not very far from the survival probability 0.055 obtained in our accurate calculations. Within the adiabatic approximation, the energy E of a photoelectron is a function of time t, $E = E_0(t) + \omega \to t = t(E)$, where $t(E)$ is the inverse function. The electrons ionized in the interval from t to $t + dt$ have energies between E and $E + dE$, where $dE = (dE_0(t)/dt)dt$. Equating the total ionization probability in this interval $P_0(t)\Gamma(t)dt$ to $C^2(E)|dE|$, where $C(E)$ is the amplitude of the photoelectron wave packet created, one finds

$$C(E) = \sqrt{P_0(t)\Gamma(t)\left|\frac{dt}{dE}\right|}\Bigg|_{t=t(E)}. \tag{6}$$

As can be seen from Fig. 2(b), for the present pulse $t(E)$ is a double valued function. Let $t_1(E)$ and $t_2(E)$ denote its two branches, and $C_1(E)$ and $C_2(E)$ denote the corresponding amplitudes defined by Eq. (6). There are two different paths for the photoelectron with the energy E to evolve from $t = t_1(E)$ to $t = t_2(E)$. The first one is to be ionized at $t_1(E)$ and then propagate until $t_2(E)$ in the scattering state. The second one is to propagate between $t_1(E)$ and $t_2(E)$ in the bound state and then to be ionized. These paths lead to the same final state, but with different phases. Summing up their contributions, the photoelectron spectrum is given by

$$P_{\mathrm{HFFT}}(E) = \left|C_1(E) + C_2(E)e^{i\Phi(E)}\right|^2, \tag{7}$$

where $\Phi(E)$ is the phase difference for the two paths, $\Phi(E) = E\left[t_2(E) - t_1(E)\right] - \int_{t_1(E)}^{t_2(E)} \left[E_0(t) + \omega\right] dt$. The results obtained using these formulas are shown in Fig. 1(b). This approximate theory yields the spectrum only in a limited energy interval from $\min[E_0(t)+\omega]$ to $\max[E_0(t)+\omega]$ shown in the figure by vertical dotted lines. Equation (7) diverges at the upper boundary of this interval because of the factor $dt(E)/dE$ in Eq. (6). It reproduces the phase of the interference substructure nicely, but the amplitude is somewhat overestimated, especially in the lower part of the spectrum. However, in spite of these limitations, it is clear that the theory correctly accounts for the mechanism responsible for the appearance of the interference substructure. This analysis confirms our qualitative interpretation of the dynamics.

4. Conclusion

In this paper we studied the photoelectron spectrum of H^- generated by high-frequency strong laser filed. The oscillating substructures are found in the ATI peaks. We employed the HFFT and showed that this is due to the interference between photoelectron wave packets produced in the rising and falling parts of the pulse.

References

[1] V. Ayvazyan *et al.*, Euro. Phys. J. D **37**, 297 (2006).
[2] R. Moshammer *et al.*, *Phys. Rev. Lett.* **98**, 203001 (2007).
[3] M. Nagasono *et al.*, *Phys. Rev.* A **75**, 051406(R) (2007).
[4] J. H. Eberly and K. C. Kulander, Science **262**, 1229 (1993).
[5] N. B. Delone and V. P. Krainov, Usp. Fiz. Nauk **165**, 1295 (1995) [Sov. Phys. Usp. **38**, 1247 (1995)].
[6] K. Toyota, O. I. Tolstikhin, T. Morishita, and S. Watanabe, *Phys. Rev.* A **76**, 043418 (2007).
[7] O. I. Tolstikhin, *Phys. Rev.* A **77**, 032712 (2008).
[8] M. Gavrila and J. Z. Kaminski, *Phys. Rev. Lett.* **52**, 613 (1984).

The 8th Asian International Seminar on Atomic and Molecular Physics IOP Publishing
Journal of Physics: Conference Series **185** (2009) 012051 doi:10.1088/1742-6596/185/1/012051

Correlated positron-electron emission from LiF(100)

GA van Riessen, FO Schumann, M Birke, C Winkler and J Kirschner

E-mail: riessen@mpi-halle.mpg.de
Max-Planck Institute of Microstructure Physics, Weinberg 2, 06120 Halle, Germany

Abstract. Using two hemispherical energy analyzers with transfer lenses we have observed positron-electron pair emission from a surface upon low energy positron impact. An 85 eV positron beam from the NEPOMUC positron source was incident upon a LiF(100) surface and emitted electrons were detected in coincidence with the positrons. Energy analysis of the positron-electron pair confirms that electrons are emitted from the valence band upon positron impact. Electron-electron pairs emitted upon electron and positron impact are also discussed. The information obtainable from these experiments has the potential to provide important insights into electron-electron correlation in solids.

1. Introduction

In the past several decades, extensive application of electron-electron coincidence experiments (e^-,e^-e^-) on atoms and molecules have contributed to a deeper understanding of electron-electron correlation effects. In the condensed phase, electron-electron correlation effects, which underlie phenomena such as magnetism and superconductivity, continue to present a formidable challenge to theory. It follows that electron-electron ($e^- - e^-$) coincidence experiments on solids have received burgeoning interest in recent years despite the experimental challenges. It has been demonstrated that information about the correlation between a pair of electrons in the solid can be recovered from the observed momenta of the pair of electrons emitted from the surface upon photon or electron impact [1, 2, 3, 4, 5, 6].

Many theoretical studies have been devoted to the various aspects of the so-called exchange-correlation hole, a depletion of the average electron density around an individual electron in the condensed phase (see, e.g., refs. [7, 8, 9]) that is a direct manifestation of Coulomb and exchange interactions. Its extent in momentum space is a measure of the electron pair-correlation function which is a central quantity in the theory of many-body systems. It has been shown that with $e^- - e^-$pair emission spectroscopy it is possible to directly probe the exchange-correlation hole [1, 2]. Separating the contributions of Coulomb and exchange interactions to the exchange-correlation hole is not straightforward. It is the indistinguishability of two electrons that leads to characteristic exchange effects in $e^- - e^-$interactions. These are absent for positron-electron ($e^+ - e^-$) interactions. It follows that by employing a low energy positron beam and observing in time coincidence an electron and positron emitted from a surface after their interaction in the solid may provide a route by which the role of (attractive) Coulomb interactions can be probed without direct influence of exchange interactions.

An understanding of $e^+ - e^-$correlations is also central to understanding positron properties in condensed matter systems. Positron annihilation methods provide a sensitive probe to the electronic structure of solids [10] but interpretation of the information obtained requires a detailed understanding of $e^+ - e^-$correlation. Despite considerable theoretical investigation, there remains some unexplained discrepancies between theory and experiment, even for simple metals [11, 12].

Recently we reported that correlated e^+-e^-pairs emitted from a LiF(100) surface could be detected and that the essential characteristics of their energy distribution could be recognized from simple considerations of the wide band gap of LiF [13]. In this article we present additional experimental detail and present a preliminary analysis of the differences observed between (e^+,e^+e^-) , (e^+,e^-e^-) and (e^-,e^-e^-) energy spectra.

2. Experiment

The experiment essentially consisted of the detection of particles (positrons or electrons) emitted upon impact with positrons incident upon a LiF(100) crystal in an arrangement illustrated in Figure 1. The LiF(100) crystal was heated constantly to 150° Celsius to minimize contamination and to mitigate electrostatic charging of the surface. The positron source was a beamline of NEPOMUC located at the research reactor FRM-II in Garching [14]. An 85 eV beam of moderated positrons was magnetically guided in a longitudinal magnetic field of 10 mT to the end of the beamline where it was extracted from the magnetic field through an 8 mm aperture in a shield made from magnetically conductive ARMCO iron and focused to a 1 mm spot by electrostatic optics onto the target. One of seven electrodes was formed by the inner part of the iron shield. A 90% transparent grid was placed at the entrance to the first electrode (at ground potential). The positron beam was accelerated between the grid and the iron shield where the magnetic field lines turn abruptly from the axial to the transverse direction as they are guided outwards through the shield. With sufficient acceleration nonadiabatic transmission through the rapidly varying magnetic field is possible and the beam divergence is small, as was demonstrated by Shi *et al* [15]. Finite element modeling with the Opera3D (Vector Fields, USA) code confirmed that with appropriately optimized potentials applied the electrodes, positron beam of initial width 7 mm, total kinetic energy of 85 eV and a Gaussian distribution of transverse kinetic energy of FWHM of 2 eV in the solenoid could be completely transmitted and focused to 1mm at the sample as illustrated in Figure b. The residual magnetic field at the crystal was negligible. Transmission was found to be poorer than expected due mostly to the large beam size and energy spread in the solenoid. The primary positron flux was estimated to be 5×10^4 s^{-1} from the measured rate of positron annihilation at the sample.

Two 200 mm hemispherical analyzer's (Scienta R4000) with 2D detectors (multichannel plates (MCP) with resistive anode) were used to detect positrons and electrons and to record energy spectra. The analyzers, sample and positron optics were mounted in a UHV chamber. Either positrons or electrons could be analyzed by reversing the polarity of all the voltages applied to the lens, and analyser components, and by appropriately changing the potential between the front MCP and a grid in front it. The lenses and analyzers were configured to accept electrons or positrons with a mean kinetic energy of 30 eV with an analyser pass energy of 300 eV. Positrons (electrons) with energy E_{e^+} (E_{e^-}) between 15 and 45 eV were measured in parallel. The total experimental energy resolution was estimated from the width of the elastic positron peak to be approximately 4 eV. Coincidence timing logic was employed applied to signals originating from the MCPs in order to record the impact parameters at each detector only when they both detected a particle and to measured the time between the detection of the two particles [13]. Data were acquired for $e^+-e^-$and e^--e^-pair emission upon positron impact for a total of 62 hours and 11 hours, respectively.

3. Results

Analysis of the time-interval distribution for the detected e^+-e^-pairs showed clearly that correlated e^+-e^- pairs were emitted from the LiF(100) surface in the energy range observed [13]. The detected intensity of correlated e^+-e^-pairs increased rapidly from sum energy $E_{\rm sum}$ =

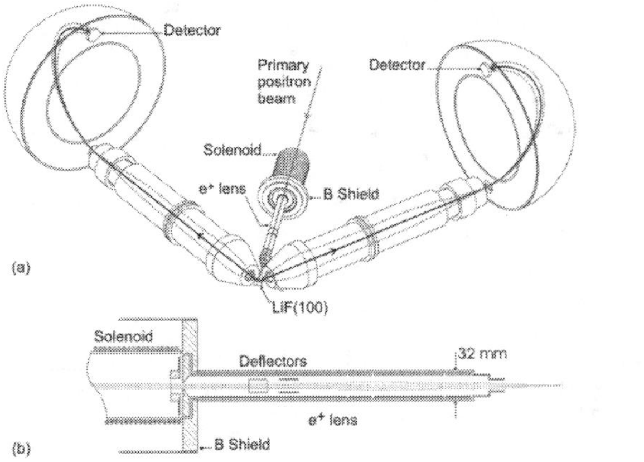

Figure 1. (a) Simplified diagram of the experiment. (b) Schematic illustration of the magnetic shield and primary positron optics.

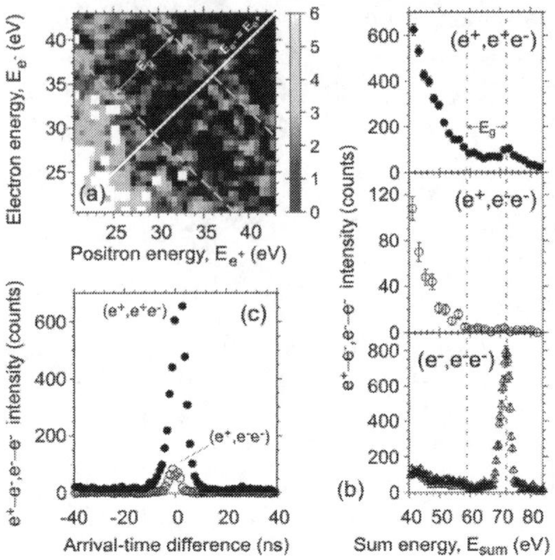

Figure 2. (a) Energy distribution of e^+-e^-pairs emitted from LiF(100) upon 85 eV e^+ impact. (b) The e^+-e^-pair sum energy ($E_{\rm sum}$) distribution obtained from integrated intensity along a 14 eV wide strip centred along the line $E_{e^+} = E_{e^-}$ shown in (a), together with positron- and high energy resolution electron-excited e^--e^-pair sum energy distributions. (c) The number of detected pairs as a function of the time interval between the arrival of each particle at the detectors.

$E_{e^+} + E_{e^-}$ of approximately 60 eV toward lower sum energy. We focus our attention to the higher energy region where a weak ridge can be seen that along $E_{\rm sum} \approx 72$ eV (indicated by a dashed white line in Figure 2a). It is revealed more clearly in Figure 2b which shows the integrated intensity along a 14 eV wide strip centred along the line $E_{e^+} = E_{e^-}$ against the sum energy $E_{\rm sum} = E_{e^+} + E_{e^-}$. A histogram of the time intervals between the detection of the positrons and electrons in this energy range is shown in Figure 2c. The area of the peak corresponds to the number of correlated pairs, or true coincidences, that arise from single scattering event and the area of the uniform background beneath it gives the number of uncorrelated, or accidental coincidences. Only events corresponding to the area under the peak are included for energy analysis. The contribution from accidental coincidences is clearly very small ($< 5\%$).

The position of the peak in the (e^+,e^+e^-) energy distribution corresponds to the maximum possible value of $E_{\rm sum}$ for a correlated e^+-e^-pair which is given by the difference between the incident positron kinetic energy, E_i and the energy required to transfer an electron from the top of the valence band to the vacuum level, E_B. Because the LiF conduction band is close to the vacuum level, E_B is approximately equal to the band gap $E_g = 13.0 \pm 0.4$ eV [3]. The peak can therefore be attributed to exclusively elastic scattering processes involving the incident positron and an electron from the top of the valence band. A corresponding feature is observed in the sum energy distribution of electron pairs obtained with better energy resolution using an electron (Figure 2b). The wide band gap of LiF prohibits an electron emitted from the valence band from losing a continuous range of energy by electronic excitations below E_g (indicated in Figure 2). This leads to a region of low inelastic contributions in the electron-excited e^--e^-pair spectrum that allows pair emission from the valence band to be clearly distinquished [3]. A corresponding valley is observed in the e^+-e^-pair energy spectrum below 72 eV. At lower e^+-e^-pair energy the relative intensity of inelastic contributions increases rapidly. Differences in the electron excited $e^--e^-$and e^+-e^-pair distributions in this region may arise from differences in the inelastic scattering cross-sections for positrons and electrons, but reliable comparison requires similar energy resolution.

Correlated $e^- - e^-$pairs emitted upon 85 eV positron impact were also observed and their energy distribution measured (Figure 2). In contrast to the (e^+,e^+e^-) pair energy distribution, pair intensity increases continuously below about 59 eV. This is consistent with energy being conserved betweeen the $e^- - e^-$pair and the undetected positron, with the maximum energy available to the $e^- - e^-$pair in this case given by $E_i - 2E_b \sim 59$ eV.

Our angle-integrated data do not allow direct pair emission mechanisms to be distinguished from indirect ones, i.e. multiple scattering. However we can speculate that the mechanism of $e^- - e^-$pair emission upon positron impact is an indirect process whereby the incident positron transfers kinetic energy to a valence electron that subsequently interacts with a second electron causing the emission of both. In the case of positron-excited $e^+ - e^-$pair emission, we expect that with improved energy resolution and counting statistics, asymmetries in the energy distribution would be revealed, but to first order the mechanism by which the $e^+ - e^-$pair are emitted is similar the multi-step scattering processes by which an $e^- - e^-$pair is emitted upon electron impact. We refer to Refs. [16, 17] for discussion of these mechanisms. Angle-resolved experiments will require considerably more time to obtain data with statistics adequate for detailed comparison of the $e^- - e^-$and $e^+ - e^-$scattering dynamics but could still be completed in time typical of most $e^- - e^-$coincidence experiments.

4. Summary

We have shown that upon impact with 85 eV positrons, $e^+ - e^-$pairs were emitted from a LiF(100) surface. Structure in the pair energy distribution is attributed to the primary positron scattering with an electron at top of the valence band. Comparison of $e^- - e^-$scattering data obtained upon impact with electrons and positrons supports this interpretation. The data indicate that $e^+ - e^-$pair emission spectroscopy with energy and angular resolution can be used to probe electronic correlations and mechanisms of electronic excitation in solids by low energy positrons.

Acknowledgments

We are grateful to Dr. C. Hugenschmidt, J. Mayer, P. Pikart and C. Piochacz for their advice and assistance at NEPOMUC.

References

[1] Schumann F O, Kirschner J and Berakdar J 2005 *Phys. Rev. Lett.* **95** 117601
[2] Schumann F O, Winkler C and Kirschner J 2007 *Phys. Rev. Lett.* **98** 257604
[3] Samarin S, Artamonov O M, Suvorova A A, Sergeant A D and Williams J F 2004 *Sol. Stat. Commun.* **129** 389
[4] Feder R, Gollisch H, Meinert D, Scheunemann T, Artamonov O M, Samarin S N and Kirschner J 1998 *Phys. Rev. B* **58** 16418
[5] Berakdar J, Gollisch H and Feder R 1999 *Sol. Stat. Commun.* **112** 587
[6] Fominykh N, Henk J, Berakdar J, Bruno P, Gollisch H and Feder R 2000 *Sol. Stat. Commun.* **113** 665
[7] Ortiz G, Souza I and Martin R M 1998 *Phys. Rev. Lett.* **80** 353
[8] Nekovee M, Foulkes W M C and Needs R J 2001 *Phys. Rev. Lett.* **87** 036401
[9] Perdew J P, Burke K and Wang Y 1996 *Phys. Rev. B* **54** 16533
[10] Puska M J and Nieminen R M 1994 *Rev. Mod. Phys.* **66** 841
[11] Stachowiak H and Boroński E 2005 *Phys. Rev. B* **71** 245107
[12] Boronski E and Stachowiak H 2005 *Physica B: Cond. Matt.* **366** 168
[13] Van Riessen G, Schumann F O, Birke M, Winkler C and Kirschner J 2008 *J. Phys.: Condens. Matt.* **20** 442001
[14] Hugenschmidt C, Schreckenbach K, Stadlbauer M and Strafler B 2005 *Nucl. Inst. Meth. Phys. Res. A* **554**
[15] Shi M, Gerola D, Waeber W B and Zimmermann U 1995 *Appl. Surf. Sci.* **85** 143
[16] Berakdar J and Das M P 1997 *Phys. Rev. A* **56** 1403
[17] Berakdar J 2000 *Nucl. Inst. Meth. Phys. Res. B* **171** 204

Journal of Physics: Conference Series **185** (2009) 012052 doi:10.1088/1742-6596/185/1/012052

Low energy K_α X-ray Satellites in Chlorine

Koteswara Rao Veeranki[1], B. Seetharami Reddy[1], M. V. R. Murti[2], S. S. Raju[1] and L.S. Mombasawala[3]

[1]Department of Nuclear Physics, Andhra University, Visakhapatnam – 530003, India.

[2]CSRE, Indian Institute of Technology Bombay, Powai, Mumbai - 400076, India.

[3]SAIF, Indian Institute of Technology Bombay, Powai, Mumbai - 400076, India.

E-mail: dr.vkoteswararao@yahoo.com

Abstract. $K\alpha$ X-ray low energy satellite spectra of chlorine generated by photon excitation with sodium chloride as target was studied using the Wavelength Dispersive X-Ray Spectrometer (Philips 2404) employing LiF 220 cystal. The measured energies and energy shifts of these X-ray satellites are compared with theoretical values available and ascribed appropriate transitions.

1. Introduction

When a multiply ionized atom de-excites, X-ray satellites are produced. The characteristic X-ray $K_{\alpha1,2}$, also called diagram line arises from a singly ionized atom out of 2P–1S transition. On either side of this, weaker lines appear which can be observed only by high-resolution spectrometers, usually crystal spectrometers. The X-ray lines appearing on the low energy side are called low energy satellites. These were first interpreted [1] as KLL Radiative Auger emissions, subsequently other modes of decay resulting in the emission of low energy satellites like RER (Radiative Electron Rearrangement), XRS (X-ray Raman Scattering) were reported, though KLL RAE is the predominant mode of decay. A brief description of these low energy satellites is given here under.

1.1. Radiative Auger Effect

A positively ionized atom with a single or more inner shell holes usually decays either through a radiative transition emitting X-rays are non-radiative transition emitting electrons by Auger effect. But as early as in 1935 Bloch [2] suggested that such an atom may emit an X-ray photon and a valence electron simultaneously and called that phenomenon "Radiative Auger Effect". Subsequently, Aberg and Utrianien [1] observed yet another possible decay mode, in which an X-ray photon and an inner-shell electron and not a valence electron as envisaged by Bloch, are emitted simultaneously. They continued the same nomenclature of Radiative Auger Effect and thus are usually credited with the discovery of this phenomenon of Radiative Auger Effect referred by the abbreviated form RAE. They also remarked that it may be called the "atomic internal Compton effect" in analogy to gamma ray emissions accompanied by ejection of atomic electrons.

If the atom initially has a hole in the K-shell, the de-excitation may occur in several ways. If the k-vacancy is filled by the jumping of an L_i electron, with an L_j electron going to the continuum or an unoccupied atomic bound state simultaneously, the transition is termed as KL_iL_j RAE (Radiative Auger Emission). If the electron is ejected to a continuum it is called shake off process while as if it fills an unoccupied atomic orbital it is called a shake- up process. In the final state of KLL RAE X-ray emission, two holes are left in the L-shell and the electron movement is similar to the KLL Auger electron transition. However, the physical process of RAE is entirely different from the non- radiative auger emission. The former occurs due to the sudden change in the potential of the atom before and after 2P –1S transition. But the non-radiative auger electron emission takes place due to the coulomb scattering. Where RAE is due to an electron dipole transition governed by the angular momentum selection rule in the auger electron emission, no selection rules are involved in non radiative auger emission. Applying the law of conservation of energy to RAE phenomenon

$$h\nu + E_k(L_j) = E(K L_i L_j)$$

Where, hν is the energy of the photon emitted, $E_k(L_j)$ is the kinetic energy of the ejected electron and $E(KL_i L_j)$ is the energy which the auger electron would possess in the case of a non-radiative auger emission. As $E_k(L_j)$ varies from zero to $E(KL_iL_j)$, this results in a continuous range of photon energies extending from an edge at $h\nu = E(KL_i L_j)$ towards lower energies and hence KLL RAE is observed usually not as a sharp peak but as a continuum. There may be more than one such band; different bands corresponding to different energy terms of the final doubly ionized atom, each band with an edge called RAE edge that corresponds to an energy term.

1.2. Radiative Electron Rearrangement

Another low energy X-ray satellite of interest is Radiative Electron Rearrangement usually referred as RER [3]. In this process an X-ray is emitted with two 2s electrons simultaneously undergoing rearrangement. While one of the 2s electrons jumps down to fill up the initial 1s vacancy, the other 2s electron gets excited to 2p sub shell. Hence this process requires the presence of at least one initial 2p vacancy in addition to the 1s vacancy in the atom. When there is only 1p vacancy in addition to 1s vacancy initially, the satellite resulting from transition from such initial state will be lowest energy RER satellite. The addition of each L-shell vacancy increases the energy of the RER transition, so that an RER satellite sequence is predicted [3] especially when the excitation energy is strong enough like in the case of ion bombardment. These lines occur between KLL RAE peaks and the K_α diagram line and were first observed by Richard et al [4] in a study by oxygen ion (30MeV) bombardment. They could not offer any definitive explanation for these peaks. Jamison et al [3] interpreted them as RER satellites. The RER satellite energies measured by them for Mg, Al and Si agreed well with the Hartree - Fock calculations for the energies of X-rays resulting from $(1s)^{-1}(2p)^{-n} \rightarrow (2s)^{-2}(2p)^{-n+1}$ transition, thus lending support to this interpretation [5].

1.3. X-Ray Raman Scattering

This form of low energy K_α X - ray satellite which appears very close to the K_α diagram line results from the inelastic scattering of the K_α diagram line by the bound electrons [6]. The energy of this X-ray is the energy of K_α diagram line less the binding energy of the scattering electron. The intensity of this satellite is very small relative to that of the K_α diagram line. Work on low energy satellites is rather scarce. The aim of the present work is to study low energy X-ray satellite spectra of chlorine generated by photon excitation for which no results are reported so far.

2. Experimental details

In the present work a Philips 2404 Wavelength dispersive plane crystal spectrometer was used. An X-ray tube with Rhodium anode was the source of photon excitation and this was operated at a suitable potential and current (30 - 50 kV, 40 - 60 mA) to get good intensities. Sodium Chloride in the form of pure powders (99.5% purity) was pressed into pellets of 50mm diameter and 5mm thicknesses were

Journal of Physics: Conference Series **185** (2009) 012052 doi:10.1088/1742-6596/185/1/012052

used as targets. Two fine collimators, one between the sample and the crystal and the second one between the crystal and the detector were used. The basic principle involved and the experimental procedure are described elsewhere [7]. LiF 220 crystal 2d = 6.53 , is well suited for taking the first order K_{α} spectrum of chlorine as the wavelength of its $K_{\alpha 1,2}$ line is 4.729 . Spectra were taken by recording the intensities in 2θ steps of 0.010^0. Good statistics were maintained in counting.

3. Results and Conclusions

The collected data were first smoothed using sliding least square fitting and then the spectra were corrected for background using 'CORSPC' programme developed by us. After smoothening of data and subtraction of background, peak fitting and deconvolution of different peaks was carried out by 'PEAK-FIT' software version 4.11 of Systat Software Inc [8]. The energies of the X-ray satellite lines were determined after calibration using the standard $K_{\alpha 1,2}$ X – ray energies [9] of potassium, chlorine and phosphorous.

Six low energy X-ray satellites albeit with low intensity (which is not uncommon in the spectra of low energy satellites) were observed in the spectrum (fig-1). Three of them are interpreted as KLL RAE satellites. Comparing their energies (Table-1) with the Semi-Empirical values [10], these X-ray satellites are assigned $K–L_2L_3$, $K – L_1L_{2,3}$, $K – L_1, L_2$ RAE transitions. The peaks appearing close to the K_{α} diagram line on the low energy side are interpreted as two of the possible RER (Radiative Electron Rearrangement) lines. The measured energy shifts of these lines relative to the K_{α} diagram line are in agreement with in 5eV with the theoretical energy shifts predicted by HF formalism [11] for RER – 1 and RER – 2 lines and hence are ascribed the transitions $(1S)^{-1}(2p)^{-1} \rightarrow (2s)^{-2}(2p)^{0}$ for RER -1 line and $(1S)^{-1}(2p)^{-2} \rightarrow (2s)^{-2}(2p)^{-1}$ for RER-2 line.

Fig -1: Low Energy K_{α} X-Ray Satellite Spectrum of Chlorine

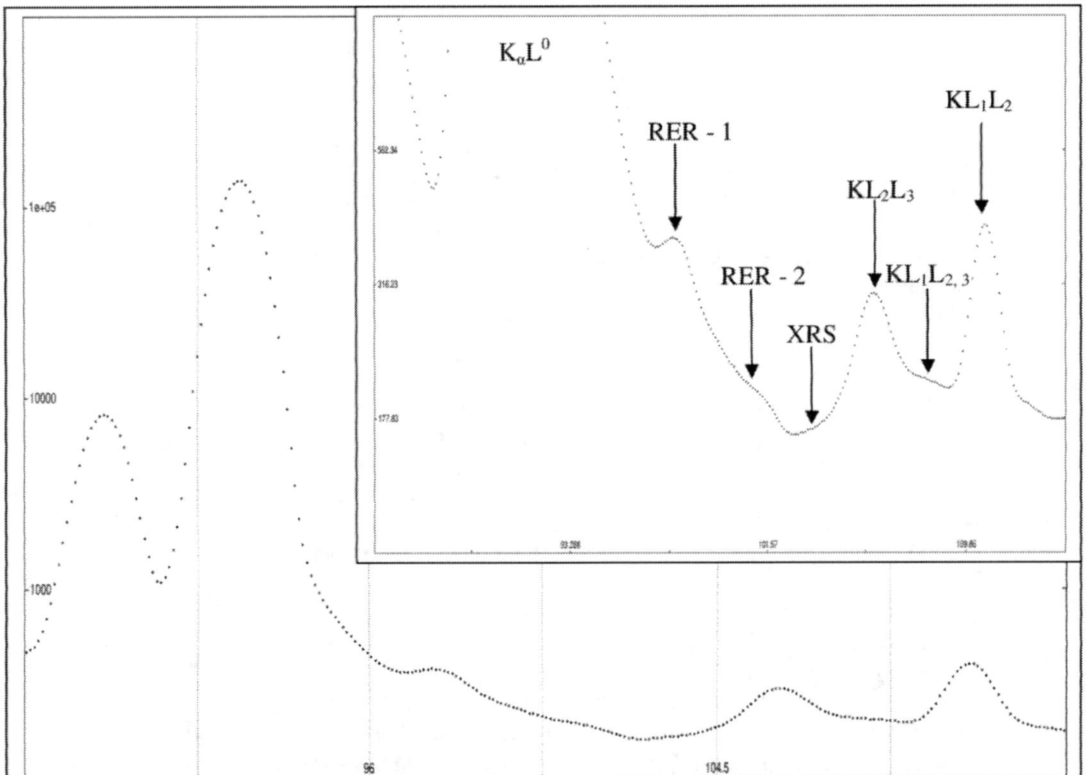

The energy of the next peak appearing with very low intensity is measured to be 2622.5 ± 1eV. This energy is found to be equal to the energy of K_α diagram line less the binding energy of the L_2 shell electron of chlorine [10]. Hence as explained in the introduction it is interpreted as XRS line.

Table-1: Energy of KLL RAE lines (eV)

Element	$K\text{-}L_1L_2$		$K\text{-}L_1L_{2,3}$		$K\text{-}L_2L_3$	
	EXPT.	SEMI EMP.[10]	EXPT.	SEMI EMP.[10]	EXPT.	SEMI EMP.[10]
Chlorine	2308.6±2	2300	2345.0±2	2330	2377.1±2	2376

Table- 2: Energy shifts of RER lines with respect to $K_\alpha L^0$(eV)

Element	ΔE (RER-1)		ΔE (RER-2)	
	EXPT.	THEORY HF[11]	EXPT.	THEORY HF[11]
Chlorine	125.8±3	121	98.6±3	91

Table: 3 – Energy of XRS line (eV)

Element	XRS	
	EXPT.	THEORY*
Chlorine	2420.1**±2**	2420.9**±1**

* E (XRS) = E ($K_\alpha L^0$) – B.E of bound electron [10]
Energy of $K_\alpha L^0$ (Cl) = 2622.5 ± 1 eV,
Energy of Scattering bound electron (L_2), B_e = 201.6 eV

4. Summary

K_α X-ray low energy satellite spectrum of chlorine generated by photon excitation with sodium chloride as target was studied using the wavelength dispersive X-ray spectrometer (Philips 2404) employing LiF 220 crystal. Three KLL RAE peaks, a sequence of two RER lines and an XRS line were observed. Three RAE peaks are assigned $K– L_2L_3$, $K – L_1L_{2,3}$, $K – L_1, L_2$ transitions in view of the agreement between their experimental and semi-empirical energies. The two observed RER lines are ascribed $(1S)^{-1}(2p)^{-1} \rightarrow (2s)^{-2}(2p)^0$ and $(1S)^{-1}(2p)^{-2} \rightarrow (2s)^{-2}(2p)^{-1}$ transitions as their experimental energy shifts relative to the K_α diagram line agree with the theoretical energy shifts for RER -1 and RER -2 lines. Another peak is interpreted as XRS line as its experimental energy agrees well with the predicted energy.

References

[1] Aberg T and Utriainen J 1969 *Phys. Rev.* **22** 1346.
[2] Bloch F 1935 *Phys. Rev.* **48** 187.
[3] Jamison K A, Hall J M and Richard P 1975 *J Phys.* **138** L458.
[4] Richard P, Moore C F and Olsen D K 1973 *Phys. Lett.* **43A** 519.
[5] Jamison K A, Hall J M, Oltjen J, Woods C W, Kauffman R L, Gray T J and Richard P 1976 *Phys. Rev A* **14** 937-945.
[6] Miyerno Y and Ohmura Y 1967 *J. Phys. Soc. Jpn.* **22** 445.
[7] Parthasaradhi K, Ramesh Babu G, Murthy V R K, Murti M V R and Rao K S 1987 *Nucl.Istrum. Meth. A* **54** 255.
[8] Peak Fit 4.11, *SYSTAT Software Inc.*, USA, www.systat.com.
[9] Storm E and Israel H I 1970 *Atomic and Nuclear Data Tables* **5** 70.
[10] Larkins F P 1977 *Atomic Data and Nuclear Data Tables* **20** 311.
[11] Forese Fisher C 1972 *Computational Physics communications* **4** 107.

Journal of Physics: Conference Series **185** (2009) 012053 doi:10.1088/1742-6596/185/1/012053

Electronic structure of artificial atoms

Jingbo Wang and **Chris Hines**

School of Physics, The University of Western Australia, Australia

E-mail: wang@physics.uwa.edu.au

Abstract. Quantum dots are artificially fabricated "atoms and molecules", in which charge carriers are confined in all three dimensions similar to the electrons in real atoms and molecules. Consequently, they exhibit similar properties normally associated with real atoms and molecules, such as quantized energy levels and shell structures. The detailed electronic structure of quantum dot systems depends on many of their physical properties such as material, size and geometry. Therefore, they can be readily engineered and manipulated to test fundamental concepts in quantum mechanics, to build more efficient and precisely controlled lasers and electronic devices, and to process vast amount of quantum information. This opens up a wide range of possibilities and areas for exploring new physics and new applications. In this paper, we present *ab initio* calculations on quantum dot systems by solving the corresponding Schrödinger's equations.

1. Introduction

Quantum mechanics is one of the most successful theories ever developed. It has provided extraordinarily accurate predictions of many physical systems, with a wide range of practical applications such as lasers and electronic devices. However the scalability of quantum calculations and the sheer resources necessary to make an accurate practical calculation limits the ability to study nano-structured devices of reasonable size using first principles of quantum mechanics.

This paper is concerned with approximations and computational schemes developed in order to perform quantum mechanical calculations with limited resources. In particular, we will describe an iterative refinement configuration interaction method (IRCI) and a spin adapted configuration interaction method (SACI). Both methods take into account full electron and spin correlation effects, giving a level of accuracy not available in alternative approaches such as the density functional theory or the self-consistent Hartree-Fock method. They are also advantageous over the standard configuration interaction method, in that a much smaller basis is required by choosing the most appropriate set of Slater determinants through an iterative process and by using spin eigenfunctions already possessing the Fermionic symmetry.

2. Theory and Computational Schemes

Journal of Physics: Conference Series **185** (2009) 012053 doi:10.1088/1742-6596/185/1/012053

The Hartree-Fock method and the density functional theory are both self-consistent mean-field models, although they are fundamentally different in the way of treating exchange and correlation effects [1,2]. In the Hartree-Fock method, exchange is considered exactly by the proper anti-symmetrization of wavefunctions using a Slater determinant, while correlation in the motion of the electrons is neglected since only a time-averaged effective potential is used in this formulation. In the density functional theory, both exchange and correlation effects are included exactly in principle, but only approximately in practice in most situations. Perhaps worse still, there is no clear route for the density functional theory to provide convergent results.

A more accurate formalism is to use a linear combination of many Slater determinants, each of which describes a different configuration, namely the configuration interaction (CI) method [3]. This method provides a convergent route to obtain numerically exact solutions of multi-electron systems. The main difficulty of the CI method lies in the rapid explosion of the number of Slater determinants to be included in the expansion as the number of spin-orbitals increases. For example, to represent a six-electron quantum dot with 10 available spin-orbitals we need 44,100 Slater determinants in the expansion, while with 20 spin-orbitals the number of Slater determinants to be included is 1,502,337,600. As a result, a truncated CI approach is almost inevitable in practice. In the following we describe two ways to significantly reduce the basis size while maintain high numerical accuracy.

2.1. Iterative refinement configuration interaction method (IRCI)

The number of Slater determinants increases drastically as the number of spin-orbitals increases:

$$N_D = \binom{n_b}{n_\uparrow}\binom{n_b}{n_\downarrow}.$$

However, upon examining the configuration interaction coefficients, some Slater determinants do not contribute significantly to the final solution. If these determinants are excluded, the accuracy of the solution can be improved by substituting more significant determinants.

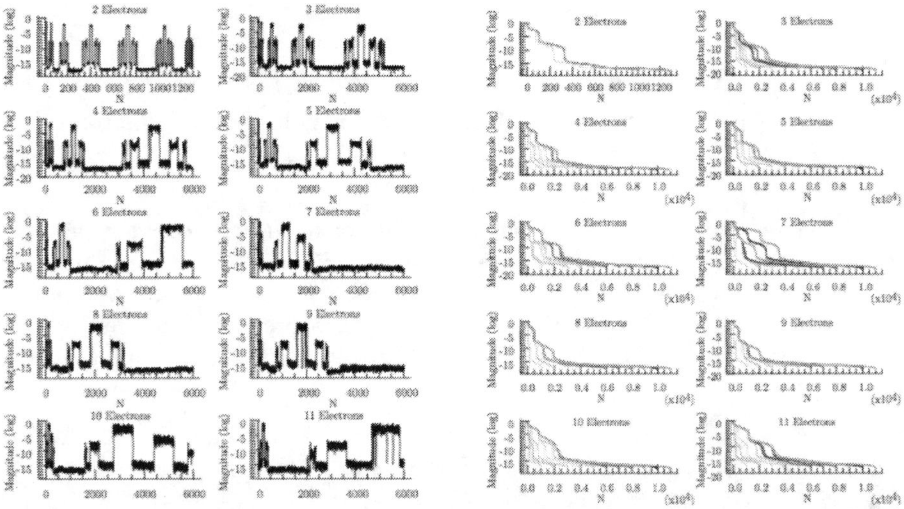

Figure 1. Left panel: original CI coefficients; Right panel: sorted CI coefficients.

The IRCI scheme starts with a large list of determinants, from which a subset is selected and a CI calculation is performed. Determinants with CI coefficients smaller than a certain threshold are replaced with new determinants that may contribute more to the solution, as shown in figure 1. This process iterates until the entire list is exhausted.

Journal of Physics: Conference Series **185** (2009) 012053 doi:10.1088/1742-6596/185/1/012053

2.2. Spin adapted configuration interaction method (SACI)

The spin-adapted CI method expands the multi-electron wavefunction as a linear combination of antisymmetrised products of basis spatial wavefunctions and spin eigenfunctions. This method has a distinct advantage over standard CI, in that a much smaller basis is needed since the spin-adapted orbitals already possess the required Fermionic symmetry [4,5,6].

Elementary Spin functions :

$$\theta_k(m_S) = \sigma(1)\sigma(2)\cdots\sigma(N)$$

where $\sigma(i) = \begin{cases}\alpha(i) \\ \beta(i)\end{cases}$ represents spin up or down of ith electron.

Note $\theta_k(m_S)$ are eigenfunctions of $\hat{S}_z$ but not $\hat{S}^2$.

Spin basis functions :

$$X(S, m_S) = \sum_k C_k \theta_k(m_S)$$

where C_k are determined by $\hat{S}^2 X(S, m_S) = S(S+1)\, X(S, m_S)$.

Spatial basis function :

$$\Phi = \phi_1 \phi_1 \cdots \phi_N$$

where ϕ_i are single electron spatial wavefunctions.

Spin-adapted orbital :

Finally, we combine both spin and spatial functions to form an orthonormal and properly antisymmetrised basis:

$$\mathcal{A}\, \Phi(1, 2 \cdots N)\; X(S, m_S)$$

where $\mathcal{A} = \sum_P (-1)^P P$ sums over all possible permutations. The symmetry properties of the spin eigenfunctions dictate with which spatial functions it can be combined.

3. Results

The electron densities for up to 11 electrons in a triangular dot and a square dot are shown in figures 2 and 3, respectively. For each system, density distributions for spin up and spin down electrons are also shown. Furthermore, we studied coupled quantum dots, such as the Quantum Cellular Automata (QCA) systems, which can be used to have been proposed to process both classical and quantum information. The first lowest electronic states for 12 and 16 coupled quantum dots are shown in figures 4 and 5, respectively.

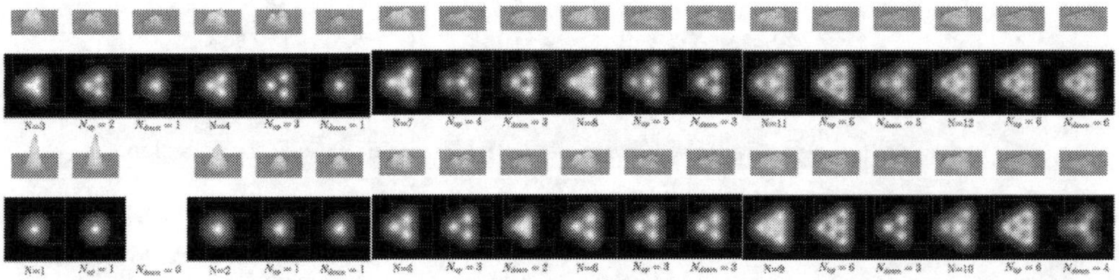

Figure 2. Total, spin up, and spin down density functions for the ground state of a triangular quantum dot (up to 12 electrons).

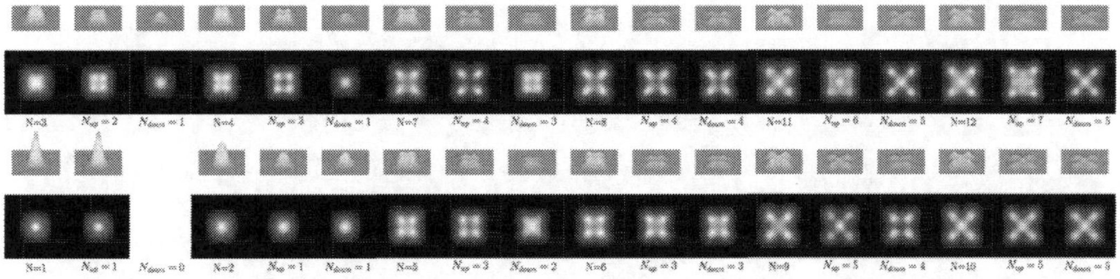

Figure 3. Total, spin up, and spin down electron density distributions for the ground state of a square quantum dot (up to 12 electrons).

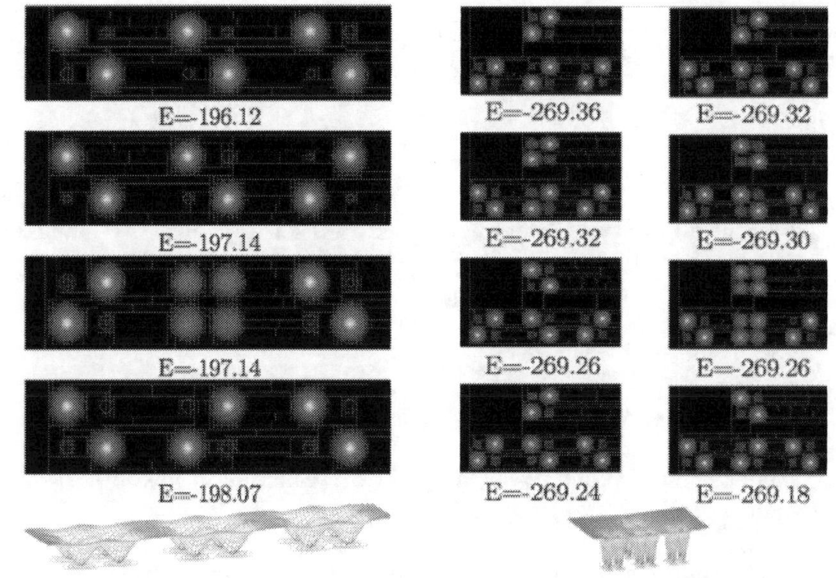

Figure 4. The lowest electronic states of coupled quantum dots.

4. Conclusions

Numerical calculations provide detailed and often very accurate information about nano-structured systems. With further development of theoretical models and advanced computational algorithms, many novel applications as well as potential problems which previously could only be speculated on can now be addressed in great detail.

References

[1] S. McCarthy, J. Wang, and P. Abbott, Computer Physics Communications 141, 175 (2001).
[2] J.B. Wang, C. Hines and R.D. Muhandiramge, Handbook of Theoretical and Computational Nanotechnology (chapter 10, p545-604, American Scientific Publishers, 2006).
[3] S. Reimann and M. Manninen, Reviews of Modern Physics 74, 1283 (2002).
[4] R. Pauncz, The Construction of Spin Eigenfunctions (Kluwer Academic Publishers, New York, 2000).
[5] R.D. Muhandiramge and J.B. Wang, International Journal of Quantum Chemistry, 106, 27–53 (2006)
[6] J.B. Wang and R.D. Muhandiramge, New Quantum Chemistry Research (Nova Science Publishers, 2007)

Acknowledgment

The authors would like to thank Ranga Muhandiramge for his earlier work on the SACI scheme and Brandon Lau for testing the code on various quantum dot systems.

Journal of Physics: Conference Series **185** (2009) 012054 doi:10.1088/1742-6596/185/1/012054

X-ray absorption near-edge structure of chromium ions in α-Al_2O_3

W Wongkokua[1,2], S Pongkrapan[1,3], P Dararutana[1,4], J T-Thienprasert[5]
P Wathanakul[1,3,6]

[1]Gemmology and Mineral Sciences Special Research Unit, Faculty of Science, Kasetsart University, Bangkok 10900, Thailand

[2]Department of Physics, Faculty of Science, Kasetsart University, Bangkok 10900, Thailand

[3]Department of Earth Sciences, Faculty of Science, Kasetsart University, Bangkok 10900, Thailand

[4]The Royal Thai Army Chemical Department, Bangkok 10900, Thailand

[5]School of Physics, Suranaree University of Technology, Nakorn-Ratchasima 30000, Thailand

[6]The Gem and Jewelry Institute of Thailand (GIT), Bangkok 10330, Thailand

pwathanakul@gmail.com

Abstract. Both synthetic and natural α-Al_2O_3 samples with different Cr concentrations were investigated by XANES. The Cr contents were analyzed using LA-ICP-MS technique prior to this experiment. XANES spectra combined with first principle calculations showed the transformation from α-Al_2O_3:Cr^{3+} to α-Al_2O_3:Cr^{3+},Cr^{3+} or α-Cr_2O_3 as the Cr-Cr content increased. The absorption transformation could be resulted from the change of environment around Cr^{3+} ions, i.e., the Al-O-Cr bonds decreased while the Cr-O-Cr bonds increased. Significant differences in XANES line shapes observed for synthetic and natural α-Al_2O_3 samples suggested the differences in local environments around Cr ions in both samples. The XANES line shape of α-Al_2O_3:Cr^{3+},Cr^{3+} or Cr ion pairs corresponded to the Cr content, which was also confirmed by the intensity ratios between N_2-line and sideband S of R-lines of photoluminescence spectra.

1. Introduction

Both synthetic α-Al_2O_3 and natural ones, known as corundum, have been used for jewelry. While pure α-Al_2O_3 is colorless, the presence of certain trace elements or impurities in the corundum structure can induce coloration, e.g., red by Cr, blue by Fe and Ti, and yellow by Fe. The amount of impurities ranges from a few ppm to one percent or more. Low Cr content (α-Al_2O_3:Cr^{3+}) gives rise to pink sapphire, higher Cr content (α-Al_2O_3:Cr^{3+},Cr^{3+}) produces ruby whereas Cr^{3+} in α-Cr_2O_3 is responsible for the green coloration. The different local environments of Cr ions were expected between the low and high Cr contents.

X-ray absorption near edge structure (XANES) spectrometry is a powerful technique in study of electronic structure around the absorbing atom. The peak positions and line shape of XANES spectrum are strongly sensitive to valence state and symmetry of surrounding/neighboring atoms or local environments. A comparative study of two Indian natural ruby samples and a flame fusion synthetic ruby showed similar peak positions and line shape in XANES spectra [1]. With variations in Cr contents, the Cr-O distances of α-Cr_2O_3 to α-$Cr_{0.07}Al_{1.93}O_3$ measured by extended X-ray absorption fine structure (EXAFS) were estimated to be of small significance (1%) [2].

The Cr-Cr coupling was expected to play a role in XANES line shape because of the change of symmetry around Cr^{3+}. For 1s shell absorption of Cr^{3+}, the photo-electrons ended up in the O 2p empty states by the selection rule $\Delta l=+1$. The Cr^{3+} unfilled 3d states took part in the absorption processes by hybridizing with those O 2p states. The pd hybridization highly depended on the local coordination environment.

The luminescence spectrum of dilute ruby (α-Al_2O_3:Cr^{3+}) was dominated by the two sharp R-lines. The Cr exchange-coupling could be observed from the existence of weaker sharp lines (N-lines) [3]. These N-lines increased in intensity relative to that of the R-lines as the concentration of chromium was increased [4].

In this study, the X-ray absorption spectra of two synthetic and two natural samples of α-Al_2O_3 (from Mong Hsu, Myanmar) with different Cr concentrations were measured by XANES technique. The increase in Cr-Cr coupling corresponding to the Cr concentrations was also investigated using the intensity ratios between the luminescence N- and R-lines.

2. Experiment

The samples consisting of two synthetic and two natural α-Al_2O_3 (Mong Hsu, Myanmar) with various Ti, Cr and Fe concentrations are listed in table 1. Their trace element contents were analyzed using laser ablation inductively coupled plasma mass spectrometry (LA-ICP-MS) technique prior to this study. NIST-612 multi-element glass was used as the standard for chemical composition calculation. It must be noted that content inhomogeneity are to be expected, especially, in natural samples. Thus, for each one of the samples, three different laser ablated sites were measured and averaged out for a representative value. Samples were polished in one side perpendicular to the c-axis to ensure the orientation was uniformed for all measurements. α-Cr_2O_3 powder was prepared for comparative XANES study.

Table 1. Contents of Ti, Cr and Fe in synthetic and natural α-Al_2O_3 samples analyzed by LA-ICP-MS.

	Synthetic samples		Natural samples	
	SYNC30	SYNC60	GMMSA	GMMSB
^{47}Ti	12.9	40.2	967.4	460.5
^{53}Cr	3518.6	6054.0	4339.3	3712.0
^{57}Fe	7.7	12.0	11.4	10.6

XANES measurements were performed at the X-ray absorption spectroscopy beamline (BL-8) of the Synchrotron Light Research Institute, Thailand, using a radiation beam from a bending magnet. The beamline details and performance have been reported elsewhere [5]. The XANES spectra were recorded at Cr-K-edge. A pure chromium foil was used as the standard for energy calibration, which was 5989 eV at the absorption edge. The detection was carried out in fluorescence mode using a 13-element Ultra LEGe detector. Flat surface of α-Al_2O_3 samples were arranged at about 45 degrees to the light source and detectors.

Photoluminescence (PL) measurements were carried out at room temperature (25°C) using a Renishaw inVia Raman microscope. The argon laser excitation wavelength was 514.5 nm. The system was calibrated using single crystal silicon with spectral resolution of 2.0 cm^{-1}.

3. Results and discussion

Since all XANES measurements were carried out at the same angular orientation with respect to the light source, the detectors, and the c-axis of samples, the polarization dependence of the X-ray absorption cross section was presumably identical.

Figure 1(a) shows comparison of synthetic samples with two different Cr concentrations. The spectra showed the transition from α-Al_2O_3:Cr^{3+} to α-Al_2O_3:Cr^{3+},Cr^{3+} and α-Cr_2O_3. The spectrum of α-Al_2O_3:Cr^{3+} was calculated using first principle multiple-scattering code FEFF8.2 [6]. Feature A, a pre-edge peak of the calculated spectrum, was displaced to higher energy as previously reported [7]. Features B, C, and E were slightly shifted to higher energy whereas Feature D was first shifted to higher energy before being shifted back to lower energy which suggested Cr-Cr coupling, i.e., the increase in Cr concentration.

Figure 1(b) shows XANES spectra comparison of two Mong Hsu samples (GMMSA and GMMSB) with different Cr concentrations. The spectra showed transitions in similar manner to the synthetic samples except at Feature C, and not directly dependent upon the increase in Cr contents. Feature C was clearly pronounced in GMMSA while Features B and D showed small significances. This could be explained by the coupling of Cr with other trace elements, especially, high content of Ti in natural samples. The development of Feature E in α-Cr_2O_3 was rather specific to eskolaite. The similarity in line shapes of SYNC60 synthetic and GMMSB natural samples suggested the possibility of them having equal contents of Cr-Cr pairs which influenced their line shapes accordingly

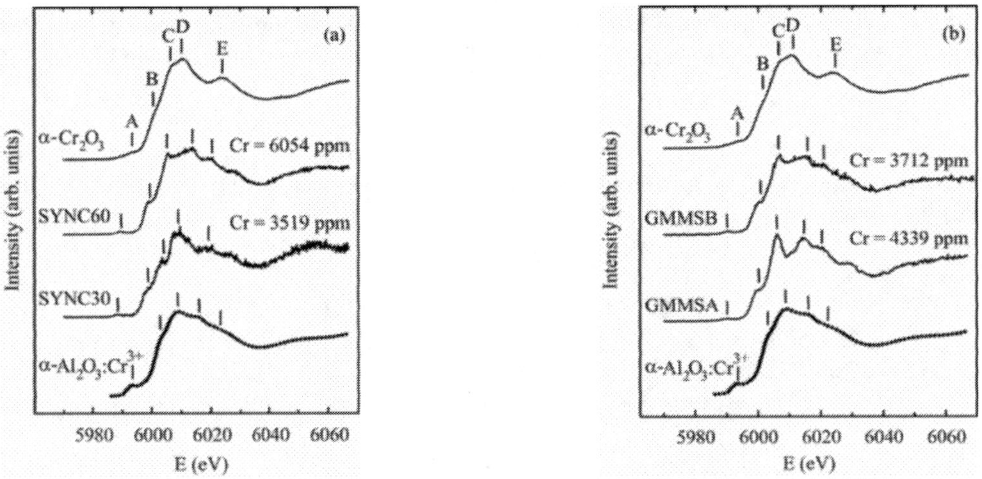

Figure 1. XANES spectra of α-Al_2O_3:Cr^{3+} (from calculation) and pure α-Cr_2O_3 (from experiment) were displayed as references. Different Cr contents of synthetic and natural samples are noted above their spectra. Features A, B, C, D, and E represent the transitions discussed in the text.

The photoluminescence of Cr was not correlated to Cr content in α-Al_2O_3 in a simple manner. The intensity ratios of the N_2 and the sideband S of R_1 and R_2 correlated almost linearly with Cr_2O_3 contents [8]. The linear relation was fitted and used as the Cr-Cr content calibration in this study. The PL spectrum of Cr in α-Al_2O_3 was measured and demonstrated in figure 2(a). The intensity ratio of N_2/S versus Cr-Cr contents were plotted as shown in figure 2(b). This suggested that, for synthetic α-Al_2O_3 samples, the higher the Cr concentration, the greater number of Cr-Cr pairs.

However, for natural samples, this may also depend on the geological environments, e.g., high concentration of Ti in this case. GMMSB with less Cr content compared to GMMSA (shown in table

1) exhibited more Cr-Cr pairs as shown in figure 1(b). Unsurprisingly, GMMSA with more Cr concentration consisted of less Cr-Cr pairs.

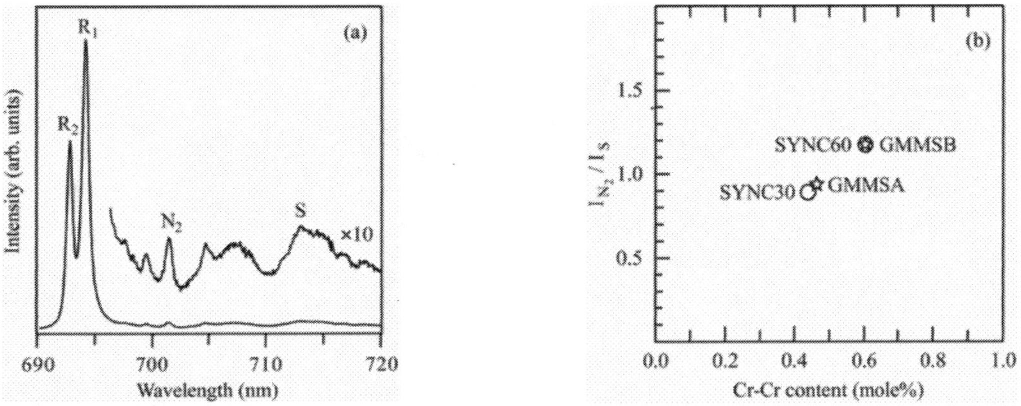

Figure 2. (a) Photoluminescence spectrum of an α-Al_2O_3 and their magnified N-lines and sidebands, and (b) a plot of intensity ratios of N_2-line (701.37 nm) and its sideband S (713.19 nm) of the synthetic (open circles) and natural samples (open stars) showing linear correlation.

4. Conclusion

Different contents of Cr in synthetic and natural α-Al_2O_3 were studied by XANES. The spectra line shapes suggested the transition from α-Al_2O_3:Cr^{3+} to α-Al_2O_3:Cr^{3+},Cr^{3+} and α-Cr_2O_3 with respect to the increase in Cr-Cr contents. XANES spectra of synthetic and natural α-Al_2O_3 (Mong Hsu ruby, Myanmar) samples differed significantly because of the different variations of Cr^{3+} local environments. Trace element contents were measured by LA-ICP-MS prior to the XANES measurements. The Cr-Cr coupling corresponded to the Cr content was confirmed by intensity ratios of R- and N- lines of photoluminescence spectra.

Acknowledgements

The Synchrotron Light Research Institute (SLRI) is acknowledged for GRANT 1-2550/PS03. The Gem and Jewelry Institute of Thailand (GIT) provided PL facility. LA-ICP-MS analyses were carried out at the GEMOC Key Centre, Macquarie University, Sydney, Australia. Prof. Lutz Nasdala from Institute of Mineralogy and Crystallography, University of Vienna, Austria is acknowledged for a fruitful discussion during his visit to Kasetsart University, Bangkok, Thailand. Funding supports were provided in parts by the Faculty of Science, Kasetsart University, Bangkok, Thailand and GIT.

References

[1] Parikh P, Bhardwaj D M, Gupta R P, Saini N L, Fernandes S, Singhal R K, Jain D C and Garg K B 2002 *Bull. Mater. Sci.* **25** 653
[2] Gaudry É, Sainctavit P, Juillot F, Bondioli F, Ohresser P and Letard I 2006 *Phys. Chem. Miner.* **32** 710
[3] Imbusch G F 1967 *Phys. Rev.* **153** 326
[4] Schawlow A L, Wood D L and Chogston A M 1959 *Phys. Rev. Lett.* **3** 271
[5] Klysubun W, Sombunchoo P, Wongprachanukul N, Tarawarakarn P, Klinkhieo S, Chaiprapa J and Songsiriritthigul P 2007 *Nucl. Instr. and Meth.* A **582** 87
[6] Ankudinov A L, Raevl B, Rehr J J and Conradson S D 1998 *Phys. Rev.* B **58** 7565
[7] Gaudry E, Cabaret D, Sainctavit Ph, Brouder Ch, Mauri F, Rogalev A and Goulon J 2005 *Phys. Scr.* **T115** 191
[8] Häger T and Dung P T 2003 Geo- and material science on gem-minerals of Vietnam. Proc. 2nd Int. Workshop, Hanoi, Vietnam 122

Journal of Physics: Conference Series **185** (2009) 012055 doi:10.1088/1742-6596/185/1/012055

Phonon dispersion and vibronic coupling in carbon nanoribbons

Yoshihiro Yamakita[1] and Koichi Ohno
Department of Chemistry, Graduate School of Science, Tohoku University, Aramaki, Aoba-ku, Sendai 980-8578, Japan

E-mail: yy@qpcrkk.chem.tohoku.ac.jp

Abstract. The phonon dispersion curves are presented for a carbon nanoribbon, polyacene. Restricted and unrestricted density functional calculations for [*n*]oligoacenes (n = 10 and 15) predict different spin distributions in the singlet electronic state. The unrestricted states involve spin-localized, singly-occupied orbitals. Bond alternations, infrared activities and phonon dispersions are studied on the basis of these closed-shell and open-shell states, respectively. Vibronic coupling in $S_0 \leftarrow S_1$ transitions are investigated using an efficient quantum chemical model (MO/8 model). Franck-Condon and vibronic coupling transitions in oligoacenes are found to be activated commonly, and the vibronic activities are assigned to the same phonon dispersion branches of polyacene.

1. Introduction

Graphene [1], the simplest lattice with the six-fold symmetry, has attracted many researchers in the fields of surface sciences and nanotechnology. On the other hand, molecular systems called polycyclic aromatic hydrocarbons (PAHs) have long been studied both experimentally and theoretically as basic systems possessing π electrons. Nanocarbons, a class of carbon networks with nanoscale sizes lying between these two cases, can be understood within the above frameworks of solid-state physics or π-electron chemistry. Namely, the periodicity in nanocarbons would ultimately converge to the six-fold lattices, whereas the effects of finite sizes and edge structures are connected with molecular properties. However, nanoscale effects arising from the periodic, finite lattices have not been understood well.

One of the ideal cases of the nanocarbons is carbon nanoribbon. They are infinite-length, fine-width honeycomb lattices with the sp^2 hybridisation. The electronic band structures in the ribbons with zigzag or armchair edges have been studied to obtain one-dimensional dispersions [2]. However, the importance of electron-phonon coupling in electron transport strongly demands comparative studies on *phonon* dispersions. The phonon dispersions and corresponding vibronic coupling have not been fully clarified. Another important issue would be magnetic properties. The infinite carbon lattices are magnetic, in contrast to PAH molecules. Indeed, spin localization has been reported for not only graphene edges [3] but also large linear [*n*]oligoacenes with $n > 8$ [4], where n is the number of rings.

In this paper, we report on the phonon dispersions of simplest carbon nanoribbons with zigzag edges, polyacenes, in different electronic configurations. Those dispersion curves are constructed extrapolatively from the force fields of [*n*]oligoacenes with $n > 8$. Spin localization is also studied

[1]To whom any correspondence should be addressed.

employing unrestricted quantum chemical calculations. They are compared with the previous results of restricted calculations [5,6].

2. Calculations

Density functional calculations were performed with a Gaussian 03 program on Linux-based computers and/or a supercomputer (Fujitsu Primequest) at the Research Center for Computational Science of the National Institutes of Natural Sciences. Becke's three-parameter gradient corrected exchange and the Lee-Yang-Parr gradient-corrected correlation functionals (B3LYP) were employed throughout this work on the basis of the 4-31G functions. Harmonic frequencies were calculated at optimized geometries without scaling to avoid inconsistency between frequency and intensity.

The force constants for polyacene were constructed from zone folding approach for pentadecacene (n = 15). The analytic force constants were folded into the force constant matrix (F matrix) for the unit cell in the Brillouin zone by weighting the phase factor, θ, in the Cartesian-coordinate space:

$$F = \sum_{k=0}^{k_{max}} F^{(k)} \cos k\theta \, , \tag{1}$$

where $F^{(0)}$ is the internal F matrix of the repeating unit, and $F^{(k\neq 0)}$ are those for off-diagonal F matrices for interaction with the kth neighbouring unit. Vibronic bands were calculated using an efficient quantum chemical model (MO/8 model) [7,8]. Briefly, force fields were obtained by scaling bond orders and bond-bond polarisabilities with 8 fixed, transferable parameters. The transitions to the electronic excited states were calculated under the Pariser-Parr-Pople approximation for self-consistent field (SCF) configuration interaction (CI) calculations, combined with Nishimoto-Mataga approximation.

3. Results and Discussion

Unrestricted occupied spatial orbitals in the singlet state correspond to open-shell diradical states, in which the singly-occupied orbitals exist along the CCC ribbon on each side [4]. The open-shell configuration is predicted to give the ground state for decacene. These diradical states are important in the 'transition' regimes between PAH molecules and magnetic carbon networks. In this context, polyacene is a basic model system, although quantum chemical calculations for polyacene have only been directed to its aromaticity and bond alternations in the restricted states [9-13]. Bond alternation and spin localization are likely to relate to the nanoscale effects in nanocarbons [3].

The open-shell diradical state is calculated to be 20.0 kcal mol^{-1} *lower* than the restricted closed-shell state. Table 1 presents the degrees of bond alternations for pentadecacene (n = 15). The first row corresponds to the bonds in the central unit cell, and these values are expected to be more independent of the edge structures than at terminal. In both open-shell and closed-shell states, shrinking (expanding) behaviours are clearly obtained for outward (inward) CC bonds, respectively, towards the maximum alternation at terminals.

Table 1. Calculated bond distances (Å) for pentadecacene in the unrestricted open-shell and restricted closed-shell singlet states, in the order from central bonds to terminal ones.

Inward CC		Outward CC		Joint CC	
Open	Closed	Open	Closed	Open	Closed
1.405	1.404	1.405	1.403	1.461	1.469
1.405	1.405	1.406	1.403	1.462	1.469
1.403	1.406	1.407	1.402	1.463	1.469
1.400	1.408	1.410	1.401	1.463	1.468
1.399	1.411	1.412	1.398	1.461	1.466
1.407	1.419	1.410	1.394	1.456	1.463
1.430	1.437	1.397	1.385	1.450	1.460
		1.370[a]	*1.364*[a]	*1.426*	*1.434*

[a] Terminal parameters are printed in italics.

Journal of Physics: Conference Series **185** (2009) 012055 doi:10.1088/1742-6596/185/1/012055

The almost identical bond distances in central parts (given in upper rows) would ultimately converge to the periodicity of infinite lattices. The experimental bond distances of hexagonal systems of graphite and benzene are 1.42 Å and 1.397 Å, respectively. Therefore, these inward and outward CC bond distances are close to that of benzene, whereas the joint CC bonds are much longer. These tendencies are due to particular structural deviations in acenes from the regular hexagonal lattices. The C_{2h} symmetry type of bond alternations in a trans form was not obtained in this work.

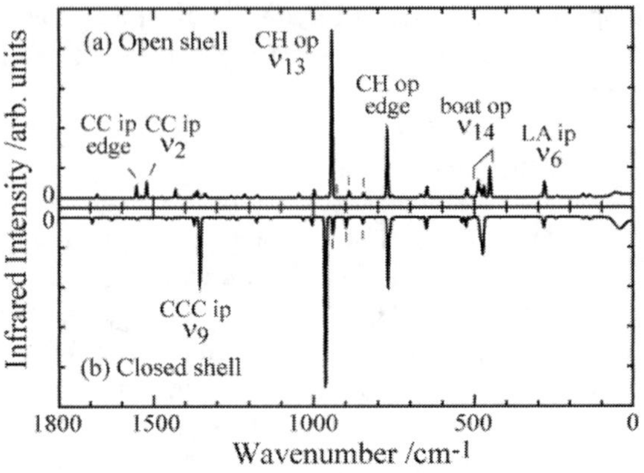

Figure 1. Infrared spectra of decacene in (a) the open-shell diradical singlet state and (b) closed-shell singlet state. ip: in-plane bend, op: out-of-plane bend, LA: longitudinal acoustic.

Figure 1 compares calculated infrared spectra between open-shell and closed-shell singlet states of decacene (n = 10). Again, the open-shell state is calculated to be 8.6 kcal mol^{-1} lower than the closed-shell state, which is in line with the previous results at BPW91/6-31G(d) and BLYP/6-31G(d) levels of theory [4]. The infrared spectra of the two states display similar bands in frequency and intensity, except for the in-plane joint CC stretching bands (CC ip edge and ν_2) obtained at ~1550 cm^{-1} in the open-shell state. The infrared intensity obtained in the short axis direction strongly suggests contribution from the singly-occupied orbitals delocalized over the CCC perimeter at each side.

Figure 2 compares vibrational dispersion relations for all modes of polyacene in the open-shell and closed-shell states. The vibrational modes of finite systems are well explained in terms of the phase difference of the infinite system, θ. For example, the band progressions calculated at ~950 cm^{-1} in Figure 1 correspond to CH out-of-plane modes with different phases of displacements. They show different phases in vibrational displacements between neighbouring CH bonds. The calculated dispersion curves are found to be very similar between the open-shell and closed-shell states, except in-plane branches ν_3 and ν_4. Hence, the frequencies of these modes can act as marker bands of localized spin states, in addition to the above mentioned infrared activities. These modes are characterised to be perimeter CC displacements in the same direction.

Next, we show electron-phonon (vibronic) coupling on the basis of molecular spectroscopy. Figure 3 depicts the calculated transition intensities for vibronic bands associated with $S_1 \leftarrow S_0$ transitions of octacene (n = 8) and decacene. Interestingly, the vibrational modes of both acenes display similar intensity patterns. Moreover, the major infrared activities are similarly obtained for acenes with n = 4, 5 (Ref. 5), 8 and 10, and their vibrational patterns are categorised in the same branches of

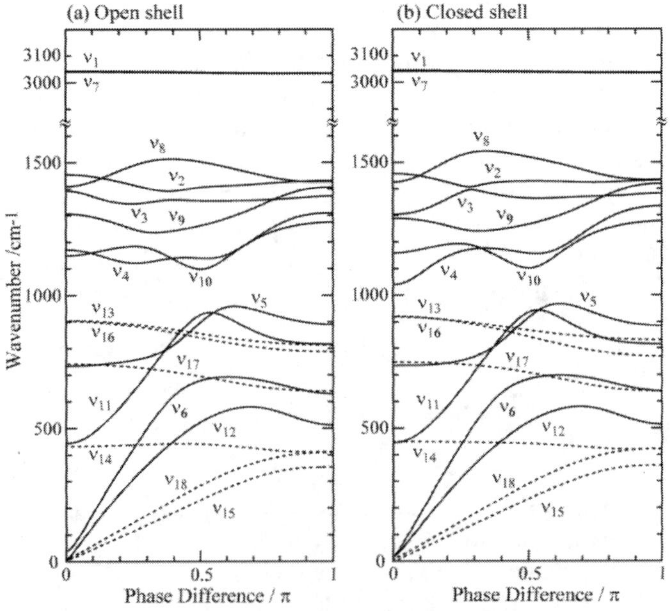

Figure 2. Phonon dispersion relations for polyacenes in (a) the open-shell singlet diradical state and (b) cloed-shell singlet state [5]. Broken lines represent out-of-plane modes.

dispersion curves in Figure 2.

The Franck-Condon type transitions are obtained for a_g modes (ν_3, ν_4 and ν_6) upon this electronic transition of the ^1L_a type, whereas the vibronic-coupling transitions are activated for b_{3g} modes through interaction with higher electronic states of the ^1B_b type. Since the activities given in Figure 3 are those for vibrational stated of the electronic ground state, these distributions correspond to resonance Raman effects. It should be noted that the scales between the two types of transitions are arbitrary. Decacene tends to display stronger intensities of the VC type relative to those of FC type than octacene.

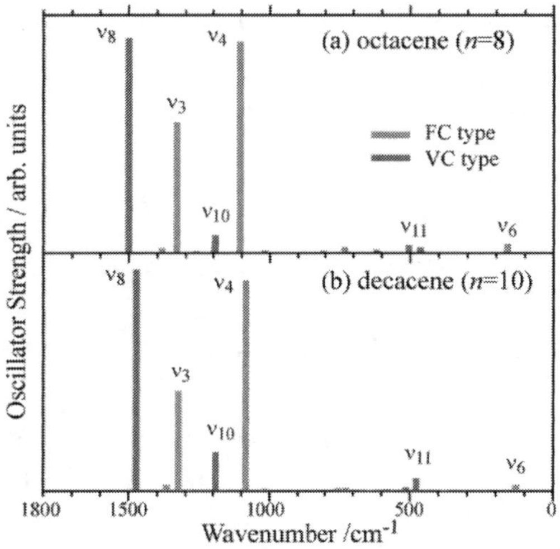

Figure 3. Calculated oscillator strength for the $S_1 \leftarrow S_0$ electronic transition in (a) octacene and (b) decacene. The Franck-Condon type and vibronic coupling type are depicted in red and blue, respectively. Scales between the two types are arbitrary.

Vibrational frequency shifts towards low wavenumber on going towards larger acenes are explained by the decreased phase difference between neighbouring units. The phase difference becomes smaller since the vibrational displacements must minimize θ to gain optical activities under the boundary conditions within the lengths of molecules. In particular, the remarkable shifts for mode ν_6 are due to the acoustic modes that converge to translational or rotational motion at $\theta = 0$. The phase difference seems to affect the transition intensities of these molecules, since only a few vibrations display vibronic activities among many vibrations with different phases. The vibrations of these non-hexagonal distorted carbon networks in the region 1000-1350 cm^{-1} with particular phase relations are expected to show strong resonance Raman effects of FC types.

4. Conclusions

The phonon dispersions of open-shell and closed-shell singlet states are similar except the perimeter symmetric CC modes. The vibrational bands of large acenes display low-wavenumber shifts systematically on going towards longer systems, whereas the vibronic activities appear very similarly. This work illustrates the possibility of molecular spectroscopic approaches to problems of nanocarbons.

References

[1] Novoselov K S, Geim A K, Morozov S V, Jiang D, Zhang Y, Dubonos S V, Grigorieva I V and Firsov A A 2004 *Science* **306** 666
[2] Fujita M, Wakabayashi K, Nakada K and Kusakabe K 1996 *J. Phys. Soc. Jpn.* **65** 1920
[3] Kobayashi Y, Fukui K, Enoki T, Kusakabe K and Kaburagi Y 2005 *Phys. Rev. B* **71** 193406
[4] Bendikov M, Duong H M, Starkey K, Houl K N, Carter E A and Wudl F 2004 *J. Am. Chem. Soc.* **126** 7416
[5] Yamakita Y, Kimura J and Ohno K 2007 *J. Chem. Phys.* **126** 064904
[6] Yamada M, Yamakita Y and Ohno K 2008 *Phys. Rev. B* **77** 054302
[7] Ohno K 1978 *J. Mol. Spectrosc.* **72** 238
[8] Ohno K 1979 *J. Mol. Spectrosc.* **77** 329
[9] Poater J, Bofill J M, Alemany P and Solà M 2005 *J. Phys. Chem. A* **109** 10629
[10] André J-M, Champagne B, Perpéte E A and Guillaume M 2001 *Int. J. Quant. Chem.* **84** 607
[11] Houk K N, Lee P S and Nendel M 2001 *J. Org. Chem.* **66** 5517
[12] Alan H, Nikolaidi A, Machado S and Humberto J 2005 *Cent. Eur. J. Chem.* **3** 361
[13] Aihara J 1994 *J. Chem. Soc. Perkin Trans.* **1994** 971

The 8th Asian International Seminar on Atomic and Molecular Physics IOP Publishing
Journal of Physics: Conference Series **185** (2009) 012056 doi:10.1088/1742-6596/185/1/012056

Quantum mechanical calculations of the S(^1D)+HD reaction dynamics on the ground electronic state

H. Yang,[1] K.-L. Han,[1] G. C. Schatz,[2] S. C. Smith[3] and M. Hankel[3]

[1]State Key Laboratory of Molecular Reaction Dynamics, Dalian Institute of Chemical Physics, Chinese Academy of Sciences, Dalian 116023, China

[2]Department of Chemistry, Northwestern University, Evanston IL 60208-3113, USA

[3]Centre for Computational Molecular Science, Australian Institute for Bioengineering and Nanotechnology, The University of Queensland, QLD 4072, Australia

E-mail: m.hankel@uq.edu.au

Abstract. We present reaction probabilities and branching ratios for the title reaction calculated employing the real wavepacket approach as implemented in the DIFFREALWAVE code. Calculations for zero total angular momentum ($J = 0$) and also for some non-zero values, $J = 10$, 30 and 40 have been carried out including all Coriolis couplings. These preliminary results predict a branching ratio greater than unity for energies above 0.5 eV total energy while more partial waves are needed to predict the ratio for energies below 0.5 eV.

The S(^1D)+H_2 reaction and its isotopic variants have received considerable interest as a prototype for the so called insertion reactions [1, 2, 3, 4, 5, 6, 7, 8, 9, 10, 11, 12, 13, 14, 15, 16, 17, 18]. Experiment predicts an isotopic branching ratio, that is the product channel branching ratio (SD+H)/(SH+D), of around 1 under thermal conditions [1] and of around 0.72 from crossed molecular beam experiments [2]. Theoretical studies consistently predict a branching ratio greater than 1 [5, 6, 8, 15, 16]. By conventional wisdom one would expect a ratio greater than one from an insertion reaction and a ratio less than one for an abstraction reaction. This can be explained by the fact that the lighter hydrogen atom is more likely to escape the intermediate complex or even by a simple density of states argument. Most theoretical studies to date employed quasi classical trajectory methods, or employed J-shifting methods based on $J = 0$ quantum mechanical calculations. Some studies included $J > 0$ calculations employing the helicity decoupling approximation and J-shifting methodologies to obtain integral cross sections.

We have carried out quantum scattering calculations to obtain fully converged cross sections and also for the first time differential cross sections to compare directly to experiment. All calculations have been performed using the DIFFREALWAVE (DRW) [19, 20] code and employing the $1A'$ electronic state from the surface routines by Schatz and co-workers [11] which presents the adiabatic surface by Ho *et al.* [8]. The calculations include all Coriolis couplings. Calculations for J=0, 10, 30 and 40 have been carried out with HD in its ground state ($v = 0$,$j = 0$). While all J values up to 40 and beyond will be needed to obtain converged total integral and differential cross sections which cover the experimental energies, ranging from 0.256 to 0.49 eV total energies, these preliminary results should give a good indication on the effect of including $J > 0$ values.

Figure 1 shows the total reaction probabilities for both product arrangement channels for several values of the total angular momentum J. Both, the SD+H and SH+D, $J = 0$ probability are non-zero even for low energies and both probabilities show a gradual increase with energy. For energies between 0.23 and 0.4 eV the SD+H probability is larger than the one for SH+D while above 0.6 eV the SH+D probability is larger. For the SH+D channel the $J > 0$ probabilities are smaller than the one for $J = 0$ especially for higher energies. They also show less structure. The same is true for the SD+H channel but here the $J > 0$ probabilities are larger. However, the magnitudes for the higher J value probabilities is very consistent.

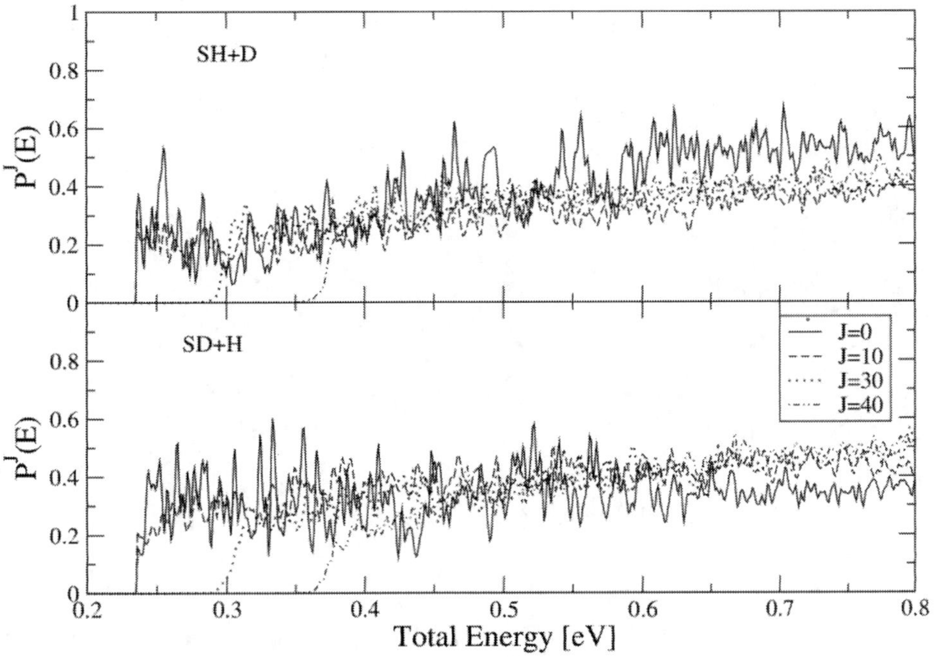

Figure 1. This figure shows the reaction probabilities for both product arrangement channels for different values of total angular momenta versus total energy in eV. Results for J=0, 10, 30 and 40 are shown.

The behaviour of the reaction probabilities is reflected in the branching ratio, shown in Figure 2. We show the ratio of SD+H/SH+D in this work. The dashed line is drawn at SD/SH=1 as a guide. As can be seen the ratio for $J = 0$ is larger than one for energies below 0.4 eV. For energies up to 0.6 eV the ratio shows sharp structures which are due to the sharp peaks in the individual product arrangement channel $J = 0$ reaction probabilities. Above 0.6 eV the ratio becomes relatively flat and nearly constant at around 0.6 reflecting the result from the previous figure that the SH+D channel is preferred for higher energies.

This difference between $J = 0$ and $J > 0$ reaction probabilities can also be seen in the branching ratios. The $J > 0$ branching ratios show very narrow and high peaks at low energies but become much more flat for higher energies. The branching ratio obtained from the $J = 0$ data is smaller than the ones obtained for $J > 0$, for energies above 0.4 eV. The $J = 0$ branching ratio is less than 1 while the $J > 0$ branching ratios are larger than 1. The energy range covered by experiment is from 0.256 to 0.49 eV. For these energies the branching ratios from the $J > 0$ calculations are larger than 1.

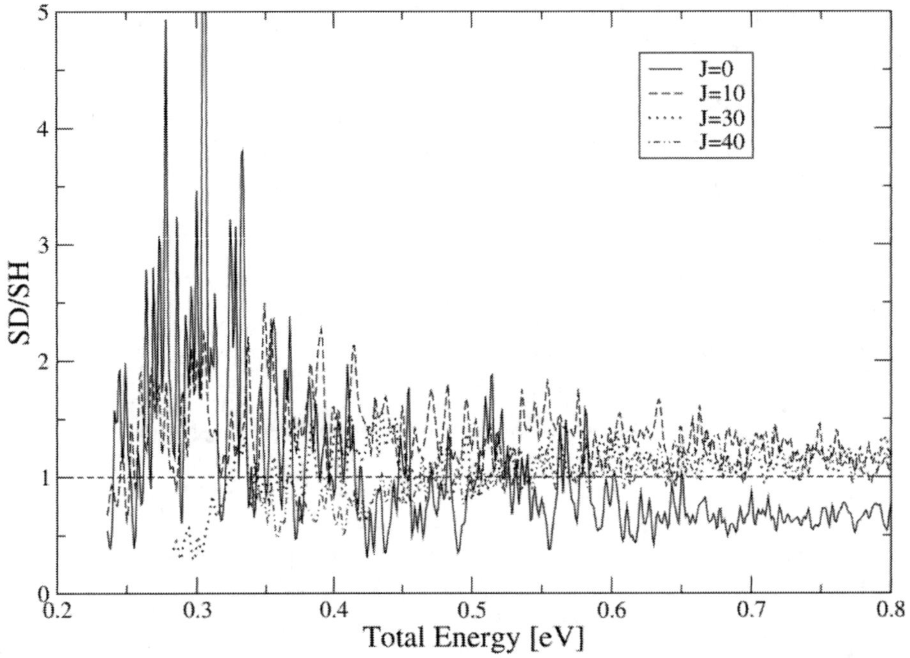

Figure 2. This figure shows the branching ratio, SD/SH, for different values of total angular momenta versus total energy in eV. Results for J=0, 10, 30 and 40 are shown.

The preliminary results from our calculations seem to indicate that the obtained branching ratio might be larger than unity, similar to most other theoretical studies carried out so far, especially for energies above 0.5 eV. To give a definite predictions for lower energies at present more partial waves are needed. A branching ratio larger than one can be expected from dynamics on the ground electronic state alone. However, the experimental branching ratio, which is around 0.72 seems to suggest that a simple density of states argument or that the lighter hydrogen atom can more easily escape the intermediate complex cannot explain the behaviour of the branching ratio. Studies including the triplet electronic states showed that the lower branching ratio is most likely to be due to quenching via non-adiabatic transitions to the triplet electronic states [11, 17].

The results for $J = 40$ show that converged results for energies up to 0.35 eV can be achieved from calculations up to this J value. Calculations for all J values up to 40 are now in progress to obtain fully converged integral and differential cross sections.

Acknowledgments
Computational resources used in this work were provided by the University of Queensland (Centre for Computational Molecular Science) and the Australian Research Council (LIEF

grant LE0882357: A Computational Facility for Multiscale Modelling in Computational Bio and Nanotechnology). MH would like to thank Sun Microsystems and the University of Queensland for funding. HY acknowledges financial support from The University of Queensland during a research visit to the Centre for Computational Molecular Science, May - July 2008. GCS was supported by AFOSR Grant FA9550-07-1-0095.

References

[1] Inagaki Y, Shamsuddin S M, Matsumi Y and Kawasaki M 1994 *Laser Chem.* **14** 235
[2] Lee S H and Liu K 1998 *Chem. Phys. Lett.* **290** 323
[3] Lee S H and Liu K 1998 *J. Phys. Chem. A* **102** 8637
[4] Lee S H and Liu K 2000 *Appl. Phys. B* **71** 627
[5] Chang A H H and Lin S H 2000 *Chem. Phys. Lett.* **320** 161
[6] Chao S D and Skodje R T 2001 *J. Phys. Chem. A* **105** 2474
[7] Zyubin A S, Mebel A M, Chao S D and Skodje R T 2001 *J. Chem. Phys.* **114** 320
[8] Ho T S, Hollebeek T, Rabitz H, Chao S D, Skodje R T, Zyubin A S and Mebel A M 2002 *J. Chem. Phys.* **116** 4124
[9] Zhang H and Smith S C 2002 *J. Phys. Chem. A* **106** 6137
[10] Honvault P and Launay J M 2003 *Chem. Phys. Lett.* **370** 371
[11] Maiti B, Schatz G C and Lendvay G 2004 *J. Phys. Chem. A* **108** 8772
[12] Mouret L, Launay J M, Terao-Dunseath M and Dunseath K 2004 *Phys. Chem. Chem. Phys.* **6** 4105
[13] Bañares L, Aoiz F J, Honvault P and Launay J M 2004 *J. Phys. Chem. A* **108** 1616
[14] Göğtas F, Bulut N and Akpinar S 2005 *J. Mol. Struct. (THEOCHEM)* **723** 189
[15] Bañares L, Castillo J F, Honvault P and Launay J M 2005 *Phys. Chem. Chem. Phys.* **7** 627
[16] Lin S Y and Guo H 2005 *J. Chem. Phys.* **122** 074304
[17] Chu T S, Han K L and Schatz G C 2007 *J. Phys. Chem. A* **111** 8286
[18] Aoiz F J, González-Lezana T and Sáez Rábanos V 2008 *J. Chem. Phys.* **129** 094305
[19] Hankel M, Smith S C, Allan R J, Gray S K and Balint-Kurti G G 2006 *J. Chem. Phys.* **125** 164303
[20] Hankel M, Smith S C, Gray S K and Balint-Kurti G G 2008 *Comp. Phys. Commun.* **179** 569

Journal of Physics: Conference Series **185** (2009) 012057 doi:10.1088/1742-6596/185/1/012057

Preparation of intense multi-element metal cluster ions with single composition

H Yasumatsu[a], M Fuyuki[b], T Hayakawa[b] and T Kondow[a1]

[a]Cluster Research Laboratory, Toyota Technological Institute:

In [b]East Tokyo Laboratory, Genesis Research Institute, Inc.

717-86 Futamata, Ichikawa, Chiba 272-0001, Japan

kondow@clusterlab.jp

Abstract A source of composition-selected multi-element cluster ions has been developed toward investigation of chemical reactivity of the clusters supported on a solid surface. The cluster ions are produced in a gas aggregation cell equipped with several magnetron sputtering devices, and their composition is selected by a quadrupole mass filter. The translational and internal kinetic energies of the single-composition cluster ions are reduced by collision with cold helium to achieve cluster-impact deposition onto the surface at a low collision energy. It has been succeeded to obtain single-composition silver-copper bimetal cluster ions more intense than several tens pA. A typical translational energy width is 0.5 eV per cluster.

1. Introduction

Chemical and physical properties of clusters change significantly and characteristically with a number of constituent atoms of the cluster (cluster size) due to specific changes of the geometric and electronic structures of the clusters. In this extension, it is straightforward to see that multi-element clusters have larger variety and novelties in their properties because of inhomogeneous electronic interaction between atoms of dfferent elements as well as many possible geometric isomers even at a given composition. Furthermore, multi-element clusters can be a model system of cocatalysis to understand effects of the second (and third, …) element(s) on the catalytic activity. In these relations, several studies on bi-element clusters have been reported [1-6]. However, these studies do not show production of single-composition clusters with practical intensities. Furthermore, no report has been made on tri- or more-element metal clusters. In this report, preparation of intense multi-element metal cluster ions with a single-composition is described, aiming at investigation of chemical reactivities of multi-element clusters supported on a solid surface. Therefore, the cluster ions are required to be as intense as several tens pA in order for the preparation of the supported clusters by cluster impact [7-9] in a reasonable time. In addition, the translational kinetic-energy width of the cluster ions should be narrower than ~1 eV in order not only to achieve a low-energy impact for avoiding the cluster dissociation at the cluster impact but to obtain a uniform geometry of the clusters on the surface, as the geometry is determined by the impact energy as well as the cluster-surface interaction.

[1] To whom any correspondence should be addressed.

2. Experimental apparatus

Figure 1 shows a sectional top plan of the apparatus developed. The machine consists of a multi-element cluster-ion source, octopole ion guides, a quadrupole deflector, a quadrupole mass filter for the cluster ions, a gas-collision cell, a cluster-deposition equipment and a quadrupole mass filter for the reaction products desorbed from the supported clusters. In this report, the description is concentrated on the production and the cooling of the cluster ions.

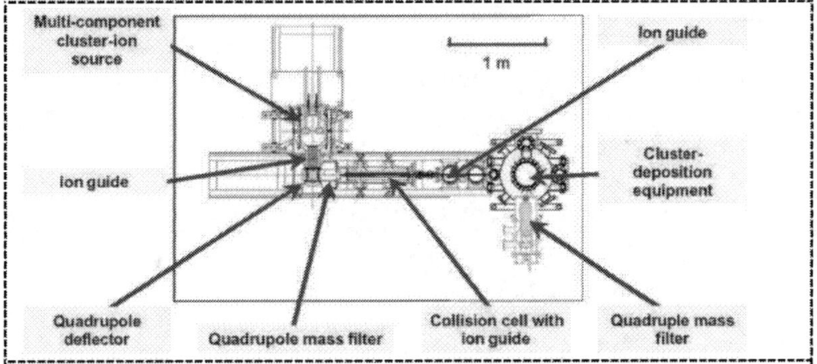

Figure 1. Sectional top plan of an apparatus for production and surface deposition of single-composition multi-element clusters.

The multi-element cluster ions are produced in a gas-aggregation cell equipped with magnetron sputtering sources [8] having a target of a different element for each placed in a parallel and off-centered arrangement; five magnetrons are attachable in this source. In simultaneous operation of the magnetrons at the powers of 20-100 W, atoms and ions of different elements are ejected from the targets by argon-ion impact, and they are allowed to aggregate into cluster ions in collision with helium and argon in the aggregation cell cooled by liquid nitrogen filled in a jacket surrounding the cell. The flow rates of the argon supplied to each magnetron and the helium are regulated independently with mass-flow controllers; typical flow rates are 15 and 60 cm^3 min^{-1} (sccm) for the argon and the helium, respectively, so that the pressure of the aggregation cell is ~10 Pa. The distance between each magnetron and an exit nozzle (7 mm of inner diameter and 10-mm long) of the aggregation cell is optimized externally and independently (typically 120 mm).

The cluster ions thus produced are admitted into the first ion guide (290-mm long) placed in a jacket filled with liquid nitrogen, in which the diameter of the ion beam is reduced by collision with helium so as to increase ion transmittances of the ion optics placed downstream. The ion-guide is supplied with an rf voltage (500 KHz, 300 V_{p-p}) from a home-built RF oscillator driven by a FET. The rf frequency was optimized by adjusting an inductance and a capacitance placed in parallel to the ion guide. In the quadrupole deflector, only cations are deflected by 90 degrees toward the quadrupole mass filter so as to keep the mass filter from contamination with intense neutral species. Simultaneously, this deflector acts as an energy filter so as to obtain a narrow translational energy width of the cluster ions for a high ion-transmittance and resolution through the mass filter. The ions are injected into the mass filter (Extrel MEXM-9000, mass range of 25-9000 amu) through two cylindrical electrostatic lenses in order for the composition selection of the cluster ions.

The composition-selected cluster ions are admitted into the collision cell filled with helium leaked through a needle valve. The collision cell and then the helium are cooled by liquid nitrogen filled in a jacked surrounding the cell, where the helium is pre-cooled by passing through a copper tube (~10 m) wound around the outer wall of the jacket. The cluster ions are guided by an octopole ion-guide (1-m long, 1 MHz, 300 V_{p-p}) equipped in the collision cell in order to prevent the ions from being scattered out during the collisional cooling.

The intensity of the cluster ions is measured as an ion current, I, hitting on a metal plate at any place (e.g. rods of the ion guide, the electrostatic lenses, etc.) through a home-built current-voltage converter having a conversion factor of 10^8 VA^{-1} and a serial home-built passive low-pass filter with cut-off frequencies of 3 and 8 Hz. The translational energy distribution of the cluster ions is measured by a retarding-potential method, i.e. a numerical difference of I with respect to a bias voltage, V_{bias}, applied to the detector plotted as a function of V_{bias}.

3. Results and Discussion

3.1. Production of multi-element cluster ions

Figure 2 shows a typical mass spectrum of bimetal cluster cations of silver and copper, $Ag_NCu_M^+$, detected at the third ion-guide rods; $Ag_NCu_2^+$ are also discernible as shoulders of Ag_N^+ at their heavier sides. We have succeeded in producing the bimetal cluster ions in a wide size range with intensities more than several tens pA. The numbers of the Ag and Cu atoms in a cluster ion are controllable coarsely by the electric powers supplied to the magnetrons. It is easily expected to produce tri- or more-element cluster ions by operating more magnetrons in a similar manner.

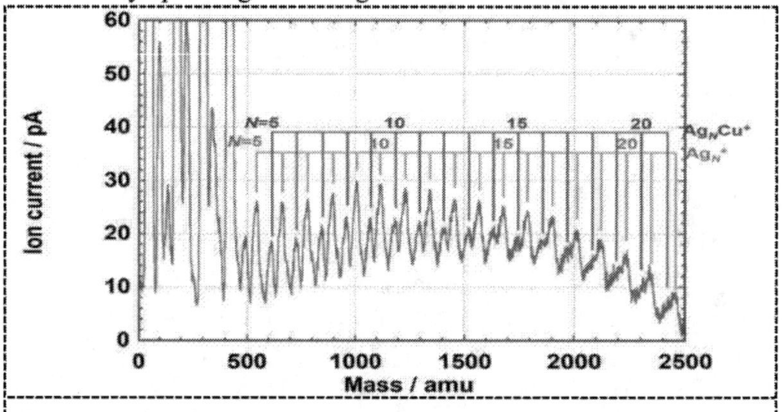

Figure 2. Mass spectrum of silver-copper bimetal cluster cations.

3.2. Cooling of cluster ions

Figure 3 shows a translational energy distribution of Ag_{44}^+ without and with introducing helium in the collision cell, the pressure of which increased from 10^{-5} to 10^{-1} Pa ranges by the helium introduction. The best-fit curves give FWHM of 5 and 0.5 eV without and with the cooling, respectively. Therefore, one can decrease the collision energy as low as 0.5 eV, which is one order of magnitude lower than the

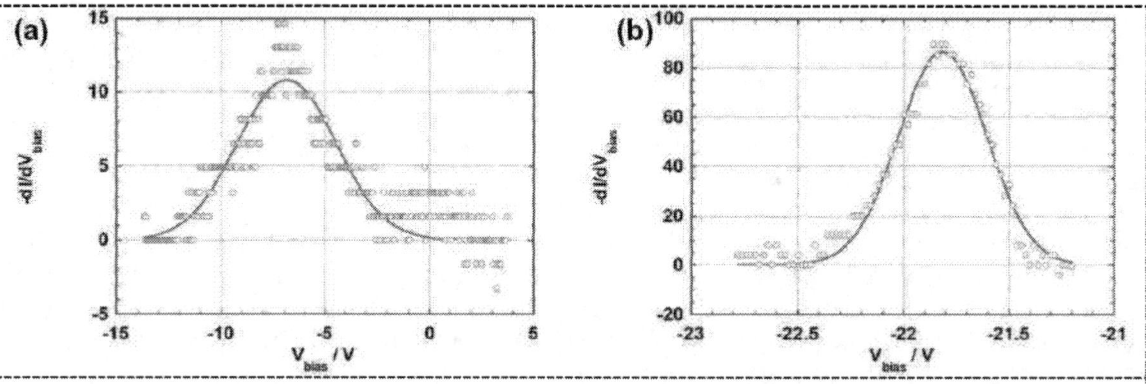

Figure 3. Translational energy distributions of Ag_{44}^+ (a) without and (b) with cooling by cold helium. The open circle shows a difference of the ion current, I, of Ag_{44}^+ with respect to the bias voltage, V_{bias}, applied to the detection plate, while the solid curves are best-fit ones obtained by fitting the measured data with a Gaussian function. The I difference was obtained by simple moving average of five adjacent measured data.

dissociation energy of Ag_{44}^+, so that the cluster is hardly dissociated by the impact on the surface. Furthermore, one can make the cluster impact at precisely-controlled collision energy to achieve a uniform geometry of the clusters on the surface. It was confirmed by mass analysis of the ions passed through the collision cell that the ions are not dissociated after the experience of the collision with the helium.

3.3. Optimization of frequency of rf voltage applied to ion guide

Figure 4 shows the transmittance of Ag_N^+ through the first ion guide at various frequencies of the rf voltage applied to the first ion guide. These results indicate that the heavier cluster ions have the higher transmittance through the ion guide with the higher frequency. In this manner, the optimized frequencies were applied to the three ion guides depending on the mass of the cluster ions of interest.

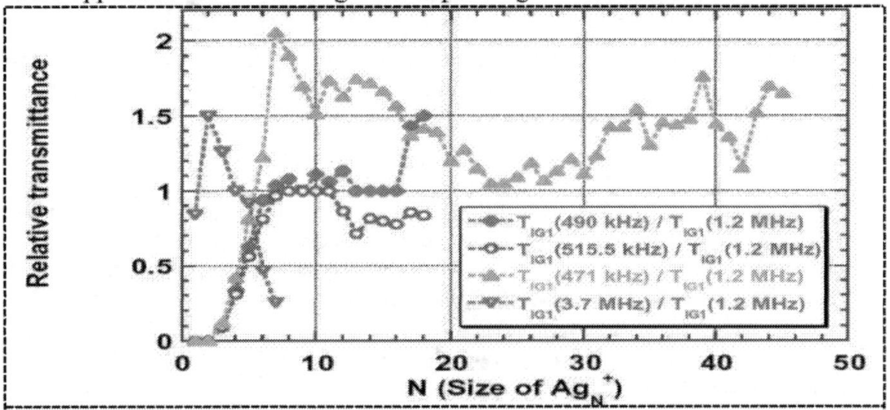

Figure 4. Transmittance, T_{IG1}, of silver cluster cations, Ag_N^+, through the first ion guide as a function of N at various rf frequencies applied to the ion guide. The relative transmittance is shown with respect to T_{IG1} at 1.2 MHz.

Acknowledgements

The authors are grateful to Professor J. M. Lisy (University of Illinois) for providing us an electronic circuit of the rf oscillator. Thanks are also due to Dr. G. Santambrogio (Genesis Research Institute, Inc.) for the fruitful discussion on the operation of the ion guide. This research was supported by the Special Cluster Research Project of Genesis Research Institute, Inc.

References

[1] Akutsu M, Koyasu K, Atobe J, Hosoya N, Miyajima K, Mitusi M and Nakajima A 2006 *J. Phys. Chem.* A **110** 12073
[2] Neukermans S, Wang X, Veldeman N, Janssens E, Silverance R E and Lievens P 2006 *Int. J. Mass Spectrometry* **252** 145
[3] Yin S, Xu X, Moro R and de Heer W A 2005 *Phys. Rev.* B **72** 174410
[4] Breaux G A, Hillman D A, Neal C M and Jarrold M F 2005 *J. Phys. Chem.* A **109** 8755
[5] Sumiyama K, Hihara T, Peng D L and Katoh R 2005 *Sci. Technol. Adv. Mat.* **6** 18
[6] Du X, Inokuchi M and Toshima N 2006 *J. Mag. Mag. Mat.* **299** 21
[7] Yasumatsu H and Kondow T 2003 *Rep. Prog. Phys.* **66** 1783
[8] Yasumatsu H, Hayakawa T, Koizumi S and Kondow T 2005 *J. Chem. Phys.* **123** 124709
[9] Hayakawa T, Yasumatsu H and Kondow T 2009 *Eur. J. Phys D* in press

Journal of Physics: Conference Series **185** (2009) 012058 doi:10.1088/1742-6596/185/1/012058

Studies of laser induced-breakdown spectroscopy of holly leaves

D.C. Zhang[1] , X. Ma[1], W.Q. Wen[1,2], H.P. Liu[1], P.J. Zhang[1,2]

1 Institute of Modern Physics, Chinese Academy of Sciences, Lanzhou 730000 China
2 Graduate University of Chinese Academy of Sciences, Beijing 100049 China

E-mail: dch-zhang@impcas.ac.cn

Abstract. The LIBS spectra of holly leaves produced by Nd:YAG nanosecond laser has been investigated. More than 20 elements and molecules were identified from the spectra. The influence of laser wavelength upon analysis of plant samples via LIBS technique has been investigated. The comparison of LIBS spectra shows that the spectrum for 1064 nm irradiation is more intense than the one for 266 nm irradiation by the same pulse energy in the near-UV region. Investigation of the influence of laser wavelength for LIBS can help to improve the analytical capabilities of LIBS.

1. Introduction

The laser-induced breakdown spectroscopy (LIBS) is a laser-based versatile elemental and molecular analysis tool. It has attracted substantial attention nowadays for a wide range of applications such as measuring of the microelements in solids, liquids and gases and for discrimination of microorganisms and bacteria[1-4]. It is a relatively new task to use the LIBS technique in plant samples[5-7].

In this paper, we present some experimental results on analysis of LIBS spectra of holly leaves. The interactions between the incoming irradiation and the sample depend upon numerous variables such as laser wavelength, pulse energy and the surrounding atmosphere, etc [8]. We also show the influence of laser wavelength for LIBS spectra of plant samples.

2. Experimental technique

The experiments were carried out with a Nd:YAG laser (Quanta-Ray, USA) that can deliver fundamental wavelength of 1,064 nm (IR) with a pulse duration of ~10ns and a repetition rate of 10Hz. Forth harmonic 266 nm (UV) with maximum pulse energy of 100mJ was acquired by KDP (Potassium Dihydrogen Phosphate) crystal. The energy of the laser pulse was monitored by an energy meter (3sigma, Coherent) during experiment. The beam was focused on the sample using a quartz lens (f=100mm). Fig.1 shows a schematic drawing of the LIBS system. All of the experiments were carried out in air without any control of the surrounding atmosphere. Plasma emission was collected by two plano-convex lenses to bifurcated fiber cable. The latter was mounted onto the entrance slits of a four-channel fiber optic spectrometer (Avantes, Nethlans). In this work the master channel (230~343nm) and first slave channel (340~438nm) were used. The CCD detector of the spectrometer was triggered with the laser lamp signal which was delay for 500ns and the gate for acquisition was set to 2ms for IR and UV laser. Before measurement, the sample leaves were cleaned by distilled water to remove the dust on their surface and dried. The leaf sample of holly was mounted on a stage with precision movements and continuously moved across the laser beam in order to provide a fresh surface for each shot. This improves the reproducibility of mass ablation by avoiding the formation of deep craters by successive ablation[9].

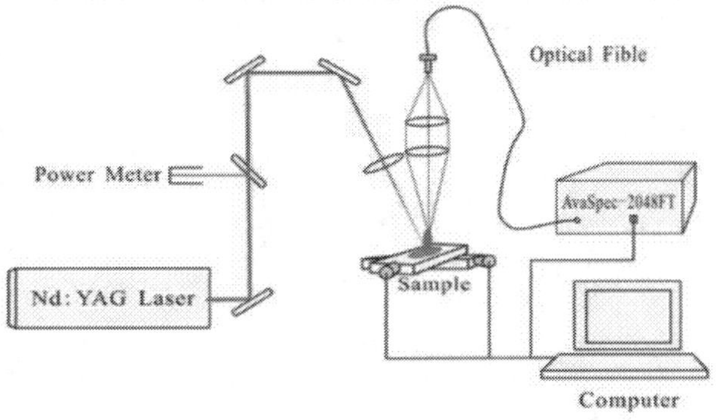

Figure 1. The schematic setup of the LIBS system for plant sample.

3. Results and discussions

3.1. Spectra lines identification

In the experiment, more than 430 lines in the range of 230~438 nm were found and more than 20 elements and molecules were identified from spectra of holly leaves samples according to the spectroscopy data of NIST[10]. A typical spectrum of the holly leaves sample is shown in Fig.1. The lines from different metal elements such as K, Ca, Na, Mg, Fe, Al, Mn, Ti, Mo, Ni, Cr, Co, Zn, Cu, V and nonmetal elements such as C, O, Si were identified according to their wavelengths. Two weak Pb lines were also found which may be used to analyze air pollution. Some molecular spectra from C_2 and CN were observed. Normally, C_2 is emitted from the leaves surface or formed in its vicinity, while part of CN is due to carbon recombination with atmospheric nitrogen in the plasma through the reaction $C_2 + N_2 \rightleftharpoons 2CN$ [11, 12]. Some species observed in the holly leaves samples are listed in table 1.

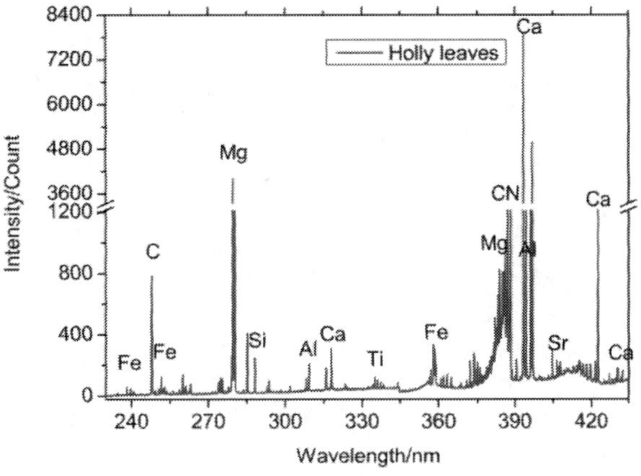

Figure 2. The LIBS spectrum of holly leaves.

Table 1. Some peaks identification of species observed in the holly leaves samples.

Wavelength (nm)	*Species*	Wavelength (nm)	*Species*	Wavelength (nm)	*Species*	Wavelength (nm)	*Species*
232.74	Fe II	281.34	Fe I	313.064	Ti II	383.491	Fe I
234.39	Fe II	281.66	Al II	317.964	Ca II	387.166	CN
239.59	Fe II	282.83	Ti II	322.815	Mn I	387.889	Fe I
243.27	Fe I	284.39	Fe I	323.436	Ti II	388.372	CN
243.51	Si I	285.19	Mg I	327.401	Cu I	388.675	Fe I
247.89	C I	286.95	Fe I	330.27	Na I	390.354	Mo I
249.34	Na II	288.18	Si I	334.204	Zn I	396.214	Al I
254.92	Al II	297.91	Na II	334.487	Ca I	396.876	Ca II
257.61	Mn II	298.36	Fe I	334.931	Ti II	402.474	Ti I
259.31	Fe II	298.49	Fe II	338.603	Ti I	403.094	Mn I
259.93	Fe II	298.77	Si I	346.377	Co I	405.789	Pb I
266.06	Al I	299.73	Ca I	347.499	Fe I	407.76	Sr II
268.463	Fe II	300.12	Fe I	350.34	Co I	416.689	CN
271.44	Fe II	300.27	Ni I	356.448	Fe I	418.027	CN
272.12	Fe I	304.77	Fe I	357.296	Pb I	422.636	Ca I
272.72	Fe II	305.89	Fe I	357.824	Cr I	423.559	Fe I
275.55	Fe II	306.21	Co I	358.024	Fe I	427.465	Cr I
276.77	Fe I	306.61	V I	359.582	Ti II	430.237	Ca I
278.31	Mg I	307.55	Ti II	364.284	Ti I	436.309	C_2
279.57	Mg II	308.82	Ti II	371.701	Ti I	437.948	V I
279.82	Mn I	309.29	Al I	373.762	Fe I	438.387	Fe I

3.2. The characters of the spectra produced by different wavelength laser

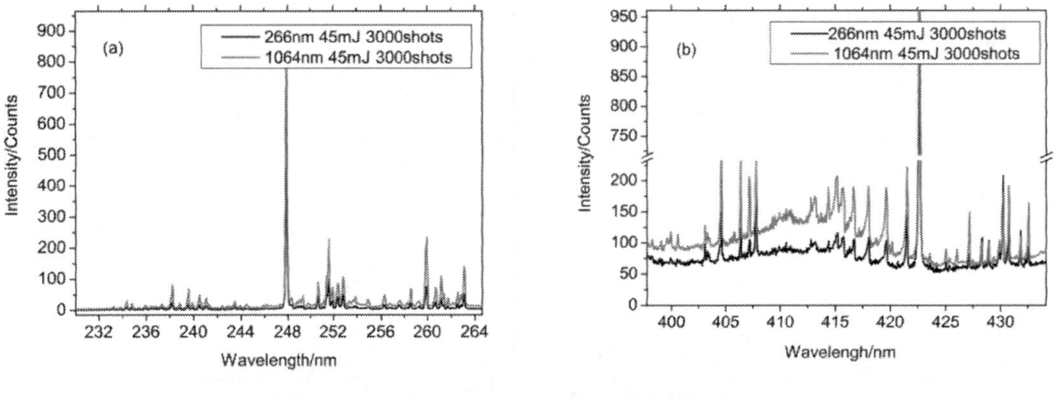

Figure 3. Parts of LIBS spectra produced by IR and UV laser.

Fig.3 shows parts of the LIBS spectrum averaged for 3,000 laser shots with 45mJ pulse energy. From plot (a), the spectrum in the near-UV region for 1,064nm irradiation is more intense than the one for 266nm irradiation. Abundant ionic lines were identified in this region. This is due to the higher absorption in the plasma caused by the inverse bremsstrahlung (IB), whose cross-section is proportional to cube of the laser excitation wavelength. Thus, the higher temperature caused by the greater absorption of the IR laser light was reached and rapidly converted into kinetic energy and ionization [8, 13, 14]. The similar phenomenon was observed in the LIBS analysis of ancient bronzes spectra [8].

Although IR laser can produce higher intensity lines which may be helpful to find weak lines in near-UV region, stronger continuum spectrum in visible region was also produced which has disadvantage to the signal-noise ratio of lines as shown in plot (b). However, one should be careful at this point, the reproducibility of LIBS spectra produced by IR laser is worse than those produced by UV laser [13]. Because the plasma shield to IR laser is much stronger.

4. Summary

The LIBS spectra of holly leaves have been investigated and more than 430 line were found and more than 20 elements and molecules were identified from spectra of holly leaves samples in the range of 230~438nm. The influence of laser wavelengths upon analysis of plant samples has been investigated. The comparison of LIBS spectra shows that the spectrum for 1,064nm irradiation is more intense than the one for 266nm irradiation by the same pulse energy in the near-UV region, and stronger continuum spectrum in visible region was also produced by 1,064nm irradiation. Thus, 266nm irradiation will be better for qualitative analysis of micro-elements in plant sample than 1,064nm irradiation for less influence of the plasma shield. Investigates of the influence of laser wavelength for LIBS can help to improve the analytical capabilities of LIBS.

Acknowledgements

The work is supported by the Natural Science Foundation of China (Grant No. 10605034)

References

[1] D.M. Diaz Pacc, C.A. D'Angelo, D.Bertuccelli, and G.Bertuccelli, Spectrochimica Acta Part B 61 (2006) 5.
[2] F. C. L. Fornarini, R. Fantoni, V. Lazic, V. Spizzicchino, Spectrochimica Acta Part B 60 (2005) 16
[3] Walid Tawfik and Y.Mohamed, optics&laser technology (2007) 9.
[4] Matthieu Baudelet, Laurent Guyon, Jin Yu, J.-P. Wolf, T. Amodeo, E. Fréjafon, and P. Laloi, Appl. Phys. Lett. 88 (2006) 3.
[5] Zhang Da-cheng, Ma Xin-wen, Zhu Xiao-long, Li Bin Zu Kai-ling, acta physica sinica 57 (2008).
[6] Myriam Bossu, HAO Zuo-Qiang, Matthieu Baudelet, YU Jin, ZHANG Zhe, and Z. Jie, CHIN.PHYS.LETT. 24 (2007) 3.
[7] A.Assion, M.Wollenhaupt, L.Haag, F.Mayorov, C.Sarpe-tudoran, M.Winter, U.Kutschera, and T.Baumert, appl. Phys. B 77 (2003) 8.
[8] L.Fornarini, V.Spizzichino, F.Colao, F. R, and V. Lazic, anal bioanal chem 385 (2006) 9.
[9] J. M. B. Le Drogoff, M. Chaker, M. Sabsabi, O. Barth´elemy , and S. L. T.W. Johnston, F. Vidal, Y. von Kaenel, Spectrochimica Acta Part B 56 (2001) 16.
[10] NIST, http://www.physics.nist.gov/PhysRefData/ASD/lines_form.html
[11] L. St-Onge, R.Sing, S. B´echard, M. Sabsabi, Appl. Phys. A 69 (1999) 4.
[12] Alan C. Samuels, J. Frank C. DeLucia, Kevin L. McNesby, and A. W. Miziolek, APPLIED OPTICS 42 (2003) 5.
[13] E. Tognoni, V. Palleschi, M. Corsi, and G. Cristoforetti, Spectrochimica Acta Part B 57 (2002) 16.
[14] C.W.NG, W.F.HO, and N.H.CHEUNG, applied spectroscopy 51 (1997) 8.

Journal of Physics: Conference Series **185** (2009) 012059 doi:10.1088/1742-6596/185/1/012059

Electron angular distributions in He single ionization impact by H_2^+ ions at 1 MeV

Shaofeng Zhang[1,2,5], J Suske[2], D Fischer[2,4], K U Kuehnel[2], S Hagmann[3], A Voitkiv[2], B Najjaril[2], A Krauss[2], R Moshammer[2], J Ullrich[2], Xinwen Ma[1,5]

[1]Institute of Modern Physics, CAS, Lanzhou 730000, China
[2]Max-Planck-Institut für KernPhysik, Saupfercheckweg 1, 69117 Heidelberg
[3]Gesellschaft für Schwerionenforschung, Planckstr. 1, 64291 Darmstadt, Germany
[4]Stockholm University, AlbaNova University Centre, 10691 Stockholm, Sweden
[5]Graduate University of Chinese Academy of Sciences, Beijing 100049 China

zhangshf@impcas.ac.cn

Abstract. For the first time we investigated in a kinematically complete experiment the ionization of helium in collisions with H_2^+-molecular ions at 1 MeV. Using two separate detectors, the orientation of the projectile H_2^+-molecular ions was determined at the instance of the collision. The electron angular distribution was measured by a "Reaction Microscope". The observed structures are found in agreement with theoretical calculations, indicating that the ionized electron of He shows a slight preferential emission direction parallel to the molecular axis.

1. Introduction

Studies of particle-induced dissociation of H_2, the simplest molecule composed of two atoms, have been reported extensively last decade. Particular attention is devoted to spatial effects in target ionization since the two centers are indistinguishable, where the contributions from each center add coherently. Early studies of collisionally induced fragmentation of H_2 mainly focused on the processes of electron capture[1, 2] and photoionization[3-5]. The following theoretical work was developed quite rapidly[6], whereas more experimental work mainly focused on the heavier molecules with synchrotron radiation[7, 8]. Up to now, little has been done for the collision experiments with H_2^+ as projectiles, which is the simplest molecular system in the nature, in despite of the advances in the collision experimental[9, 10] and theoretical work[11-13] with H_2 as a target. In a kinematically complete experiment the spatial effects in the ionization of helium induced by the molecular ion H_2^+ are investigated. In this work the orientation and internuclear distance of the molecular ion at the instance of the collision could be determined by momentum imaging technique. The phenomena resulting from the two-center potential of the molecular projectile in the dissociation channel were found.

2. Experimental setup

In our experiment, we employed a Recoil-Ion Momentum Spectroscopy (RIMS)[14] , which is suitable for the investigation of any kind of atomic reaction dynamics. The features of the experimental setup concerning molecular ion impact are presented. Briefly, the H_2^+ ion beam (1 MeV) was delivered by a linear accelerator at Max-Planck institute for nuclear physics. After collimation the ion beam entered the interaction chamber through a 0.5 mm diameter hole and crossed a cold localized gas-jet target. The outgoing projectile was charge selected by a dipole magnet, the neutral and the proton produced in the dissociation channel were delivered to different directions and detected by two separate position sensitive detectors respectively, while the primary beam was directed into a Faraday cup. The ejected electrons and the recoils produced in target ionization were extracted and accelerated by a 3 V/cm electric field from the interaction region, and then they were projected onto the detectors, respectively. In order to achieve a high acceptance together with a good resolution for both electrons and recoil-ions in coincidence, a weak homogenous magnetic field (generated by a pair of large Helmholtz coils) was superimposed along the spectrometer axis effectively confining the electron motion in space (figure 1).

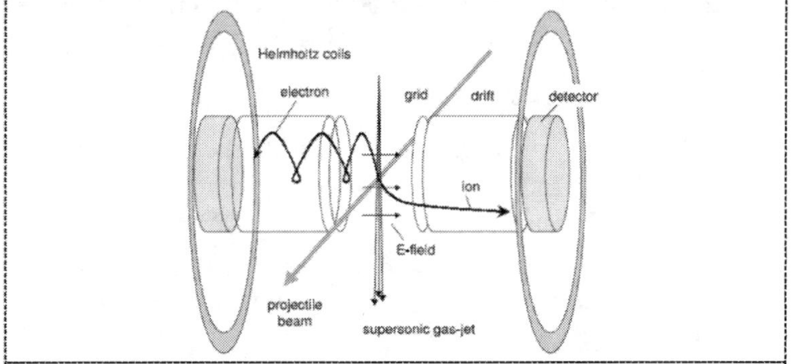

Figure 1. Schematics of an reaction microscope

The reaction channel of interest in the present experiment is the target ionization and simultaneous dissociation of the projectile molecular ion,

$$H_2^+ + He \rightarrow H + p + He^+ + e \qquad (1)$$

Here, four-fold coincidence measurements between the ionized electron, the recoil ion He^+, the projectile fragments (the neutral hydrogen atom H and the proton p) were performed, the data were recorded in event-by-event mode.

3. Results and discussion

The momenta of the four particles in eq. (1) were reconstructed according to their time and position on the detectors. The kinetic energy release (KER) of the projectile molecular H_2^+, were illustrated in the left panel in figure 2. By selecting specific KER the nuclear distance of the molecular ion can be determined. The right panel in figure 2 shows the projection of the spherical shell of the molecular ion with the KER condition around 10 eV.

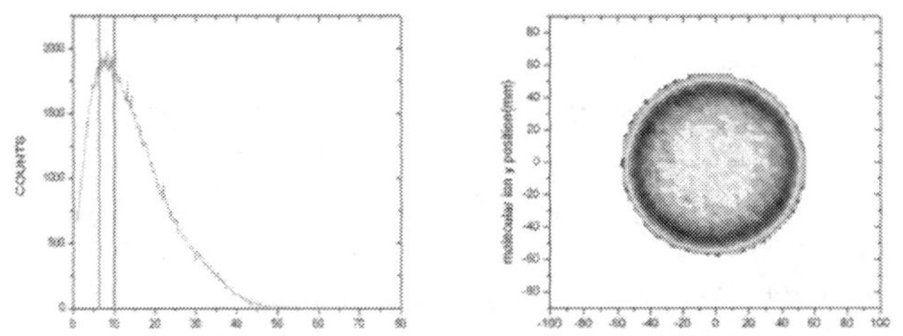

Figure 2. Left: The kinetic Energy Release (KER) spectrum of the H_2^+ ion in dissociation channel; Right: The projection of spherical shell of the fragments when the events between the two straight lines of the left plot are selected.

Journal of Physics: Conference Series **185** (2009) 012059 doi:10.1088/1742-6596/185/1/012059

The theoretical electron angular distribution with respect to the molecular axis was obtained under the axial recoil approximation[15], which holds that when a diatomic molecular ion is formed in a dissociative state, the atoms produced in the dissociation process will move outward along the straight line defined by the internuclear axis of the molecule. From the estimation below it is clear that this approximation works well in our case. If we treat the diatomic molecule as a rotating dumbbell composed of two point masses with a distance R, its rotational energy is given by

$$\frac{1}{2} I_0 \omega_0^{\ 2} = J(J+1)\frac{\hbar^2}{2I_0} \tag{2}$$

where ω_0 is the initial angular velocity of the molecular ion, and I_0 is its moment of inertia. Taking the rotational temperature of the gas to be 300 K, the final rotation angle can be written as below, by neglecting other aspects which can contribute to the rotational angle[16].

$$\Delta\varphi = \frac{8R_0\sqrt{E_\omega E_c}}{q^2} \tag{3}$$

where E_c is the coulomb potential energy, E_ω the rotational energy, R_0 the initial nuclear distance, and q the charge of the fragment. In our case, R_0=1, q=1, E_c=0.2, and E_ω=0.001 (in atomic units), we finally get $\Delta\varphi$=6.3°, which is sufficiently accurate to account for the data in our work.

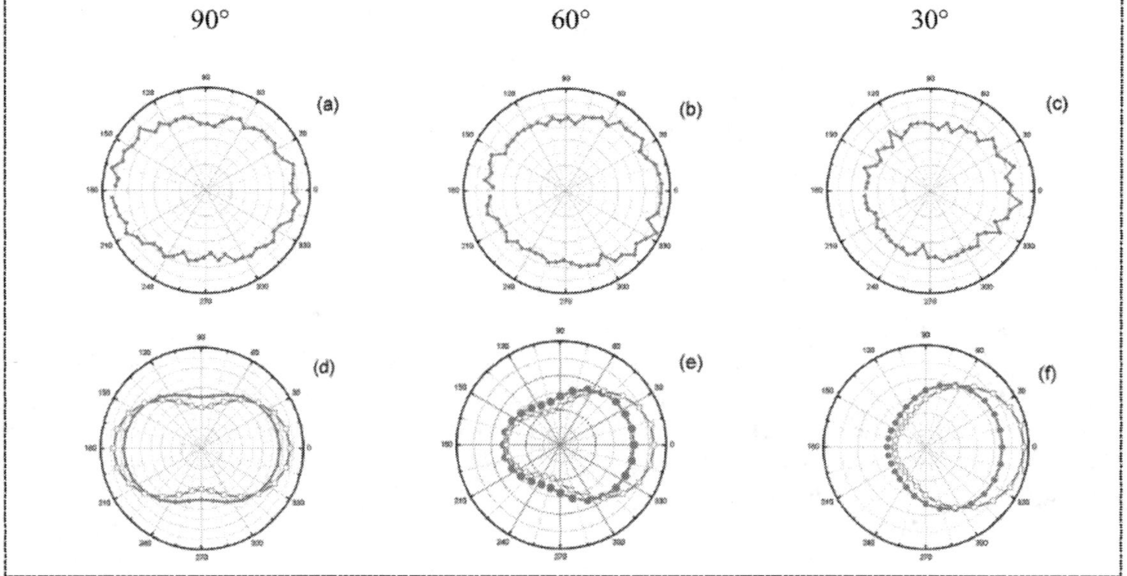

Figure 3. Electron angular distributions corresponding to three different molecular axis angles with respect to the projectile direction (from left to right the angles are 90°, 60°, and 30°, respectively). Upper row: experimental (a, b, c), lower row: theoretical (d, e, f).

Since the orientation of the H_2^+ axis can be reconstructed for each event, the electron angular distributions with respect to molecular axis can be obtained. Both experimental (upper row) and theoretical (lower row) results of the electron angular distributions are shown in figure 3. The projection plane in figure 3 is perpendicular to the beam propagation direction (the propagation direction is looking down into the figure). Plot (a) shows the electron angular distribution when the molecular axis is perpendicular to the beam propagation and plot (d) shows the theoretical results under the same situation. Plot (b) and (e) as well as plot (c) and (f) present the experimental and theoretical electron angular distributions for the molecular axis angle 60° and 30° with respect to projectile direction, respectively.

The experimental data show that the emission of the ionized electron from the target has a slight preference along the molecular axis. Theoretical calculations from the quantum plane-wave Born approximation show more pronounced structure variation while the direction of the molecular axis changes. This phenomenon indicates the structured projectile does have some impacts on the collision reaction.

4. Summary

The first successful experiment for projectile-orientation-reconstruction in high energies (1 MeV) collisions was performed. The experiment was precisely organized to study the spatial effects in ionization of helium by the hydrogen molecular ion impact in dissociation channel: $H_2^+ + He \rightarrow H + p + He^+ + e$. The experimental results show less pronounced structure effects than the quantum plane-wave Born approximation predictions.

Acknowledgement

This work has been supported by the Doctoral Training Program between Chinese Academy of Sciences and the Max-Planck Society. Zhang and Ma were also supported by NSFC (No. 10434100).

References

[1] Tuan T F and Gerjuoy E 1960 *Physical Review* **117** 756
[2] Joulakian B, Hanssen J, Rivarola R and Motassim A 1996 *Physical Review A* **54** 1473
[3] Cohen H D and Fano U 1966 *Physical Review* **150** 30
[4] Dörner R, Bräuning H, Jagutzki O, Mergel V, Achler M, Moshammer R, Feagin J M, Osipov T, Bräuning-Demian A, Spielberger L, McGuire J H, Prior M H, Berrah N, Bozek J D, Cocke C L and Schmidt-Böcking H 1998 *Physical Review Letters* **81** 5776
[5] Walter M and Briggs J 1999 *Journal of Physics B: Atomic, Molecular and Optical Physics* **32** 2487-2501
[6] Corchs S E, Busnengo H F and Rivarola R D 1999 *Nuclear Instruments and Methods in Physics Research Section B: Beam Interactions with Materials and Atoms* **149** 247-256
[7] Heiser F, Geßner O, Viefhaus J, Wieliczek K, Hentges R and Becker U 1997 *Physical Review Letters* **79** 2435
[8] Mills J D, Sheehy J A, Ferrett T A, Southworth S H, Mayer R, Lindle D W and Langhoff P W 1997 *Physical Review Letters* **79** 383
[9] Stolterfoht N, Sulik B, Gulyás L, Skogvall B, Chesnel J Y, Frémont F, Hennecart D, Cassimi A, Adoui L, Hossain S and Tanis J A 2003 *Physical Review A* **67** 030702
[10] Stolterfoht N, Sulik B, Hoffmann V, Skogvall B, Chesnel J Y, Rangama J, Frémont F, Hennecart D, Cassimi A, Husson X, Landers A L, Tanis J A, Galassi M E and Rivarola R D 2001 *Physical Review Letters* **87** 023201
[11] Galassi M E, Rivarola R D, Fainstein P D and Stolterfoht N 2002 *Physical Review A* **66** 052705
[12] Nagy L, Kocbach L, Pora K and Hansen J P 2002 *Journal of Physics B: Atomic, Molecular and Optical Physics* **35** L453-L459
[13] Sarkadi L 2003 *Journal of Physics B: Atomic, Molecular and Optical Physics* **36** 2153-2163
[14] Ullrich J, Moshammer R, Dorn A, Dörner R, Schmidt L P H and Schmidt-Böcking H 2003 *Reports on Progress in Physics* **66** 1463-1545
[15] Zare R N 1967 *The Journal of Chemical Physics* **47** 204-215
[16] Wood R M, Zheng Q, Edwards A K and Mangan M A 1997 *Review of Scientific Instruments* **68** 1382-1386

Journal of Physics: Conference Series **185** (2009) 012060 doi:10.1088/1742-6596/185/1/012060

The potential of an EBIT in assisting plasma diagnostics and progress at the Shanghai EBIT

Y Zou and R Hutton
The Key Laboratory of Applied Ion Beam Physics, Educational Ministry, China
Shanghai EBIT Laboratory, Modern Physics Institute, Fudan University, Shanghai 200433, China

E-mail: zouym@fudan.edu.cn, rhutton@fudan.edu.cn

Abstract. In this paper, the potential of Electron Beam Ion Traps (EBIT) in assisting plasma diagnostics is reviewed. Examples of how EBITs have contributed to plasma diagnostic studies, including both magnetic fusion and inertial fusion plasmas, are shown. The progress of the instrumentation development at the Shanghai EBIT and ongoing research in the Shanghai EBIT laboratory are reported.

1. Introduction

Electron Beam Ion Traps (EBIT), initially developed at Lawrence Livermore National Laboratory (LLNL) [1], are sophisticated devices capable of acting both as highly charged ion (HCI) light sources and ion sources. As a HCI light source, they can basically provide light from emission states of any charge state of any element in the periodic table, hence almost unique for spectroscopic research. When an EBIT is operating, an electron beam is emitted by the cathode of an electron gun. This beam is accelerated by a high voltage, applied between the cathode and a drift tube assembly. The drift tube assembly provides a trap region for positively charged ions. On the way towards the trap region, the electron beam is compressed by a strong magnetic field provided by a pair of Helmholtz coils. The compression is made to increase the electron beam density, as the rates for all electron-ion collision processes are very dependent on the beam density. A strong compression, leading to three orders of magnitudes increase in the beam density, can be achieved by using superconducting Helmholtz coils. A cryogenic system is needed to maintain the superconduction, this also helps to produce a ultra high vacuum in the trap region. Such EBITs fall into the category of Cryogenic EBITs.

In an EBIT, a tenuous plasma can be formed from basically any element, as long as the element can be injected into the trap region. The way an EBIT works gifts it with the property of a tuneable mono-energetic electron beam. Unfortunately there is a small energy spread caused by the temperature of the emitting cathode and the electron beam space charge. The range the electron beam energy can be tuned can cover a few hundreds of eV to several tens of keV, in some cases even up to around 200 keV [2]. This property makes it possible to use EBITs for detailed studies of processes in hot plasmas, i.e. disentanglement studies of the various processes in hot plasmas and to assist plasma diagnostics for temperature, density, electromagnetic field, as well as ion motion.

After the groundbreaking developments at LLNL, EBITs have been set up in Oxford [3], NIST [4], Tokyo [5], Berlin [6], Freiburg [7] (now moved to Heidelberg), Dresden [8], Vancouver [9], Stockholm [10], Belfast [11] and Darmstadt [12]. To promote the above mentioned studies in China, the Shanghai EBIT project was launched in January of 2002, under a collaboration between Fudan

University and the Shanghai Institute of Applied Physics. The design of the Shanghai EBIT puts it well into the category of so-called super EBITs, which means the electron beam energy should reach 100 keV or higher. There were already three members of the super EBIT category before the Shanghai EBIT was setup, they are the LLNL EBIT (super EBIT), the Tokyo EBIT (YEBISU) and the Heidelberg EBIT.

2. EBITs in assisting plasma diagnostics

Many laboratory and astrophysical plasmas involve a Maxwell-Boltzmann energy distribution of electrons colliding with an ensemble of ions. Understanding the behaviour of plasmas from their emission spectra needs accurate electron-ion collision rate coefficients, which include cross sections for various processes, such as excitations, ionizations, recombinations etc. The cross-sections are needed in the whole energy range at a certain plasma temperature. Theoretical calculations need to make approximations in almost all the cases. and may not always lead to accurate results. Experimental studies using laboratory plasma sources, like Tokamaks and θ pinch devices, often suffer from limited accuracies caused by complications such as density effects, radiative transfer, ion abundance, field gradients and temperature effects etc. and combinations of all these effects [13].

At LLNL, a special setup was devised to simulate a Maxwell-Boltzmann electron distribution for a clean and optically thin EBIT plasma, in order to study collisions between ions of specific charge states with electrons of a Maxwell-Boltzmann energy distribution [14]. In their work, they developed a system to sweep the electron beam in energy and current as a function of time to simulate conditions in a hot plasma. The current is swept to keep the electron density constant during the energy sweep, so as to minimize differences caused by differing space charge distributions While the energy sweeping is done by converting the number of electrons under the distribution to a time distribution, for example, for a higher fraction of electrons at a given energy the time duration at that energy was set to a longer value according to a Maxwell-Boltzmann distribution. The energy sweep covered most of the range of the electron energies at a given temperature, with a minimum and a maximum energy cut off according to their EBIT operation conditions. Because EBIT operating parameters can be easily changed, a wide range of temperatures can be simulated. Ion density effects are generally unimportant, as the EBIT plasma is optically thin. By measuring the emission photons at the same time as sweeping the electron beam energy and current, they could automatically integrate the collision cross sections with the desired electron distribution. The observed properties of ions in such experiments are directly dependent on the relevant rate coefficients [14].

Although it has been emphasized that X ray spectroscopy is important for ion temperature measurements in Tokamaks, the spectra of highly charged ions, in particular, He-like and Li-like ions, contain much more information than just the ion temperatures obtained from Doppler line broadening. The spectra show, in addition to the strong resonance lines, numerous satellites which are produced by dielectronic recombination and inner shell excitations. Since the intensity ratios between the resonance lines and their dielectronic satellites depend only on the electron velocity distribution, they are sensitive measures of electron temperature. In addition, the line intensity of an inner shell excited satellite is proportional to the density of the neighboring charge state. These satellites then provide information of charge balance and impurity ion transport. From the experimental point of view, the satellite spectra of He-like ions of medium Z elements are very attractive, because the separation between the resonance line and satellites are sufficiently large to allow for clean identifications. At the same time, the wavelength range of the spectrum is small enough, so that the sensitivity of the instrument is constant over the entire range. The measurements of the satellites-to-resonance line rations are therefore reliable.

In a work done at TEXTOR-94 [15], satellite spectra of He-like Ar from a discharge with auxiliary heating by neutral-beam injection was measured, as shown in Fig.2 in [14]. In that work, the intensity ratio of the line q, from the Li-like $1s^2 2s\ ^2S_{1/2}$ - $1s2s2p\ ^2P_{1/2}$ transition, over the line w, from the He-like $1s^2\ ^1S_0$ - $1s2p\ ^1P_1$ transition, was used to deduce the charge state ratio of Ar^{15+}/Ar^{16+}, believing that the satellite line q was produced by impact excitation of the ground state of Li-like Ar, $1s^2 2s\ ^2S_{1/2}$

.However, the result was about four times higher than theoretical predictions. A few years later the n=2-1 spectral emission pattern of He-like Ar, together with the associated satellite emission originating from Li-like Ar was measured with high-resolution x-ray spectroscopy at the Berlin EBIT [16]. In the Berlin work, a mono-energetic electron beam was ramped in energy between 1.9- and 3.9-keV (from below the resonance energy for dielectronic satellites to above the direct excitation threshold for n=1 to 2) to examine the He-like ions. In this way all the radiating processes of recombination and excitation could be studied with high-resolution spectroscopy. Hence, the satellites caused by dielectronic resonance are separated from those caused by direct excitations, as they appear at different electron energies. The results show that the line q could be caused by both direct excitation of the ground state of Li-like Ar and by dielectronic recombination to the ground state of He-like Ar. This questions the applicability of using the q/w line ratio to deduce the charge state ratio. X-ray emissions from He- and Li-like Ar ions are also used in diagnostics for inertial confinement fusion plasmas. Ar is usually added as a dopant to the solid fuel. The electron temperature is deduced from the intensity ratio of transitions between n=3 and 1 levels of H-like and He-like Ar, $I_{H\beta}/I_{He\beta}$; while the density is derived from Stark broadening of the Kβ line. But the satellites from Li-like 1s3lnl'-1s^{2}nl' transitions could distort the profile of the Kβ line. They can cause asymmetry of the line profile and broadening of the line width. This would mislead the temperature and density interpretation, if the satellite contributions were not properly taken into account. These contributions were usually estimated by theoretical models, with only the n=2 and 3 satellites being taken into consideration [17,18]. In 1995, the contributions of higher-n satellites to the Kβ line of He-like Ar were measured [19], using a high resolution crystal spectrometer to analyze the X-ray emission of the Ar ions trapped and excited in the LLNL EBIT. The electron beam energy was swept to hit the KLM, KMM, KMN and KMO dielectronic resonances, as well as through the excitation threshold of n=1 to 3. Their results showed that the resonance strength of n=4 satellites was larger than that of n=3, while the resonance strength of n=5 was nearly equal to that of n=3. That means the contribution of the satellites from n$\geqslant$ 4 was considerably larger than expected from standard scaling procedure. This result has lead to reassessment of line profile calculations of the He-like Kβ line used in density diagnostics in Laser fusion.

In the past few decades, EBITs have been producing definite values of electron impact excitation, ionization and resonance excitation cross sections, as well as dielectronic resonance strengths. Some results have demonstrated shortcomings of theoretical data. For example, systematic measurements of n= 3 to 1 line emission from He-like ions at the LLNL EBIT [20] showed significant disagreement between the measured and the calculated ratios of the 1s3p $^3P_1$ – 1s^2 $^1S_0$ intercombination and the 1s3p $^1P_1$ – 1s^2 $^1S_0$ resonance line for low and medium Z ions. In [20], systematic disagreement was also reported for the measured and the calculated ratios of the 2p^{5}3d $^3D_1$ -2p^6 $^1S_0$ intercombination and 2p^{5}3d $^1P_1$ -2p^6 $^1S_0$ resonance line in the Ne iso-electronic sequence.

Table 1. Design parameters of the Shanghai EBIT.

Highest electron beam energy	200 keV
Highest electron beam current	200 mA
Electron beam radius	35 μm
Highest magnetic field	5 T
liquid He consumption	0.5 L/h

Hence EBITs have played an important role in producing accurate data for testing atomic theory, as well as in disentangling plasma processes and assisting plasma diagnostics.

3. Progress at the Shanghai EBIT

The Shanghai EBIT projected started at the beginning of the year of 2002. The conceptual design was done in around 6 months. The characteristic design parameters are listed in Table 1. The engineering design was finished by the end of 2003 while manufacturing the components took another year. The installation of the Shanghai EBIT took another half year and was completed by the end of 2004 [21]. In the May of 2005, the first X-ray spectrum was obtained. The electron beam energy was pushed to 100 keV by the end of 2005. Presently the Shanghai EBIT can be operated at electron beam energies spanning from 0.8 keV to 130 keV. The electron beam current depends on the beam energy, however at 130 keV the current has reached 160 milliamps. A picture of the main body of the Shanghai EBIT is shown in Fig.1. For machine diagnostics, a portable slit imaging system, together with sophisticated software for imaging processing was developed [22]. Through detailed imaging studies of the ion clouds in the Shanghai EBIT, upper limits for the electron beam diameter were obtained under various operating conditions. Table 2 lists some electron beam properties at 81 keV and for a 3 T magnetic field.

Figure 1. A picture of the main part of the Shanghai EBIT.

A metal vapour vacuum ion source, MEVVA, and a gas injection system have been installed for metal and gas element (also gas compound) injection. To give the Shanghai EBIT move flexibility in its variety of usable elements, a laser ion source and an atomic oven are under development.

For spectroscopic studies of HCIs, we have set up three spectrometers covering the wavelength region of around 1 Å to above 10000 Å: a 1- meter normal incidence (McPherson 225) spectrometer for the wavelength region of 300 – 10000 Å, a flat-field grazing incidence spectrometer (home made) for the wavelength region of 30-300 Å, and a flat crystal spectrometer (home made) for 1–30 Å. A high purity Germanium detector is constantly attached to the Shanghai EBIT for X-rays measurements and monitoring.

Table 2. Electron beam widths and electron densities at the beam energy of 81 keV and magnetic field of 3 T.

I(mA)	FWHM(μm)	Electron density(cm^{-3})
120	76.2	7.1×10^{11}
100	74.8	6.5×10^{11}
80	68.5	6.8×10^{11}

While the diagnostic equipment was being developed, other highly related research projects were initiated at the Shanghai EBIT laboratory. These studies include: numerical simulation of the charge state distribution and temperature evolution of an EBIT plasma, angular distribution and polarization of photon emission from radiative recombination, studies of hyperfine interaction dependence of metastable state lifetimes, disentangle studies of hot plasma processes using EBIT, electron correlation and QED effects on atomic level structure. Some results from the above mentioned ongoing research projects are reviewed briefly in the following.

The work on numerical simulation of the charge state distribution and temperature evolution of an EBIT plasma was based on the work of Margolis [23], but substantial modifications and improvements were made. We assumed a Boltzmann distribution instead of a homogeneous radial distribution for the ion density and a Gaussian shape instead of a homogeneous radial distribution for the electron beam intensity. We also considered multiple charge exchange, up to four-electron transfer, instead of only single-electron transfer as in previous works. In this work, relativistic expressions for the electron energy and electron impact cross sections are employed, as the electron velocity is already a few tenths of light velocity even at modest electron beam energy (i.e. several tens of keV). Also in this work, the electron beam radius is calculated instead of being input as a fixed parameter. The computer code in our work was developed to handle two different experimental situations, one is for pulse injection and the other is for continuous injection [24]. In the project for hyperfine interaction effects on metastable lifetime, we have made theoretical studies along the iso-electronic sequences of Ni-, Zn-, Be- and Mg-like ions. For the Ni-like ions, interest was on the lifetime of the $3d^9 4s\ ^3D_3$ level. The focus of this work was to evaluate the importance of the hyperfine induced electric quadrupole channel of the $3d^{10}\ ^1S_0$–$3d^9 4s\ ^3D_3$ transition for isotopes with nuclear spin. Result shows that hyperfine quenching is important along the whole iso-electronic sequence. At the low-Z end the hyperfine induced electric quadrupole transition channel dominates, being several orders of magnitude larger than the rate for the magnetic octupole transition channel, while for higher Z the two rates are of comparable size [25]. For Zn-like ions, our interest was on the lifetimes of the $4s4p\ ^3P_2$ level. Our research reveals that the hyperfine-induced electric dipole transition contributes to the $4s^2\ ^1S_0$-$4s4p\ ^3P_2$ transition at about the same level as, or even higher than, the M2 transition whenever the nucleus has a spin. This is true all along the Zn-like sequence [26, 27]. The results of Be-like and Mg-like ions are under preparation for publication.

For experimental disentanglement studies of hot plasma processes, much effort was spent in studying KLL dielectronic recombination processes, both resonance strengths and resonance energies, for Xe ions [28, 29]. In this project, a stable and precise high voltage divider [30] was developed to meet the need to accurately determine the resonance energies. We could compare calculations by different theoretical methods at the precision level of 0.03 percent, on average, because of our accurately measured resonance energies. Under the project concerning electron correlation and QED effects on atomic structure, we studied the QED energy shift in the $1s^2 2s_{1/2}$–$1s^2 2p_{3/2}$ x-ray transition in Li-like $^{208}Pb^{79+}$ as part of a collaboration with the Tokyo EBIT [31]. The contribution of QED effects to the transition was determined to be -24.99 ± 0.10 eV. Data from the experiments on electron correlation studies are under analysing.

4. Conclusions

In this contribution, the application of EBITs in assisting plasma diagnostics is reviewed. The progress of machine development at the Shanghai EBIT, the related experimental setup, and the ongoing research projects together with some of their results are reported.

5. Acknowledgements

The work was jointly supported by the National Natural Science Foundation of China under Grant No. 10434050, Shanghai Leading Academic Discipline Project No. B107, and '985' research platform setting up project reimbursed by the Chinese Educational Ministry.

References

[1] Levine M A, Marrs R E, Henderson J R, Knapp D A and Schneider M B 1988 *Phys.Scr.* T **22** 157
[2] Marrs, R E., Levine M A, Knapp D A and Henderson J R, 1988 Phys. Rev. Lett. **60** 1715
[3] Silver J D, Varney A J, Margolis H S, Barid P G E, Grant I P, Groves P D, Hallet W A, Handfort A T, Hirst P J, Holmes A R, *et al* 1994 *Rev. Sci. Instrum.* **65** 1072
[4] Morgan C A, Serpa F G, Takacs E, Meyer E S, Gillaspy J D, Sugar J, Roberts J R, Brown C M and Feldman U 1994 *Phys. Rev. Lett.* **74** 1072.
[5] Currell F J, Asada J, Ishii K, Minoh A, Motohashi K, Nakamura N, Nishizawa K, Ohtani S, Okazaki K, Sakurai M, *et al*, 1996 *J. Phys. Soc. Japan* **65** 3186.
[6] Biedermann C, Forster A, Fussmann G and Radtke R 1997 *Phys. Scr.* **T 73** 360.
[7] Crespo J R, Dorn A, Moshammer R, and Ullrich J 1999 *Physica Scripta*, **T 80** 502.
[8] Ovsyannikov V P, Lornack G Z 1999 *Rev. Sci. Instrum.* **70** 2646.
[9] Froese1 M, Champagne C, Crespo López-Urrutia J R, Epp S, Gwinner G, Lapierre A, Pfister J, Sikler G, Ullrich J and Dilling J 2006 *Hyperfine Interactions* **173** 1572
[10] Bohm S, Enulescu A, Fritioff T, Orban I, Tashenov S and Schuch R 2007 J. Phys: *Conference Series* **58** 303
[11] Watanabe H, and Currell F 2004 *J. Phys.: Conf. Ser.* **2** 182
[12] Geyer S, Kester O, Pfister J, Sokolov A *et al*, *GSI report: ATOMIC-PHYSICS-19* 290
[13] Neill P, Harris C, Safranova A S, Hamasha S, Hansen S, Safranova U andBeiersdorfer P 2004 *Can. J. Phys.* **82** 931
[14] Savin D W, Beiersdorfer P, Kahn S M, Beck B R, Brown G V, Gu M F, Liedahl D A, and Scofield J H 2000 *Rev.Sci.Instrum.* **71** 3362
[15] Bertschinger G, Biel A, the TEXTOR-94 team, Herzog O, Weinheimer J, Kunze H-J and Bitter M 1999 Physica Scripta **T83** 132
[16] Biedermann C and Radtke R 2002 *Phys. Rev.* **E66** 066404
[17] Mancini R C, Hooper C F, et al, 1992 *Rev. Sci. Instrumt.* **63** 5119
[18] Hammel B A, Keane C J, Cable M D, et al, 1993 *Phys.Rev. Lett.* **70** 1263
[19] Smith A J, Beiersdorfer P, Decaux V, Widmann K, Reed K J, and Chen M H 1996 *Phys.Rev.* **A 54** 462
[20] Beiersdorfer P, Brown G V, Gu M F, Harris C L, Kahn S M, Kim S H, Neill P A, Savin D W, Smith A J, Utter S B, and Wong K L 2000 *Proceedings of the International Seminar of Atomic Processes in Plasmas, Nagoya, Japan* 25
[21] Zhu X F, Liu Y, Wang X *et al.* 2005 *Nucl. Instr. & Metho.* **B 235** 509
[22] Yang Y, Lu D, Fu Y, Yao K, Chen W, Xiao J, Geng Z, Hutton R and Zou Y, submitted 2009 *Rev. Sci. Instrum.*
[23] Margolis H S 1994 DPhil Thesis (Oxford: Pembroke College)
[24] Liu Y F, Yao K, Hutton R, and Zou Y 2005 *J. Phys.* **B 38** 1
[25] Andersson M, Yao K, Hutton R, Zou Y, Chen C Y 2008 *Phys. Rev.* **A 77** 042509
[26] Andersson M, Liu Y, Chen C Y, Hutton R, Zou Y 2008 *Phys. Rev.* **A 78** 062505
[27] Yao K, Andersson M, Brage T, Hutton R, Jo¨nsson P, and Zou Y, 2007 *Phys. Rev. Lett.* **98**, 269304, *Phys. Rev. Lett.* **98** 269903(E)
[28] Chen W D, Hu W, Fu Y Q, Xiao J, Liu Y, Meng F C, Shen T M, Chen C Y, Wu S, Wei B, Hutton R, and Zou Y 2007 *Phys. Plasm.* **14** 103302
[29] Chen W D, Xiao J, Shen Y, Fu Y Q, Meng F C, Chen C Y, Zhang B H, Tang Y J, Hutton R, and Zou Y 2008 *Phys. Plasm.* **15** 083301
[30] Chen W D, Xiao J, Shen Y, Fu Y Q, Meng F C, Chen C Y, Hutton R, and Zou Y 2008 *Rev. Sci. Instrum.* **79** 123304
[31] Zhang X, Nakamura N, Chen C Y, Andersson M, Liu Y and Ohtani S 2008 *Phys. Rev.* **A 78** 032504

AUTHOR INDEX

AUTHOR INDEX

Institute of Physics
Dirac House, Temple Back
Bristol BS1 6BE UK

CURRAN ASSOCIATES INC
proceedings
.com